To Eric
the ultimate geological polymath
with kind regards
Don   November 2019

DEAR SAFFRON

Don Hallett

Little Henry Publishing

Published in Great Britain by

Little Henry Publishing
13 York House, Courtlands
Sheen Road
Richmond TW10 5BD, UK

Content and illustrations copyright © Don Hallett, 2018
All rights reserved

No portion of this book may be reproduced, stored in a retrieval system or transmitted at any time or by any means, mechanical, electronic, photocopy, recording or otherwise without the prior written permission of the publisher.

The right of Don Hallett to be identified as the author of this work has been asserted by him in accordance with the Copyright Designs and Patents Act, 1988

A CIP record of this book is available from the British Library

Published 2018
Reprinted with corrections 2019

Cover design, lay out and type-setting by *Could You Just...?*
Kingswood, Maidstone, UK

Printed on acid-free paper by Biddles Books Ltd
King's Lynn, UK

ISBN 978-1-5272-2273-1

Letter 1
To the casual visitor (and there are not too many of them), Halifax is a rather depressing place. A typical north of England industrial town which has seen better days. The buildings are mostly old, undistinguished and provincial. Terrace houses, built for working-class artisans during the nineteenth century spread row upon row up through the town. An elevated road, opened in 1973 to reduce congestion, cuts brutally through the centre. There are dozens of former mills, factories and warehouses, relics of the industrial past, converted to mundane uses as carpet showrooms, antique emporia, supermarkets and DIY stores. The largest mill complex of all, Dean Clough, which once produced top-quality carpets, is now home to numerous small craft shops, studios, keep-fit classes, a theatre group, and small cafes. The inhabitants are an uneasy mix of old established locals and a large number of immigrants, mostly from Pakistan, who came to work in the textile mills, and more recently Poles, eastern Europeans and refugees. Halifax, like most other northern industrial towns, is struggling to come to terms with its industrial heritage. What employment is available for those who used to work in the dozens of textile mills, the engineering works, the confectionary trade? It is difficult to say. It seems we have reverted to being a nation of shopkeepers.

But Halifax is where I was born and brought up, and to me it represents my childhood, a magical place of security, adventure and discovery, the place which moulded my character and shaped my outlook. Halifax in those days was a very different place, thriving, bustling, proud of its manufacturing industries, and its achievements. There were literally dozens of factories and mills scattered through the town, producing carpets, textiles, and confectionary (Mackintosh's toffees were famous). Most of them had mill chimneys which spewed out smoke, forming a pall over the town, which turned the honey-coloured stone from which it was built to a uniform black. But no one cared. The town was prosperous; it built itself a magnificent town hall, designed by Charles Barry (who designed the Houses of Parliament in London) in the Italian Renaissance style, a monument to the town's civic pride. It was also home to the Halifax Building Society, the largest and most respected building society in England. Its mill owners became rich, and in the tradition of many Victorian industrialists they were also philanthropists. The Crossley brothers invented a steam driven loom for weaving velvet and carpets, which reduced the price of carpets to a level most people could afford, and on the proceeds built the enormous Dean Clough mill complex which was the largest carpet manufactory in the world. They left their mansions to the town, and built alms houses, a school and an orphanage, plus a park designed by Sir Joseph Paxton (who built the Crystal Palace for the Great Exhibition in London), which in true public-spirited style was named the People's Park, Edward Akroyd, who owned a rival mill complex, developed a model village for his workers, built a lavish French medieval style church (where my parents were married), and after his death his mansion and park passed to the town.

Halifax is located on a slope with a river, in fact not much more than a large stream, in the valley bottom. The oldest buildings, including a medieval moot hall, scandalously demolished in 1956, and the parish church (recently elevated to the status of a minster) were built near the river, and as the town grew it spread up the slope. Dozens of terraces of working class houses were erected during the Victorian era, amongst which industrial buildings were located almost at random. At the end of our street was a textile mill, and nearby there was a steam laundry, an engineering works and a bobbin factory. In contrast to nearby places like Bradford and Leeds, all the houses were built of stone which was locally abundant. This was considered to be greatly superior to the red brick of neighbouring areas, and a source of pride to the inhabitants.

Halifax's wealth was based on wool, and a sheep features on the town's coat-of-arms. Incidentally, the coat of arms also bears the image of the severed head of John the Baptist, in the mistaken belief that the name Halifax is derived from the Old English *halig feax*, the holy hair (of John the Baptist). The town has a long history, extending back to Norman times, and wool was always the principal industry. There is a reason for this which is connected to geology. The town is located on a formation called the Millstone Grit, which extends westwards and forms an area of exposed, peaty moorland up to the crest of the Pennine hills and over into Lancashire. The moorland is fit for nothing but sheep grazing, but equally importantly the water from these moorland areas is soft, and ideal for washing the grease out of sheep fleeces. In medieval times, isolated farmsteads kept sheep which would be hand sheared and the wool cleaned, spun and woven at home on hand looms into 'pieces'. Some of the older houses still have abnormally large windows on the upper floor where the hand looms were located to obtain the maximum amount of light. The pieces would then be taken out to dry in a tenter field where, in order to prevent the woven cloth from shrinking, they were placed on a wooden tenter frame studded with hooks to keep the cloth in a state of tension while it dried, which gives rise to the phrase 'on tenterhooks', meaning a state of painful tension. The finished pieces were then taken to market by packhorse.

Halifax became the principal market for the wool trade, and the woven cloth was sold from stalls in the streets (one of which is still called Woolshops), or in the yards of local inns. As the trade flourished so did the town which, until the Industrial Revolution, held a dominant position in the area. In the 1770s the town authorities decided that instead of random trading in the streets, they would build a dedicated market where traders could display their wares and conduct their business. This was called The Piece Hall, the place where the hand-loom weavers sold their pieces. It was, and still is, an impressive building. Constructed around the four sides of a large courtyard it consisted of over 300 individual rooms on three floors with galleries, which were rented to individual traders. Unfortunately, it came too late. The Industrial Revolution made the local clothier and hand-loom weaver redundant. Machines could do the work of dozens of spinners and weavers, and large textile mills quickly replaced the former cottage industry. Halifax's dominant position was lost to towns like Bradford and Leeds which had access to coal and easier transport links and soon outstripped Halifax in importance. The Piece Hall eventually became a wholesale fruit and vegetable market, but has since been nicely restored, and the rooms are now home to small boutiques and studios.

Halifax has another claim to fame. It had a gibbet, a sort of guillotine which was used for the summary execution of convicted criminals. In medieval times the lord of the manor had the right to hold trials and to conduct executions, a relic of an Anglo-Saxon system of justice known as *infangthief and outfangthief.* The executions could be, and often were, carried out, not only for capital offences but for thieves who stole goods 'over the value of thirteen pence'. This system remained in use in Halifax later than any other town in England. Over sixty known executions took place there, with the last one in 1650, giving rise to the prayer 'from Hell, Hull and Halifax, good Lord deliver us'. (Hull had a notorious gaol). A curious feature of these executions was that no one was keen to act as executioner, so a horse was often employed to pull the pin which released the blade.

A steep escarpment rises to the east of the town and the crest is named Beacon Hill, the site since Tudor times of a beacon which could be lit to warn the surrounding area of some major threat, or to signal a military or naval victory. It is still used to celebrate events such as the Queen's Jubilee. In the eighteenth century, public executions took place on Beacon Hill, the gibbet having been long abandoned, and the bodies were left hanging in chains, *pour encourager les autres*.

With the coming of the canals and then the railway Halifax began to diversify its economy. Textile mills, carpet factories and dye works took the place of the cottage-based industry. Other industries – particularly confectionary, machine tools, boiler making, and metal-working – created jobs for hundreds of workers, and the population grew rapidly through the nineteenth century. It was during this time that large fortunes were made by the mill owners, but for the workers, like my grandparents, it was a very hard, hand-to-mouth existence, with long hours, low pay and harsh working conditions. This environment encouraged the growth of the Cooperative Movement, which originated in Rochdale in 1844 and spread rapidly. It was based on the principle of common ownership of stores (mostly food and clothing) of which anyone could become a member for a small fee. Each year the profits were distributed among the members in proportion to the amount of money they had spent. It was an idea that appealed to thrifty Yorkshire folk, and there were nine or ten such 'co-ops' in Halifax.

Yorkshire folk are also renowned for their canny individualism. The joke is that a Yorkshireman can buy from a Jew and sell to a Scot and still make a profit. This trait produced a people who were independent, industrious, and thrifty, (some would say bolshie, mean and penny pinching), and who were in consequence particularly receptive to the ideas of nonconformism, especially when presented by firebrand preachers of the 1770s like Oliver Heywood and John Wesley. By 1800 there were a dozen or so nonconformist chapels in the town – Unitarians, Baptists, Congregationalists, Methodist and Quaker - existing alongside the traditional Anglican parish church. Catholicism on the other hand never had much of a presence in Halifax. The spirit of individualism was encouraged by the chapels, along with sobriety, plain speaking, self-respect and self-reliance. These qualities in turn induced a desire among the people for betterment and self-improvement. Libraries were established, workingmen's clubs were set up, lectures were held and musical concerts arranged – attracting such illustrious figures as Paganini, Johann Strauss and Liszt. My family were a product of this social framework.

The other notable feature of Halifax was its strong regional accent – homely, distinctive and quaint - particularly to southerners. The Yorkshire accent uses short, rounded vowels. I remember being ribbed when I first went to London for my pronunciation of bath bun which down there is pronounced barth ban. Yorkshire folk also favour shortened words like allus for always, owt for anything, nobbut for nothing but, hasta for have you, geroff for get off, yerwot for what did you say? and mi for my. Aitches are dropped, so you have 'enry's 'ouse, and there is a solecism of using a plural verb with a singular pronoun, giving rise to expressions like 'he wer going' instead of 'he was going' or 'it wer a col' neet' for 'it was a cold night'. 'The' is shortened to th' or t', so that 'up the hill' and 'in the oven' become 'up th'ill' and 'in th'oven', or as in the quaint expression 'it's 'ot enough to crack t'flags' (flags are flagstones used for paving). There is also the curious habit of hyperbaton, inverting the word order in a sentence, producing for example 'he's a good batsman, is Geoffrey'. Yorkshire-speak also includes dialect words, many derived from Old Norse, like addle, gawp, reek, gormless, slavver, wuthering and tyke, which indicate a Viking influence extending back more than a thousand years. This is also seen in place names like Sowerby, Ringby and Fixby, and in Halifax Northgate, Southgate and Westgate have nothing to do with gates; the word *gata* is Old Norse for street. Broad-Yorkshire can be almost incomprehensible to an outsider with expressions like 'e's allus at t'last push up (he left it very late), better keep t'band in t'nick (better keep quiet rather than annoy someone), nobbut a mention (a small amount), it caps owt (it beats anything), 'e teks a good likeness (he takes a good photograph), goin' down t'nick (heading for trouble), and 'e wer' 'ard on (he was sleeping soundly). But interestingly during the Second World War, when the BBC was trying to encourage a feeling of homely patriotism they turned to Yorkshire speakers like Wilfred Pickles from Halifax, and J.B. Priestley from Bradford as presenters and news readers.

In working class England organized sport did not exist before about 1870, but when legislation freed up Saturday afternoons for most people, sport really took off. (Sundays were sacrosanct - reserved for church or chapel. On Sundays there was no sport, the mills lay idle, all the shops were closed, and public transport was greatly reduced). A rugby club was formed in 1873 which later became the Halifax Rugby League Club, a club of some renown, which for many years was among the top five or six in the country. They won the league championship four times and the Challenge Cup five times, and in 1954 a crowd of 102,000 spectators, the largest crowd ever to watch a rugby game, saw Halifax play Warrington in a replay of the Challenge Cup final - Halifax unfortunately lost. A rugby union club (a different game, usually associated with the middle and upper classes) was formed in 1919, but was never popular in what was a predominantly working-class town. Halifax formed a soccer club (called football in England) in 1911, but it never had much success, except for one glorious occasion in 1971 when they beat Manchester United in a cup qualifying match. Cricket was also a very popular sport, and Halifax had a dozen or so local or work-related clubs, but real cricket *aficionados* would go to watch Yorkshire play at Leeds, Bradford or Huddersfield. And as anyone will tell you Yorkshire are the best. They have won the county championship more times than any other county, have provided more England players, and have produced some of the most famous cricketers in the history of the game. I exaggerate, of course, but you get the idea.

Politically Halifax has tended towards liberalism, but this is obscured by the fact that universal voting rights were not established for men until 1918 and for women until 1928. Before these dates the franchise depended on property ownership, which greatly limited the number of people eligible to vote. In 1895, for instance, only one person in seven in Halifax qualified to vote, and these people were by definition the better-off, - landowners, mill owners, doctors, lawyers or those with inherited wealth. Canvassing before the twentieth century tended to be a rough and tumble affair, and bribery of one sort or another was a common means of securing votes. Things changed with the twentieth century and the rise of the Labour party, and Halifax has elected more Labour MP's than any other party since the end of World War II, which is entirely consistent with its predominantly working-class population.

And what of Halifax today? I haven't lived in Halifax for many years, but I have been back on numerous occasions. The cinemas we used to go to are closed, the park near our house has been vandalised and the copper plaque on the war memorial with the names of the dead has been stolen, the immigrants have gradually spread up through the town, and the old chapels have given way to mosques. (it is a curious thing to hear people of obvious Pakistani origin speaking English with a broad Yorkshire accent). The buses are no longer green, orange and white (the choice of a former transport manager who had a previous connection with Glasgow) and excursion trains to Blackpool no longer run from the old station. The rugby club sold its ground to a developer, and the prestigious headquarters building of the Halifax Building Society, once the pride of the town, is the headquarters no longer. The 'Halifax' became one of the many casualties of Thatcher's deregulation of the banking industry. It made a reckless dash for growth, merged with the Bank of Scotland, and in 2006 disappeared as a separate entity, handing over its assets to Bank of Scotland, which in turn was acquired by Lloyds in the bloodbath of 2008. So now the former headquarters is surmounted by the black horse, and memories of the famous building society are but a distant echo. On the other hand, the mill chimneys are long gone, the air is clean and the soot-coated buildings have now been restored to their original honey colour. The moors are still purple with heather in autumn, the seventeenth and eighteenth-century houses in the surrounding villages are still largely intact, and people no longer have to work for eight hours a day in a noisy textile mill. The terrace houses have been modernised, and virtually every house has a car parked outside. Life is much easier than it was, the cost of living is relatively low, the people are friendly, and commuting is easy. Halifax has changed a lot in fifty years, sometimes for better, often for worse, but for me it is still the dirty, bustling, proud old town of my childhood.

Letter 2
My grandparents were a product of their time, as we all are. They adopted the beliefs, the attitudes and the morals of late Victorian and early twentieth century England. My paternal grandfather, Alfred, was ten when Queen Victoria died and 24 at the start of the First World War, my maternal grandfather, Walter, was eight years older. Unusually neither of them was called up for war service; Walter worked as a mechanic in a factory producing armaments, and Alfred was an engine driver on the railway, jobs which were regarded as essential to the war effort. Both sets of grandparents came from very humble backgrounds. Alfred's family were originally agricultural workers near York, but with industrialisation and the coming of the railways they moved to Leeds, and my great grandfather obtained work as a waggon inspector for the Lancashire and Yorkshire Railway. Alfred followed in his footsteps and became a locomotive fireman and later a driver, operating mostly on the route from Leeds and Bradford through the Pennines to Manchester and Liverpool. This route has many tunnels and in the days of coal-fired engines the smoke and fumes were a real health hazard, and as a result he suffered badly from asthma. He married my grandmother Mary Hannah in Leeds in 1912 and they moved to Sowerby Bridge, not far from Halifax, shortly afterwards. They had two children, my aunt Rhoda, and my dad, George.

I did not know my paternal grandparents very well. Mary died when I was six and Alfred when I was nine. I remember my grandmother as a kind and gentle woman. She had been very attractive in her youth and retained her good looks well into middle age. She died of cancer, or perhaps of a broken heart after my father was killed. They lived in a modest brick terrace house at Sowerby, a hilltop village, a mile or so from the engine shed at Sowerby Bridge from which my granddad worked, and three or four miles from our house. There was a steep climb from the valley bottom up to the village, past an imposing eighteenth century mansion called White Windows. Their house faced onto open fields and the pastoral Ryburn Valley, and there were delightful walks through the woods and fields. My abiding memory is of wandering along paths through fields fenced with iron railings, kicking up autumn leaves, and arriving at Sowerby's surprisingly elegant Georgian church, where my grandparents were eventually buried. Sowerby was the birthplace of John Tillotson, the son of a Puritan clothier who became Archbishop of Canterbury. He was a moderate man in an intolerant age, saying of Roman Catholics that 'Papists, I doubt not, are made like other men'. He was a tutor to the son of Edmund Prideaux, Oliver Cromwell's Attorney-General, and married a niece of Cromwell, but he survived the Restoration, prospered under the protection of Princess Anne and William and Mary, and served as Archbishop of Canterbury from 1691 to 1693. Of my grandmother I have only the faintest memory. She always wore a wrap-around pinafore, she was neat and tidy, house-proud and welcoming. I remember my grandfather as a man in a brown suit and tie, and smelling of pipe tobacco. He seemed like a contented man, but with his asthma and ailing wife and having lost his son, life must have been difficult for him. I have no memory of his interests, his politics or his religion, but a couple of incidents give a clue to the kind of man he was.

During the war German bombers used to fly over the area *en route* to Manchester and Liverpool. Incidentally the authorities constructed starfish decoys on the Pennine moors. These were simulated buildings made of wood and combustible material which could be set alight in an effort to deceive the bomber crews into believing that this was the place to drop their bombs. Occasionally, if visibility was bad, the planes would jettison their bombs, and on this particular occasion they dropped their bombs not far from our house. Our area was visible from Sowerby and when my grandfather saw the fires burning he immediately set out on foot in the middle of the night to walk the three or four miles to our place, involving some arduous climbs, to check that we were safe.

Another example is more revealing. My mother's family included three farmer/butchers, and during the war the slaughter of animals was closely controlled as food was strictly rationed. One of her butcher cousins illegally slaughtered a pig, and gave some of the meat to my mother. When Alfred heard about this he was horrified, and forced my dad to take the meat back to the butcher. I don't know whether this was based on moral principle or the fear that it might compromise my dad who by then was a policeman.

After my grandmother died my granddad became a rather lonely figure, and he struck up a friendship with a lady known as Miss Clegg. She fleeced him for a loan of £700 which she never repaid. It rather tarnished his image as a sensible, prudent and practical man. Just before he died I remember him urging me to 'take up languages', on the basis that with a knowledge of languages you would never be out of a job. Unfortunately, I was to discover that I had little aptitude for 'languages'.

My maternal grandparents, Walter and Ada, had a much greater influence on me. They lived until I was about twenty, so I knew them much better. They came from an even poorer background than my other grandparents. I doubt if in their entire life they owned more than £200. During the twenties and thirties my granny had to go out and deliver milk for her farmer-cousins, in order to make ends meet and, just to be on the safe side, she kept her money in a Spanish pocket underneath her dress. She would occasionally look after the neighbours' children, and one of her charges was John Christie who became infamous as a notorious serial killer in London during the Second World War. Both sets of grandparents had much in common, they did not drink alcohol, they were all committed to the very Victorian principles of self-help and improvement and the importance of education, and they were honest (more or less), hard-working (when they had to be) and God-fearing (though they rarely went to church). This is not to say they were without faults. Walter once disgraced himself by losing a significant sum on the horses, which he could ill afford, but he never did it again.

Walter's family came from the Dewsbury area where they were weavers and dyers, but they moved the fifteen miles to Halifax in the 1860s, presumably because of better work prospects. My grandmother Ada came from a family of carpet weavers, poultry keepers and farm labourers who lived in the area between Halifax and Bradford. When she was young she lived across the street from the home of the Shaw family. One of the boys,

Percy, was a mechanic with a taste for invention. There are several versions of the story, but the one I like best is that one night he was returning from Bradford by car along an unlit twisting road with a steep drop on one side when his headlights reflected the eyes of a cat. This gave him the idea of reflecting road studs which could be placed either in the middle of the road or along the edge. He patented the idea and experimented until he found a suitable design of small reflecting lenses embedded in rubber with a space for water to collect, so that when a vehicle passed over them they became self-cleaning. They became a best seller and his workshop produced over a million cat's eyes a year with exports around the globe, and he was rewarded with an OBE for his invention. He was highly eccentric; he never married, he removed all the carpets and curtains from his house because they collected dust, and he had four televisions constantly running, but with the sound turned off. He became something of a Halifax celebrity.

My maternal grandparents lived in a rented house which I found endlessly fascinating, but was in fact so primitive that it was demolished not long after they died. It was called a back-to-back, which meant that the back of their house also formed the back of a similar house in the street behind. There were four rooms one above the other, a living room, the main bedroom an attic and a cellar. There was no bathroom; the toilet was twenty yards along the street, and there were no toilet rolls; you used cut-up pieces of newspaper. There was no electricity; lighting was by gas, and there was no running hot water; the water had to be heated in a range. And this was home to four people, Walter and Ada, my mother and my aunt Eva, at least until the girls got married. But what a treasure house! In the living room they had a wonderful rocking chair, an elegant chaise longue, a mahogany sideboard which almost reached the ceiling, and a large gilded Victorian clock. Lighting was by gas mantle, and 'lighting-up' with a wax taper was a daily ritual. There was a large black-leaded kitchen range with an oven on one side and a water tank on the other which were heated by a coal fire in the middle. On the inside of the street door was a horseshoe for luck, but it posed a bit of a conundrum. If it was open at the top the devil could sit in it, but if it was open at the bottom their luck could run out. The bedroom was also lit by gas, and the bed had a flock mattress, filled with little pellets of wool, and there was also that other necessity - a commode. If you wanted a bath you went to the public baths which were located not far down the road.

The attic was my favourite room. It had a skylight, a bed with brass knobs which my mother and aunt had shared as girls, and a mysterious curtain behind which was my granddad's workshop. This was full of wonderful things, a last for repairing shoes, planes, spokeshaves, saws, hand drills, chisels, drill bits, awls, hammers, braces, tee-squares. It was an Aladdin's cave for a little boy. In those days you didn't throw things away, you repaired them. My granddad was particularly adept at resoling shoes. There was no lighting up there so if you went at night you had to take a candle. Also in the attic there was a very elegant glazed mahogany book case containing my granddad's books – a rather odd selection, some of which I still remember. They included *Wilson's Tales of the Borders, A History of Protestantism, As we sweep through the Deep, In Sunlight and Shadow*, volumes of poetry by Wordsworth, Longfellow and Tupper (who was Tupper, I wondered?), English and French Dictionaries, a fascinating etymological dictionary, and *Wensleydale* by Edmund Bogg, an evocative Victorian book with slightly sinister pen and ink illustrations, all shadows and moonlight, about an area which figured largely in my later life. There was also *A Springtime Saunter around Bronte Country*, to which I discovered later, my granddad had been a subscriber. My grandma and my mother were disparaging about this collection, but I rather suspect that was the spark which generated my fascination with books.

In the basement was a cellar where my granny did the laundry. She would put on her clogs, light a fire under the copper boiler, and wash the clothes by hand, and then rinse them in a peggy tub. There was an enormous mangle, a machine with two heavy wooden rollers for wringing the water out of wet clothing. It had large cast iron gear wheels and a handle, and was more satisfying to me than any of my toys. The damp clothes were dried on a creel suspended from the ceiling above the fire in the living room. I find it amazing to think that they lived like this until the day they died in 1959.

Letter 3
Walter was my maternal grandfather and he deserves a letter of his own. Walter was a good-looking man, and he knew it. He dressed as well as circumstances permitted. He always had bright, shiny shoes and a pocket watch on a gold chain. He had only the most basic education, but he obtained employment as a mechanic and machinist-turner in a machine tool factory which during both the first and second world wars was engaged in making parts for shells and guns. After the second war he

worked for the Halifax Corporation Water Works as a maintenance mechanic.

Walter was very committed to the ideal of self-help. He attended evening classes, went to lectures, and joined the local workers' education association. He frequented the reading room at the local library to read the newspapers, and to borrow books, and as a result taught himself the rudiments of French. He got himself elected onto committees for horticulture, old folks' treats, agricultural shows and the like, and in his retirement he became a cricket umpire for the local cricket league. He had a taste for music and was particularly fond of Handel's *Messiah* (for which he had a full musical score), especially when the Scottish soprano Isobel Baillie was performing. They had a piano at home which my granny, my mother, and my aunt Eva could all play. Lacking electricity, my granddad had a radio powered by a large and heavy wet battery, similar to a car battery, which had to be charged at a local shop every couple of weeks or so. This was a very temperamental affair which used to hum and whistle and fade away, and my grandfather would often be crouched over it trying to listen to a cricket commentary, and shushing everyone else to keep quiet.

My granddad had a short temper, and when my aunt Eva was beginning to spread her wings in her late teens she had several angry confrontations with him which occasionally resulted in a slapped face. To me however he was a mentor and I suppose he became a surrogate father. Each time we visited him he would give me a silver threepenny piece, a coin which had gone out of circulation before the war. I ended up with quite a collection, but where he got them from I have no idea. He taught me about cricket, and took me to watch Yorkshire play at Park Avenue in Bradford and to Headingley in Leeds, and to my first test match in which England played New Zealand at Old Trafford in Manchester. He would cut out articles for me from the newspaper with the names of cricket fielding positions, different spin actions and batting techniques. He taught me how to use a slide rule, and a drawing pantograph, and a micrometer, which could measure thicknesses to an accuracy of a thousandth of an inch. He also had a skill that I much admired. He was able to peel an apple with a penknife so that the peel came off in one continuous strip. He taught me how to play chess and bought me a chess set, which was a cause of great embarrassment. I must have been about thirteen and by then I was attending a traditional old-established grammar school. For some unaccountable reason, he took it on himself to come to the school and ask for the headmaster. I was called out of class to go to the headmaster's study where I found my grandfather with the chess set. My class mates must have been agog to know what was happening. I can only assume that my granddad either wanted to see the school or to enquire about my progress, but I was not amused.

Working for the water works my granddad had the opportunity to take me to some of the remote reservoirs on the bleak and exposed moors around Halifax – the very names are evocative - Gorple, Walshaw Dean, Widdop, Dean Head, Cold Edge, Fly Flats, Ogden. These were wild and dark locations with a sinister and forbidding atmosphere, where the dark, peat-stained water cascaded over the overflow channels. He would occasionally arrange family 'outings', usually by motor bus, which he called by the quaint old name of chara, short for charabanc. One of my favourite trips was to Castle Carr, a sort of lost domain in a fold of the moors west of Halifax. In about 1860 a wealthy local merchant, Captain Joseph Edwards, who had made his money from a blanket mill, decided to build a mock medieval castle, complete with gatehouse, tower, battlements, great hall, ballroom, picture gallery and courtyard, a quite incongruous creation for such a setting. It was one of the grandest houses built in West Yorkshire during the nineteenth century. The surrounding park was planted with trees and shrubs, and a water garden was constructed with cascades and a fifty-foot fountain. Captain Edwards did not long enjoy his creation. In 1868 both he and his son were killed in a train crash in North Wales. The castle remained in private ownership, and was only rarely opened to the public. After the war its condition steadily deteriorated, the contents were sold, and in 1962 it was demolished. It is now a romantic ruin of broken, ivy-covered walls, the park has reverted to nature and the cascades and fountains are sad and forlorn.

My granddad introduced me to other fascinating houses – High Sunderland, a strange provincial manor house on a bleak, windswept hillside built about 1600, which Emily Bronte used as the model for Wuthering Heights, with 'grotesque carvings of crumbling griffins and shameless little boys'. It fell into ruin, but I remember entering it one day and finding a Latin inscription over the fireplace which read something like *'si proprius sederit flagras, si longius alges'* which translates roughly as 'if you sit too close you burn, too far away and you freeze'. Nothing remains; the house was demolished in the 1950s and even the site is not easy to find now. In the adjacent valley Scout Hall was built at a slightly later date for a dissolute silk merchant, much given to gambling and drink.

It is a so-called 'calendar' house, having 12 bays, 52 doors and 365 window panes. The owner died at the age of 37, reputedly attempting to fly with a pair of makeshift wings from the top of a quarry behind the house. The house stood empty and semi-derelict for decades, but is now a listed building with some faint prospect of restoration. There is a quaint and crude frieze over the main door showing hunters chasing a red fox. Further down the valley is Shibden Hall, a 15th century timber-framed house which belonged in the nineteenth century to Anne Lister, a rich and independent-minded heiress, and incidentally the daughter of a soldier who fought on the British side at the battles of Lexington and Concord. She was a landowner, diarist, mountaineer and traveller who kept a diary throughout her life, some of it in code, graphically detailing her lesbian relationships. This was not decoded until the 1980s, which is just as well; my grandfather would not have approved. Anne died while on a walking and climbing tour in the Caucasus mountains of Georgia after an insect bite turned septic. Shibden Hall has fared better than the other two houses. It received a large lottery-funded grant and has been restored to the condition it was in when Anne lived there.

But I digress. My granddad had an allotment, a small plot of ground, about 30 yards by 20, on which he grew fruit and vegetables, strawberries, raspberries, gooseberries, lettuces, potatoes, carrots and the like. In summer I would gorge myself on strawberries until my skin came out in red blotches. There was a bench at the top of the allotment from which there was a view of the town spread out below. The only time the atmosphere was clear was during Wakes week when the factories closed, and the town emptied as all the workers headed off to the seaside.

Walter and Ada were born within a month of each other, and in 1959 they died within a month of each other. From my grandfather I inherited an interest in books, acquiring knowledge, of rough and rugged countryside, and of cricket. It was not a bad inheritance.

Letter 4
My mother Mary was a remarkable woman. She attended the local secondary school where she became a prefect, and dramatized Charlotte Bronte's book *Jane Eyre*. At the age of thirteen she won the Mayor's Prize which would have allowed her to progress to a commercial college for three years, but my grandparents could not afford the related expenses, and in any case needed her to work and contribute to the family budget. So she was sent off to a sweat shop where she sewed clothes for eight

hours a day for a pittance. She hated it. She was a lively, vivacious girl with an open personality. She taught herself to play the piano, which she did remarkably well, and she loved dancing. She met my dad, who was a policeman, in about 1937 on what used to be called a monkey-run, a place in town where teenagers promenaded to flirt and exchange banter. It obviously worked since before long they were 'walking out' together, and were introduced to each other's families. I am sure that Walter and Ada would have been delighted that their daughter had found a personable young man with prospects. He was, in the phrase of the day, 'a good catch'. What his family thought about the match is less clear. I would like to think that his parents recognised my mother's merits and potential, and I have no reason to doubt it, but I think my Aunt Rhoda, George's elder sister, disapproved. It is a sad reflection on human nature that people look down on others less fortunate than themselves, despite having similar origins.

George too attended the local secondary school, where he won a couple of prizes, and then followed in the footsteps of his father and grandfather by joining the railway company to work as a signal man, but he found the job unchallenging, and after a year or two he changed careers and joined the police. There the work was much more to his liking and involved 'the beat', a prescribed route to provide security at night around the town, traffic duties, and as a receiving officer in the charge office when offenders were brought in. He was based in Halifax which was some distance from home, so he would often stay at lodgings close to the police station. He had an impressive shorthand speed and owned an antique Remington typewriter dating from about the time of the First World War. He was a physical fitness enthusiast and a good swimmer, and was awarded a bronze medal in life saving. He had a punch ball and boxing gloves, Indian clubs, dumbbells and he practised ju-jitsu (which would have been handy for a policeman). He also had a bicycle, which I later inherited. He smoked, as most men did in those days, but very moderately, and he would have the occasional beer.

My mother and dad were married in 1938 at All Souls Church, the fine, wonderfully-decorated church built and given to the town by Colonel Akroyd. I was born in May of the following year in the front bedroom of 4 Westbury Terrace, and it snowed on the day I was born. Perhaps it was an omen of difficult times ahead. It was to be my mother's home for the rest of her life. The house was in a stone built terrace at the top end of the town. It was much superior to Ada and Walter's house, since it had a bathroom, electricity and hot and cold running water. My mother must have been delighted. There were two rooms downstairs, a large kitchen/living room and a sitting room, and upstairs two bedrooms and a bathroom, plus a cellar for laundry, and tiny patches of garden both back and front.

And so they started their married life, unhappily soon to be marred by the outbreak of the Second World War, which began four months after I was born, and life suddenly became much more sombre. Food was rationed and windows had to be blacked out at night so as not to attract enemy bombers. Families were issued with gas masks, and air raid shelters were prepared. At the end of our street a siren was installed to warn of the approach of German planes. This produced a chilling wailing warble which sent everyone scurrying into their cellars, and then the relief of the steady one-note signal for the all clear. I can very clearly remember seeing German bombers flying overhead *en route* to the industrial cities and ports of Lancashire, and on one occasion we had an incendiary bomb land in our garden, which fortunately did no harm, apart from crackin' t' flags.

Initially, in what was called the Phoney War, nothing much happened. The Germans were busy overrunning and occupying western Poland and the Russians were busy overrunning eastern Poland and attacking Finland. Churchill was very suspicious of the Russians, who he described as 'a riddle wrapped in a mystery inside an enigma', but in April 1940 the Nazis invaded Denmark and Norway, and in May Belgium, Holland and Luxembourg. No country could withstand the German *blitzkrieg*, and when France fell in June 1940, Britain was left as the only country still holding out against the Nazis. The outlook was bleak. Churchill, the Prime Minister made stirring patriotic speeches: 'We shall go on to the end, we shall fight on the seas and oceans. We shall fight in the air. We shall defend our island whatever the cost may be. We shall fight on the beaches. We shall fight on the landing grounds, in the fields, and in the streets. We shall fight in the hills. We shall never surrender!' But the prospects were not good, and the reality was that Britain was ill-prepared. When it became clear that Britain would not negotiate, Hitler's full wrath was directed on us. He drew up an invasion plan, preceded by a campaign of terror bombing designed to destroy the Royal Air Force and break the morale of the people. Shelters were hastily constructed and gas masks issued to everyone in case of gas attacks. Night after night, starting in July 1940, the cities of Britain were bombed: London, Glasgow, Liverpool, Southampton, Birmingham, Coventry, Manchester, Belfast, Portsmouth.

London was bombed on 57 consecutive nights. Air raid deaths in September 1940 were 6,900 and in October 6,300, and all told the blitz claimed the lives of over 40,000 civilians. But invasion depended on achieving air supremacy, and the RAF had two fighter planes, the Spitfire and the Hurricane, which inflicted a heavy toll on the German bombers during what became known as the Battle of Britain. The Germans lost five planes for every three British, and despite a final effort in May 1941 when 1,500 people were killed in London, and many famous buildings destroyed, Hitler never did achieve air superiority and the invasion plans were called off, and left as unfinished business, while he turned his attention to Russia, his erstwhile ally, by launching a surprise attack on 22nd June. It was his biggest mistake.

Halifax was not a significant target for the German bombers, but in November 1940 537 houses were destroyed or damaged and 11 people killed in a rare attack. Every factory had been converted to assist the war effort. Iron railings were cut down from public parks and from people's gardens to feed the blast furnaces to make armaments (but in fact they were never used), and even silver paper was collected for recycling. Halifax sent buses to London to replace those destroyed by the bombing, imports of food dwindled to a trickle and food rationing became ever tighter. Britain basically ran out of money, but President Roosevelt came to the rescue, forcing through Congress an act to provide Britain and other allies with food, fuel, warships, planes and arms on a lend-lease basis, meaning that after the war equipment could either be returned or paid for. In all Britain received goods to the value of 31 billion dollars, and the Soviet Union 10 billion dollars. Britain's repayments were spread over fifty years with the final payment being made on 29 December 2006. Both Britain and Russia have admitted that without this help they could not have won the war.

At sea too, the war was going badly. In the first two years of the war Britain lost six capital warships, one of which was torpedoed in the supposedly safe anchorage of Scapa Flow, and they antagonised the French by destroying the French fleet at Oran to prevent it falling into the hands of the Nazis. Britain imposed a naval blockade on Germany, but the Nazis retaliated with unrestricted U-boat attacks on Allied shipping in the Atlantic. Britain needed to import more than a million tons of material a week in order to survive, and through much of 1940 and 1941 this target was not achieved. Britain was in danger of being starved into submission. But from 1942 the invention of ASDIC (a sonar anti-submarine detection device) along with depth charges and the establishment of a convoy system slowly began to turn the tide. But it was a close-run thing. During the entire war 3,500 Allied merchant ships and 175 warships were sunk in the Battle of the Atlantic at a cost to the Germans of 783 U-boats and virtually all of their capital ships.

On land the situation was equally bleak, Britain had been forced to evacuate troops from Norway, France, Crete and Greece, and in Egypt, which was the key to the Suez Canal and Middle East, they faced attacks first from the Italians and later from the Germans. In June 1941 the Nazis invaded Russia and by September Leningrad was surrounded and subjected to a siege which was to last for 900 days. By October they were at the gates of Moscow, and by December they had reached Rostov, beyond which lay the oilfields of Baku. The whole of Europe from the North Cape to Crete and from Brittany to the Caucasus, with the sole exception of Britain and the neutral states was under the control of the Nazis. In December 1941 however the entire dynamic of the war changed when the Japanese launched a surprise attack on the United States Pacific fleet at Pearl Harbor. The entry of America into the war was a godsend to Britain, and marked a critical turning point, but it took time for the American presence to be felt. The Japanese overran south-east Asia and the western Pacific just as the Nazis had done in Europe, and Britain ignominiously surrendered Singapore to an inferior Japanese army in February 1942. This was the lowest point of the war for Britain. We had been at war for two and a half years and had precious little to show for it.

This litany of woe formed the backdrop to my parents married life together. Rationing, blackouts, the threat of invasion, bombs, a baby. It must have been a very trying time for them. Public parks were dug up to grow vegetables, there were little or no imports of fruit from overseas, shops sold goods only on production of a ration book, and the standard ration per person per week was 8 oz. sugar, 2 oz. butter, 4 oz. margarine, 2 oz. lard, 1 shillings-worth of meat, 4 oz. bacon or ham, 2 oz. tea, 2 oz. jam and 2 oz. sweets. But of course, this was princely in comparison to the situation in Leningrad or Warsaw or Shanghai. We were reduced to eating snoek and whale-meat and Spam (curious, because all these items had to be imported). People were encouraged to grow their own vegetables in a campaign called 'Dig for Victory', and this was the time at which my granddad acquired his allotment. Clothing was rationed from 1941 with coupons which initially allowed roughly one new outfit per year, but later the points were reduced to the level equivalent to only one new coat per year. Clothing was made to a reduced and inferior 'war time standard'. If you wanted a new sweater or cardigan you got out your knitting needles and made one. Petrol was reserved for official users such as the emergency services, bus companies, doctors and farmers. People were encouraged to 'make do and mend'. Envelopes were reused with economy labels. Slivers of soap were collected in a soap shaker. My mother preserved eggs in isinglass and fruit from the allotment or from the countryside in preserving jars. Rose hips and blackberries were collected from the hedgerows and made into jam. Shoes were repaired, clothes were mended or patched or dyed, and children at school were given a daily dose of cod liver oil and highly concentrated orange juice to prevent them developing rickets or scurvy. Those attitudes of not wasting anything and not throwing things away 'because they may come in useful' lingered long after the war.

It was at this nadir in Britain's fortunes that, on 6th August 1942 my dad quit the police and joined the army. He was sent to a Primary Training Wing in Harrogate and in September was assigned to the Royal Artillery as a trainee gunner at Marske in Yorkshire, where my mother was able to visit him once or twice. He applied for officer training in December and was accepted in January 1943. He underwent anti-aircraft gunnery training at Llandrindod Wells in Wales and was commissioned as a Second Lieutenant in the Royal Artillery in November 1943. My mother took me to the passing out ceremony, travelling across country in wartime to this little spa town in the middle of Wales. I clearly remember the demonstration they organised where the anti-aircraft guns fired holes in sleeves towed behind planes. It was a proud moment for both of my parents. The boy from a humble background commissioned as an artillery officer. He was posted to an anti-aircraft unit based in Gateshead near Newcastle where he served for the next five months. He would come home for brief periods of leave of 48 or 72 hours, but of course he was away most of the time. However, by this time air-raids on the Tyne shipyards were rare events and in April 1944 my dad was sent to the London District Assembly Centre for re-assignment to a more active role. He was transferred to the East Yorkshire Regiment, an infantry regiment for which, of course, his training was not appropriate. We went to see him in London, where I was thrilled to see huge barrage balloons floating above the city. The balloons were attached to steel cables which were supposed to create a hazard for incoming bombers. I have a photograph of the three of us in Hyde Park, with my dad in his uniform, and he came to King's Cross station to see us off on the train back to Yorkshire. It was the last time we saw him. He was sent by troopship to Tunis on 2nd May 1944 and on the 12th was reassigned yet again to the 11th Battalion Lancashire Fusiliers with promotion to Lieutenant. The battalion had been based in Malta for most of the war, which had suffered incessant German bombing due to the island's strategic position in the Mediterranean. He joined them in Malta on 20th May, but they had already received instructions to join the allied effort in Italy. They travelled to Italy by troopship and arrived in Naples on the 22nd to be greeted by Mount Vesuvius still smoking after a major eruption earlier in the year, which had destroyed four villages, killed 30 civilians and destroyed a squadron of ninety American bombers.

By then the tide of war had changed. The United States, after their victory at Midway in June 1942, started to roll back the Japanese in the Pacific, Britain's first major victory was won at El Alamein in November 1942, and marked the end of the threat to Egypt. Churchill remarked that 'before El Alamein we never had a victory, after it we never had a defeat'. The Russians turned the tide at Stalingrad in the winter of 1942/3. Sicily was invaded in July 1943 and mainland Italy in September whereupon the Italians surrendered. German troops were quickly moved in to replace their former allies. Naples fell in October 1943 but progress up through Italy was painfully slow and was halted at Monte Cassino for over a hundred days in early 1944. The German commander Kesselring commented, when he was finally captured, 'next time, don't begin at the bottom'. Rome finally fell (or was liberated, depending on your point of view) in June 1944, and the same month the Allies launched their long-awaited cross-Channel invasion of Europe, this time with the active participation of the Americans, and towns and cities were liberated one after another. It was now only a matter of time before the Nazi terror was annihilated. The British and the Americans bombed German cities by both night and day. The chief of Britain's Bomber Command quoted from the bible 'they sowed the wind, and now they shall reap the whirlwind'. He was stopped by the police for speeding one night, and the police officer said to him, 'you want to be careful, sir, you might kill someone', to which Bomber Harris replied, 'young man, I kill thousands of people every night'.

Meanwhile the 11th Lancashire Fusiliers were held in reserve, awaiting events to the north of Rome and shuttled around on non-combat duties between Naples, Brindisi and Bari. By August the Allies had pushed on to the outskirts of Florence in preparation for another major offensive, this time against the Gothic Line, a formidable but incomplete German defensive position in the mountains between the valleys of the Arno and the Po. The Nazis staged a strategic withdrawal, laying mines and blowing up bridges as they retreated. Despite the bitter fighting the historic heart of Florence survived more or less intact and was occupied by the Allies without opposition on 13th August. On 1st August the Lancashire Fusiliers, still in Naples, celebrated Minden Day on which every recently commissioned officer in the regiment, which included my dad, followed tradition by eating a rose from a glass of champagne, commemorating the day in 1759 when the regiment advanced to the Battle of Minden through thickets of roses, which they plucked and put in their caps. It seemed a fitting gesture, the regiment was about to be moved into the line for the assault on the Gothic Line.

The army in Italy was reorganised during August following the loss of seven divisions to Operation Anvil in southern France, at the insistence of President Roosevelt. This left the forces in Italy severely depleted, particularly the US 5th Army, which was reinforced by the transfer of a British Corps which included the 1st Infantry Division of which the Lancashire Fusiliers formed a part. The battalion was moved up from Naples, through Rome to Florence, and crossed the River Arno on 30th August and was brought into the front line. On the 31st they sent out patrols probing the German positions, and were issued with orders to form the spearhead of the attack on the following day in pursuit of the retreating Germans along route 65. The weather on 1st September, a Friday, was dull and wet, and the battalion transport was halted after just a few kilometres by destroyed bridges and mines, which also halted the advance of the armoured support. The troops advanced on foot up the steep climb around Monte Rinaldi towards the village of Trespiano, with B and D companies leading the way, gingerly picking their way through minefields. My dad led a platoon in D company and by midday they had reached the large municipal cemetery in the village. As they emerged from the cemetery into a small meadow D company came under intense mortar fire ('stonked' in the language of the day), from a German position on the slope beyond. My dad and two others were killed instantly and nine others were injured, and the rest beat a hasty retreat. My dad was the first officer fatality of the battalion.

The dreaded yellow telegram was delivered to my mother a couple of days later and her grief and disbelief are etched forever in my memory.

What is life to me without thee?
What is left if thou art dead?
What is life; life without thee?
What is life without my love?
What is life if thou art dead?

She was frantic with despair. The war was so nearly over. Italy was supposed to be the soft option compared with northern Europe. He had promised that he would keep out of harm's way. Her world collapsed. All those dreams of a happy married life, of steady progress, of increasing prosperity, lay in ruins. It devastated her life, and it probably explains why I grew up as a shy and diffident boy, lacking in self-confidence and self-esteem. She was a widow at 27, I was an orphan at five.

Letter 5

The aftermath of my father's death was traumatic. My mother received details of what had happened from my dad's C.O. and from some of his fellow officers, with the conventional messages of condolence, how well he had been liked, what a fine officer, how he had been killed instantly and had not suffered. He was temporarily buried, by monks according to some reports, near where he fell, and his body was later moved to a British military cemetery by the banks of the Arno in Florence. Bodies were not repatriated in those days. We received some grainy photographs of the temporary grave taken in the rain, with his name incorrectly spelt.

My dad's death devastated his mother too. She was already sick with cancer, and within a year she also was dead. When my mother brought the news of my father's death to my aunt Rhoda (George's sister), she said 'I already know. I saw him at the foot of my bed last week in his army uniform. He had come to say goodbye'.

The attack on the Gothic Line fizzled out in the winter storms and rain, and the Po valley was not reached until the spring of 1945. The Italian campaign cost the lives of 60,000 to 70,000 allied troops. There was a bitter and ironic song at the time called the D-Day Dodgers, a phrase used in a public speech by the British MP Lady Astor, which implied that the serious fighting was taking place in Normandy, and those in Italy were shirkers who had somehow avoided the real war. The song contains the lines

> *When you look around the mountains, through the mud and rain*
> *There you'll find the crosses, some which bear no name.*
> *Heartbreak, and toil and suffering gone*
> *The boys beneath them slumber on*
> *They were the D-Day Dodgers, who'll stay in Italy.*

The war in Europe ended in May 1945, but my mother took no part in the victory celebrations, and the peace held no pleasure for her.

> *Perhaps someday I shall not shrink in pain*
> *To see the passing of the dying year,*
> *And listen to Christmas songs again,*
> *Although You cannot hear.*
>
> *But though kind Time may many joys renew,*
> *There is one greatest joy I shall not know*
> *Again, because my heart for loss of You*
> *Was broken, long ago.*

She was a widow with a young son in a grimy northern town with little prospect of improvement. She was resentful of those who had not gone to fight, angry with Churchill for precipitating the war, and with the War Office for transferring my dad from the artillery for which he had been trained, to the infantry for which he had not. She was always upset by the annual Remembrance Day commemorations, and by popular songs such as *We'll Meet Again*. She hated Germans, she would never watch a war movie, and vowed never to visit Germany under any circumstances. When they sent back my dad's personal effects, we found among them his army ID card. On the back was printed 'On Active Service for God and King'. My dad had modified this to read 'For Mary, God, and King'. They later erected memorial plaques on which his name was recorded, at police headquarters in Halifax, and at the Methodist Church he attended in his youth – again with his name incorrectly spelt. In April 1945 my mother received a letter from the War Office listing debits and credits on my dad's Military Account which showed a deduction of seventeen shillings and six pence for pay and Mediterranean Allowance for 1st September 1944, the day he was killed, but adding that in the circumstances they would not press for settlement. I guess the principle is that if you don't complete a day's service, you don't get paid.

My aunt Eva (my mother's sister), was a medical auxiliary during the war, where she met my uncle Dick, who was recuperating at a hospital in York.

He married my aunt during a period of leave in May 1944 at a chapel just around the corner from my grandparents' house. Dick was a hearty, sociable chap who had left home in 1937 to join the army, perhaps inspired by his mother's cousin, Fred Greaves, who won a Victoria Cross at Passchendaele in 1917. (He led an attack on a German machine gun post, seized it, ejected the defenders and held the position during a German counter attack). Dick enlisted in the Coldstream Guards and saw service with the Guards Armoured Division all the way from Normandy to Berlin. He brought me fragments of mosaic from the Reichstag, and a rather nice fountain pen, war booty from some hapless German soldier, perhaps. In my childish way I asked him how many Germans he had killed, but of course that was not a question he was about to answer. My other uncle, Leslie, who was married to Rhoda, also saw military service, but in a non-combatant role.

Relief at the end of the war in Europe was short lived. Life was hard, and Britain faced an uncertain future. Churchill was kicked out in the 1945 election and a Labour Government was elected with Clement Attlee as Prime Minister. Churchill's acerbic comment on Attlee, who had served with him on the War Cabinet throughout the war, was that 'he is a modest little man, with much to be modest about'. The Government was elected on the promise of a welfare state, public ownership of basic industries, a free health service and the granting of independence to the countries of Britain's former empire. All these policies were put into effect, but Britain was crippled by huge debts, the loss of thousands of homes and factories due to wartime bombing, a run-down, bomb-damaged transport system, and the return of thousands of servicemen from overseas. It was an age of austerity, but once again the Americans came to Britain's aid, somewhat reluctantly since they did not approve of a socialist government, which they regarded as the next thing to communism, but having offered the Marshall Plan to rebuild France and Germany, they could hardly refuse aid to Britain, and in the end Britain became the largest beneficiary.

Rationing continued well into the 1950s, people who had been bombed out were housed in 'prefabs', prefabricated houses built in factories which could be assembled in days. The National Health Service was established, a state pension system was introduced, and the welfare state came into being. The Bank of England, the rail industry, road transport, coal mines, iron and steel, and gas and electricity industries were nationalised, and the 1944 Education Act was implemented which abolished fees for secondary schools. India, Pakistan, Burma, Ceylon and Malaysia were granted independence. This was a formidable achievement in the circumstances. Yet despite the austerity Britain was able to stage the Olympic Games in 1948 and organise the Festival of Britain in 1951. Britain still regarded herself as a great power, but as someone aptly pointed out, at the start of the war there were the big three Allied powers (USA, USSR and Britain), but by the end there were only two and a half.

And so my mother and I began a new life. I didn't have any brothers or sisters, and at that time only one cousin, so it was a very small family unit, dominated by females and my granddad. I suspect my mother considered the idea of remarriage, but she had been very much in love with my dad (as his letters reveal) and she had me to bring up. I think she would have regarded it as a betrayal of his memory, and in any case I don't think she had the heart for it. I can only admire how she pulled herself together and made the best of it, in the difficult conditions of post-war Britain. There was a War Orphans Committee in Halifax which organised Christmas parties for several years, and occasional excursions, but my mother resigned herself to the fact that we would be staying in our terrace house and that our financial situation was not likely to improve. She received a small pension from the army and from the police, and she took part-time employment at the junior school which I attended, as a secretary and assistant. This allowed us to live adequately, but without any luxuries. In her spare-time my mother made hand bags out of cured calf skin, which she obtained from her butcher relations. It was never intended as a commercial venture, but I think she sold one or two to friends and neighbours. She also made a duvet cover for me from large panels of silk printed with extremely detailed maps of South East Asia, which I believe were from 'escape and evasion' maps supplied to air crews during the war in the Far East. Maybe that is where my interest in maps first began.

Radio played a significant part in people's lives in those days. We had a big cabinet radio with lots of glowing valves inside, and a tuning dial showing the frequencies of strange sounding places like Hilversum, Lahti, Schenectady, Tallinn, Saarbrücken, Reykjavik, but we listened exclusively to the BBC, which had been headed in the 1930s by Lord Reith. He established three fundamental principles – the BBC would be independent and impartial, there would be no advertising, and the role of the BBC would be to inform, educate and entertain (in that order). The service that he established became respected and envied around the world. Music was divided more or less equally between classical, dance and light music, and all the presenters had to wear a suit or dress, even though they were invisible to the listeners. Other countries like Nazi Germany used broadcasting for propaganda purposes and there was an infamous character, William Joyce, known as Lord Haw-Haw, who broadcast nightly bulletins from Germany detailing German successes and Allied failures (including shipping losses unreported in Britain) which were surprisingly accurate, at least in the early part of the war. He was an American citizen, brought up in Ireland who had lived in England in the 1930s. He was arrested in 1945 and tried for treason on the basis that he had lied about his nationality in order to obtain a British passport. He was found guilty and hanged in 1946. The BBC was more subtle. It used innocuous personal greetings to send coded messages to resistance fighters in Nazi occupied Europe.

We were not too interested in the cultural side of broadcasting, but we listed to light music, comedy shows, children's programmes, and of course the news. My favourite by far was a long running serial called *Dick Barton, Special Agent*, a sort of precursor James Bond. It had a 15 minute slot every evening and always ended with our hero in a desperate, cliff-hanging predicament. In those days, cinemas put on special shows for kids on Saturday mornings, showing cartoons, Charlie Chaplin comedies and cowboy movies, and when the villain finally bit the dust all the kids would cheer and stamp and shout. Strange to think of it now, but at the end of every cinema show they played the National Anthem. People were expected to stand in respectful silence, but I can't say that it received much respect in Halifax. I didn't see my first television until I was fourteen, when my friend Bryan's parents bought one specially for the Queen's coronation. About a dozen of us crowded around and watched the event in black and white on a flickering screen about twelve inches across.

My uncle Dick came back from Germany but remained in the army. He was based initially at Regent's Park Barracks in London with my aunt Eva, where we visited them. Their flat overlooked Albany Street, and late one night a group of revellers were creating a nuisance in the street below, so my uncle emptied a slop-bucket over them. We went on holiday with them in 1946 to Bournemouth on the south coast just after the grisly Branksome Chine murder, which fascinated me no end. The next year we went to Keswick in the Lake District where I vividly remember a visit to a fairground where there were dodgem cars. Dick paid for me to go on them, not once, nor twice but four times which, in our straightened circumstances, I thought was the height of magnanimity.

On another occasion, we went to see the four-masted sailing ship *Pamir* in Shadwell Dock in London. The *Pamir* was built in Germany in 1905 and was the last large sailing ship to round Cape Horn and the last to carry commercial cargoes. It had been taken as a war prize by the New Zealand government and was paying a courtesy visit to Britain. We all stepped onto the ship which somehow began to drift away from the quay. We didn't know whether to jump or not, but Dick had the presence of mind to take control, and prevented us from doing anything stupid. In 1957, after the ship had been transferred back to German ownership, the *Pamir* was caught in a hurricane in mid-Atlantic and foundered leaving only six survivors from a crew of eighty-six.

My uncle was then sent off to Palestine to monitor the end of the British mandate, and then in 1948 to Libya which was still under allied control, but about to gain its independence. Eva went with him to Tripoli where they stayed for a year or so, and my cousin John was born there. It is hard to believe now, but they sent us food parcels from Libya with dates and food items which were still unavailable in England.

We saw my granddad Alfred occasionally and he took me to the train shed from which he had worked. That was a thrilling place, with engines ready for duty blasting off steam and eager for action, others simmering gently in the sun, and others dead, their fires out, standing forlorn and neglected. There were the ubiquitous Black Fives (do-anything, go-anywhere) used for the trans-Pennine passenger trains, large powerful austerity locos used to haul coal trains, little shunters, and old-fashioned engines from the Lancashire and Yorkshire days with tall chimneys and very basic weather protection for the driver and fireman. My granddad died not long afterwards and was buried alongside my grandmother at the little Georgian church in Sowerby.

We also saw my aunt Rhoda and her husband Leslie, who lived not far from Alfred. Rhoda had a sensitive nature and tried her hand at writing poetry. It seemed over-sentimental to me, but she had some of her poems published in a local magazine. Their house had an earth toilet with two seats side by side which I guess was fine if you wished to be sociable. We visited each other now and again, but the relationship between my mother and Rhoda can be illustrated by the example that when my granddad Alfred died, the one thing that my mother coveted was a nice roll-top desk, but my aunt said that it was going for auction, and my mother could bid for it there if she wished. As Robert Louis Stevenson said 'sooner or later we all sit down to a banquet of consequences'. And so it was in this case; this insignificant event would have consequences later in life.

Children of that period didn't have much in the way of toys, but I was fortunate in being given a few from a neighbour along the street whose boys were now grown up. In particular there were two table-top racing games dating from before the war, one for horses, which involved the navigation of your horse over a jump, and one for dogs which required some skill in jerking your dog along a stretched string. These were always popular with visitors and with my friends along the street. I had a cut-out cardboard train set, a home-made tank, a home-made, rather ingenious set of four-way traffic lights, a table tennis set, a bag of marbles, a whip and spinning top, and a money box in the shape of a negro's head and shoulders. You put a coin in his hand, pressed a lever and the coin disappeared into his mouth. And that was about it until Dinky toys came along. Dinky toys were the defining toys of the time, small die-cast cars, sports cars, vans, army vehicles, buses and so on. I slowly acquired about a dozen of them, and with my friend Henry, who also had a collection of lead soldiers, we would fight out the battles of the war among the roots of a privet hedge in his garden, which to us were the forests of the Ardennes or the jungles of Burma. We were also avid collectors – stamps, coins, cigarette cards, anything that could be made up into sets, so swapping of surplus items became a major occupation. And then there were conkers. Conkers are horse chestnuts, lovely, shiny, tactile objects which we would thread onto a string to fight against an opponent's conker. If your opponent's conker had ten victories to its credit and you were able to smash his, then you added his victories to yours. We would go to extraordinary lengths to harden our conkers – bake them in the oven, bury them in the ground, soak them in vinegar, or paint them with varnish.

I started school about the time the war ended, and my enduring memory of those years is of school broadcasts, when we would listen to programmes on *How things began* where an intrepid reporter would become esntangled with raging dinosaurs, or belligerent cave-men or Roman legionaries, and on music which introduced us to instruments, the orchestra, and a few popular classical pieces. The school had a collection of musical instruments which the teacher struggled to teach us. The problem was that no one could read music, everyone wanted to play the drums or cymbals, and no-one wanted the triangle or castanets, which were regarded as 'sissy'.

Letter 6

Westbury Terrace was (and is) a street of about fifteen houses which faced a similar row across a narrow back street. At the front it faced onto a much wider street paved with setts, rectangular stone blocks held together with tar. The houses were built about 1900 and by my time they all had inside toilets, hot and cold running water and electricity. No one on the street had a telephone, a television or a car, and very few had refrigerators. If you wanted to make a phone call you went to a cold, draughty phone box a hundred yards away where you fed coins into a slot to pay for the call. There were no facing houses at the front, but instead a row of garages, always painted dark green, one of which was used as a shoe repair shop and another as a fish and chip shop. I should explain that fish and chips were, and probably still are, the north of England's principal contribution to English cuisine, and there were dozens of them in Halifax. The fish was usually haddock or cod (unlike the south of England which at that time used 'huss' or 'rock', which was essentially dogfish or sand-shark). There is a distinction between Yorkshire fish and chips and those from Lancashire, just over the Pennines. There they use vegetable oil (originally cotton-seed oil from the cotton industry) for cooking, whereas in Yorkshire we use beef dripping (which may explain the high incidence of heart disease). Fish and chips have been elevated to cult status in Yorkshire. There is a famous place called Harry Ramsden's near Bradford which has a huge dining room seating two hundred and fifty, with crystal chandeliers, fitted carpets, oak panelled walls and waitresses dressed in black dresses, white pinafores and waitresses' headdresses, little changed since the 1930s. At Harry's the fish and chips were served with tea and bread and butter.

At the end of the street was a small Methodist chapel to which my mother sent me for Sunday school, probably to get a little peace for an hour or two. Further along was a large textile mill, always known as Courtaulds, although it changed hands several times. The mill had a distinctive square tower which could be seen from far and wide, including from my granddad's allotment three miles away. The neighbourhood was what demographers would call 'working class, mixed use'. A short distance away was a park with extensive views to the west, which provided an excellent lesson in geology, which I only came to appreciate later. It was in this park that I used to play cricket in summer and rugby in winter with my school friends.

We knew most of the people in our street. The husbands worked, usually in one of the local mills, while the women stayed at home and looked after the children and the housework. My closest friends were Henry and Margaret from No 10, who were a similar age to me. We spent a lot of time together. Henry was very keen on rugby and we would often kick a rugby ball in the street (in a game that we called 'gainings'), once neatly putting it through the window of one of our irate neighbours. We went to every home game that Halifax played, and this was during their heyday when they were regularly in the top four or five teams in the league. The ground had a capacity of 10,000 and was often completely full. We knew the team intimately, Chalkley the ponderous full-back, Daniels the flying winger, Freeman the agile centre, Kielty the wily little scrum half with Dean at stand-off, and Ackerley the hooker, who played for England several times. In later life Henry became a professional rugby league referee, and was selected to oversee the cup final at Wembley, which would have been a great honour, but that year Halifax made it to the final so he had to withdraw.

Across the back street from us lived an old chap with dementia who went for a ramble on the moors and was lost for several days before being found by the police. Next door to him was a young man who had served in the Parachute Regiment during the war and had been parachuted into France during the Normandy landings in 1944. Further along there was a bow-legged man who had played rugby for Halifax in the 1930s and beyond that was a family who upped-sticks and emigrated to Australia. But the biggest sensation happened near the end of the street when a woman came back from a routine shopping trip to find her husband burnt to death in an armchair. He had deliberately poured petrol over himself and set it alight.

Monday, by long established custom, was wash day, the idea being that you would be able to wash, dry and iron the laundry in time to use it again the following weekend. There were no washing machines, so the washing was done by hand in the cellar using a large zinc tub called a peggy tub, a posser and a rubbing board, and often using a muslin bag of Dolly Blue to make white clothes whiter. The washing was then hung out on a line in the garden to dry. If the weather was wet, it would be hung on a creel in the kitchen. No one had central heating and houses were heated by coal fires. Coal was delivered in large sacks from lorries, and the bags were emptied down a grate into the coal cellar below. This produced clouds of dust, and if a coal lorry appeared on a Monday the women would go crazy and dash to take in the washing, to prevent it being soiled by the coal dust.

Shopping too was rather primitive. There was no credit, no supermarkets, no packaging, no plastics. Instead there were neighbourhood shops where you would go with your wickerwork shopping basket or string bag to buy butter which would be cut from a large block, bacon which they would cut for you on a slicer, and fruit and vegetables which were sold loose. On the other hand, fresh milk was brought to the door by horse and cart every day in glass bottles, and newspapers were delivered early in the morning before people went to work. My first paid employment was delivering newspapers before heading off to school. We also had the rag and bone man who came around with his horse and cart collecting anything that he could recycle. Originally he collected rags which were made into paper and bones which were made into glue, but he had progressed a bit since then and now wanted metal objects, bottles, newspapers and the like. We used to get hawkers, and knife grinders and gypsies coming around, and my mother's *bête noire*, the Kleen-Easy man. He was a pro. As soon as you opened the door he would wedge his case of samples over the threshold so that you couldn't close it on him, then he would go through every damned item in his case, however many times you told him 'no'. My mother used to bake her own bread, placing the leavened dough in front of the fire to rise, before putting in the oven to bake. She enjoyed cooking, particularly things like apple pie, fruit cake, custard pie, rice pudding and parkin, all of which I liked except for the yucky skin on the rice pudding and the soggy parkin. The idea of rice or pasta or curry was totally alien in those days, and the only take-aways were fish and chips.

The people on the street were all working-class and all with a similar outlook. At that time there were no Moslems or blacks or immigrants. All, with very few exceptions, were white Anglo-Saxon protestants. Most were politically to the left, none that I can recall were drunkards or gamblers and all believed in God, although not many were regular church attenders. It was a community united by shared values and circumstances, and for the most part on friendly terms with each other. They usually had one week's paid holiday a year which they would take during Wakes Week, when all the factories and mills closed down, when they would 'set off' for places like Blackpool, Morecambe, Bridlington or Filey. This annual exodus was supplemented by weekend 'outings' to local beauty spots like Hardcastle Crags, Shibden Park or maybe as far as Haworth or Skipton.

In England the 5th of November is Bonfire night (known to us as Plot night) which is held each year to commemorate the discovery of a plot to blow up the House of Lords (and the king) in 1605. Traditionally a large bonfire is prepared and an effigy of Guy Fawkes, one of the conspirators, is placed on top, all to the accompaniment of fireworks. In Halifax each street had its own bonfire, and kids would go from door to door asking for 'a penny for the guy' or for any old clothing for the guy. We would collect as much wood as we could find, broken furniture, old armchairs, odd bits of fencing, crates, tree branches – anything that would burn. We held our bonfire in a sand and gravel yard just behind the garages, and people would bring potatoes to bake in the embers, whilst the kids let off their bangers, jumpers, rockets and Catherine wheels. Some years we built the bonfire in the street (which gives some indication of how little traffic there was), but it made rather a mess of the stone setts which cracked, and the tar binding them together, which melted.

I attended school just down the road, where we had a real dragon of a headmistress. She would check to see that we were properly dressed and woe-betide if you did not possess a handkerchief. In those days the town employed a board man, a relic from the nineteenth century board schools. If you were absent from school without having sent a note to the teacher, he would come to your house to check that you were not playing truant. They also had nit inspectors who came around to school once or twice a year to check whether any children had lice in their hair. The cradle-to-the-grave welfare state provided free milk during the morning break, but instead of issuing it fresh and cold they would heat it on the radiators, which made it quite revolting. I enjoyed school, although my grades were never great, but it was here that I first developed an interest in history and geography. I was fascinated by the story of Captain Scott's tragic death whilst returning from the south pole in 1912, and every time we had a free reading period I would go to the school encyclopaedia and read the account of his ill-fated expedition, until I knew it almost by heart. When I was nine or ten we acquired a dog, a male black Labrador who we named Major. He was a great companion, and we formed a very close attachment. Because of the continuing austerity we fed him largely on horsemeat and leftovers, a diet which these days would be looked upon with horror, but he seemed to enjoy it. He accompanied me on my paper round every morning and he would wait from me to come home from school outside the front door, and run along the street to greet me, and proudly carry my school bag home.

Towards the end of junior school my closest friend was Bryan Hartley who lived nearby. His mother and mine became great friends. Bryan was a pretty smart kid, usually a step or two ahead of me, and there was a certain amount of rivalry between us. We sometimes went to Shibden Park where you could hire a boat on the lake for sixpence an hour. Bryan got over the difficulty of paying by turning up late in the evening when the boats were being called in, and he would help the boatman to bring in the boats and get half an hours' boating for free. We played a lot of cricket together, sometimes by ourselves, sometimes with his dad, and sometimes with other kids. Occasional we would go to Bradford when Yorkshire were playing, and on one occasion during the lunch interval we were fooling around with a cricket ball which went over the boundary fence and onto the pitch. I climbed over to get it and ripped the seat out of my pants on the fence railings. I was greatly embarrassed, but Bryan thought it was a huge joke.

Bryan's dad also extricated us from a potentially awkward situation. We had gone by bike to Mytholmroyd, a village along the Calder valley, and were amusing ourselves throwing stones into a stream when an irate farmer appeared and accused us of destroying his wall by throwing stones from it into the stream (which was not the case). He took our names, but we thought no more about it, until we both received letters accusing us of vandalism and threatening legal action. Bryan's dad was an insurance agent and was able to sort things out with the farmer, who we thought was a jerk. Just up the road from Mytholmroyd is Cragg Vale, an isolated and remote area where, during the eighteenth century, a gang of coiners was active. They clipped a small amount of gold from genuine coins, re-milled the edges, melted down the clippings and cast them as new coins, using dies of French and Spanish coins which they had obtained. Two of the coiners were betrayed, and arrested by a government investigator, but the gang tracked him down in Halifax and killed him. Thirty suspected coiners were arrested and three were publicly hanged in York, and the case became something of a *cause célèbre*.

Bryan's long-suffering dad also took us for a week's holiday to the Norfolk Broads, an area of lakes and waterways in eastern England, where we hired a yacht. We were rather inefficient sailors, but it taught us something about sailing, and both of us took up more serious sailing in later years.

At the age of eleven children took an exam called the Intelligence Tests or the Eleven Plus. This was a big deal because it determined your future. If you were judged as being suitable for an academic education you might be lucky enough to get a place at a grammar school, which were now free in the post-war socialist utopia. If not, you went to a Secondary Modern school which provided more of a vocational education. There was a third option of going to a fee-paying school, but I think not one in a hundred chose that route in Halifax. My mother had the foresight to send me to a crammer before the exam for a few evening sessions to learn how to handle the tests. This paid off because I succeeded in passing the exam and being given a place at Heath Grammar School, an old and much-respected grammar school, a couple of miles away in the 'nice' part of town. I was lucky; without that start in life I would probably have ended up as a plumber or decorator or maybe followed the family tradition and worked on the railways.

Letter 7
The Brontës are a local phenomenon - a clergyman and his family living a secluded life in a remote village on the edge of the moors a few miles from Halifax, which produced two (or perhaps two and a half) of the most famous female novelists of the nineteenth century, a family with little formal education, no useful contacts and no experience of life beyond the immediate family circle, and where all of the six children died before the father. This is the stuff of legend (and big business), and many Yorkshire folk feel a special kinship with the family, since they too can roam the bleak moors, and see the wide windswept skies, and the remote isolated farmsteads on the hilltops, which form the backdrop for some of their most famous novels. I used to cycle to Haworth, the village on the hill where they lived, and visit the parsonage where you can see examples of the tiny books they wrote as children with writing so small that you need a magnifying glass to read it.

Their story is fascinating and tragic. Patrick Brontë (originally Brunty, but he thought Brontë sounded more impressive, or perhaps because bronte in Latin means thunder) came from Ireland, moved to England and was ordained as an Anglican priest in 1807. He married and had six children in quick succession before moving to the small Yorkshire village of Haworth in 1820. The family occupied the parsonage next to the church, but Patrick's wife died of cancer the following year, so her sister came to look after the young children, and stayed for the rest of her life. The two eldest girls Maria and Elizabeth were sent to boarding school at the ages of 11 and 9, where they both contracted tuberculosis and died. Charlotte and Emily, the next two sisters, had also enrolled at the school but were withdrawn following their sisters' deaths. For the next six years the remaining children, Charlotte, Branwell, Emily and Anne were educated at home by their father, but for much of the time, and especially in the evenings, they were left to their own devices. They created a make-belief world and wrote long and elaborate accounts of life and battles in the fictional kingdoms of Angria and Gondal, inspired partly by a collection of toy soldiers which had been given to Branwell, the only boy in the family, and partly by the hero of the time, the Duke of Wellington.

At the age of 16 Charlotte was sent to a girls' finishing school, to be followed by Emily a couple of years later. Emily could not stand the separation from her home and left after only three months, being replaced by her younger sister Anne. After a couple of uninspiring teaching jobs the girls came up with the idea of opening a school themselves, and to that end Charlotte and Emily enrolled at the Pensionnat Heger, a school for young ladies in Brussels, where they planned to perfect their skills in French and German. After only a few months they were called back home after the death of their aunt, but Charlotte returned to Brussels alone soon afterwards, where she developed a powerful but unrequited passion for Monsieur Heger, the proprietor of the school, which was viewed with great alarm by his wife, so Charlotte was dismissed. In spite of circulating prospectuses for their proposed school, not a single pupil enrolled. Perhaps it was the remoteness of the location which put them off.

Branwell meanwhile had aspirations of becoming a portrait painter but failed to obtain a place at the Royal Academy School in London, and failed again when he tried to set up a portrait studio in Bradford. He secured a job on the recently constructed Lancashire and Yorkshire railway but was dismissed for 'financial irregularities', and then as a tutor, but he had an affair with his employer's wife and was dismissed again. Not an auspicious career, and Branwell's life, once so full of promise, began to deteriorate as he took to drink and opium.

Taking on the role of housekeeper after her aunt's death, and with time to spare in the evenings, Emily sorted out the poems she had written as a girl, and these were discovered by Charlotte which produced a furious reaction from Emily for the invasion of her privacy. Charlotte regarded some of them as worthy of publication, and eventually persuaded Emily to allow her to include a selection, along with some of her own and of Anne and offer them for publication. Their authorship was concealed by the genderless pseudonyms of Currer Bell for Charlotte, Ellis Bell for Emily and Acton Bell for Anne. The poems were published in 1846 by a London printer, and sold just two copies. Nevertheless, the sisters were encouraged to venture into the world of writing, and with their adolescent experience of the Gondal and Angria tales, they decided to each write a novel. Charlotte submitted the manuscript of a novel called *The Professor*, based on her experiences in Belgium. It was rejected, but the publisher was encouraging, and suggested that the author (whose identity was still unknown) should write a longer novel, and she immediately set to work on *Jane Eyre*. Emily and Anne submitted their offerings, *Wuthering Heights* and *Agnes Grey* to a different publisher and both books were somewhat grudgingly accepted for publication, but at a cost to the authors of £50 for an edition of 350 copies. Charlotte's *Jane Eyre* was completed shortly afterwards and was enthusiastically received by the publishers. It was published in October 1847, and Charlotte was paid the princely sum of £500 for the book. Emily and Anne's publisher suddenly woke up to the fact that he had a book by possibly the same author and published *Wuthering Heights* and *Agnes Grey* immediately.

Neither their father nor their brother were aware of these events, and their revelation must have delighted Patrick, but stunned Branwell. Here were his spinster sisters achieving a major literary success, while he, the erstwhile leader of the troop, had nothing but failure to show for his efforts. There was great confusion about the identity of the author or authors, particularly after Anne's second novel *The Tenant of Wildfell Hall* was published in June 1848, and was mistakenly attributed to the author of *Jane Eyre*. Charlotte and Anne were obliged to travel to London to reveal themselves to George Smith, head of the publishing house of Smith, Elder and Co. They became overnight celebrities, and their books sold like hot cakes.

This marked the apogee of their lives, but not for long. From this point it plunged rapidly to the nadir. Branwell's dissolute life came to an abrupt end in September when he died of tuberculosis aggravated by his weakened state. Three months later Emily succumbed to the same disease, and five months after that Anne was gone. Charlotte was left as the only surviving child, and spent her time editing Emily and Anne's poems, and she published two more novels, *Shirley* and *Villette*, before marrying her father's curate the Rev. Arthur Bell Nicholls in 1854. Nine months later she too was dead from complications arising during childbirth, exacerbated by the ever-present tuberculosis. Patrick Bronte survived until June 1861 when he died aged 84, having witnessed the deaths of his wife, all six of his children (not one of whom reached the age of forty), and his sister-in-law. Charlotte's first novel *The Professor* was finally published two years after her death.

Between them the three sisters produced an enduring literary legacy, but for me it is Emily Brontë who is by far the most impressive. Monsieur Heger commented that she should have been a man. He wrote about 'her powerful reason and her strong imperious will, never daunted by opposition or difficulty, with a head for logic and a capability for argument unusual in a man but rarer in a woman', and on 'her stubborn tenacity which rendered her obtuse to all reasoning where her own sense of right was concerned'. She had a reclusive, secretive nature, and no close friends. She liked to walk on the wild and bleak moors with her dog Keeper, and she had an iron will. When she was bitten by a dog she cauterised the wound herself with a red hot poker, and as she lay dying she refused to see a doctor until just before the end, by which time it was too late.

Her novel *Wuthering Heights* is remarkable. Whereas Charlotte and Anne's novels are 'conventional' in the sense that they more or less reflect Victorian values and mores, Emily's novel is wild, passionate, brutal and unforgiving. How she was able to create such characters as the cold, vengeful Heathcliff, the indecisive, equivocating Catherine, and the genteel, educated Edgar, not to mention the spunky Isabella who escapes from her abusive husband by fleeing to London, and the humourless, misanthropic servant Joseph, with her limited experience of life is a mystery. The novel is set against a backdrop of the bleak, inhospitable landscape of the Pennine moors, as reflected in the word wuthering, which carries a sense of fierce, blustery, violent winds. It is a *tour de force* which has stirred the hearts of generations of Yorkshire men (and women), and many others besides.

Emily also left a large number of poems, the vast majority of which were not published in her lifetime. Seventeen of her poems, heavily modified and added to by Charlotte, appeared with the 1850 edition of *Wuthering Heights*, but the remainder were sold by Charlotte's husband in 1895, and they became widely dispersed. They were not reassembled and published until 1941. Emily's poems reveal the same unconquerable spirit as observed by Monsieur Heger all those years before, but also compassion and forgiveness particularly for such weak-willed characters as her brother Branwell.

The Reverend Nicholls, Charlotte's husband, has earned himself a good deal of obloquy from Brontë lovers by editing, destroying and selling some of the remaining manuscripts. For instance, almost nothing remains of Emily's Gondal writing, and he insisted that some events should not be mentioned in Mrs Gaskell's biography of Charlotte.

I have visited the Brontë Parsonage many times, and seen the couch on which Emily died and the kitchen where she taught herself German as she went about her housework, the tiny study where the four children spun their web of childhood and the adjacent church where seven members of the family were laid to rest. Only Anne is missing. She died in Scarborough and is buried there. It never fails to astonish me that a clutch of three young and inexperienced women from such an isolated and provincial location, could have produced such an outpouring of drama, tragedy and passion.

Letter 8
About twenty years ago we were invited to a literary *soiree* held in the apartment of very artistic woman overlooking Westminster Cathedral. The family included a poet, an actor and a conductor. Appearing on this particular evening was an actor, Stephen Oxley, who was doing a dry run for a one-man show at a London theatre the following week. He performed scenes from *Tristram Shandy* by Laurence Sterne which is often considered to be the first English novel. During the interval the hostess asked me if I had heard of Laurence Sterne and I think I rather shocked her by telling her that we had both been educated at the same school.

Heath Grammar School in Halifax was a very traditional boys-only grammar school which was determined to hold onto its long and illustrious traditions even in the egalitarian post-war world. It was founded in 1585 and still preserved an apple and pear window from the original building. The school colours were not red and yellow, but crimson and gold. By long tradition a half-day holiday was granted on Wednesday afternoon, compensated by attendance on Saturday morning. The school was ruled over by a strict disciplinarian, Walter Swale, who had held the rank of Lieutenant-Colonel during the war. Newcomers and their parents were invited to an introductory meeting before the start of term, when the masters appeared in their gowns and academic hoods in the school hall which was embellished with honours boards recording all those who had won scholarships to Oxford and Cambridge. One of the boards carried the name of Ralph Fox, who I will describe later, since he deserves a letter to himself. The school deliberately tried to emulate Britain's famous public schools. We were assigned to one of four 'houses' which was supposed to engender a sense of group loyalty. We played rugby, not the rugby league of the town but the rugby union of the establishment. The school also had a fives court, a game akin to squash, but played with gloves rather than a racquet, and a game which originated at Eton and Rugby. I was quite a good rugby player. I was fast, and a proficient kicker of goals, but in each year we had two groups, an elite group, and a second, more run-of-the-mill group. I was on the borderline between the two, but I always preferred to play in the second group where I could shine, rather than in the first group where I was rather ordinary, and I never had any ambition to play for the school. This taught me a rather profound lesson, although I didn't realise it at the time. My preference for a background role was an indication that I was not destined to become a leader of men. I lacked charisma. We had one boy however, Phil Horrocks-Taylor, two or three years ahead of me, who did make the grade and played at fly-half for England for two or three years in the mid-fifties.

We used the public school hymn book and the King James bible, and the school had memorial gates commemorating the former pupils who had 'given their lives for their country'. We were required to wear a school uniform, even a school cap wherever we went about the town (which we were expected to touch on encountering 'our betters'), and it was made clear to us that the purpose of the school was to prepare us for significant careers in the future; there was no working class ethos here. The curriculum included Latin, French, English language and literature, history (mostly British), maths, art, music and religious studies, with lesser importance assigned to geography, chemistry, physics, biology and crafts. In those days, teachers were allowed to administer corporal punishment with a gym shoe, ruler or strap. 'Six of the best' was a feared and humiliating punishment.

I can remember my first day there. School always started with assembly, a hymn, a bible reading, a prayer and an address from the head master wearing his gown and mortar board. We were then shown to our dark and rather Dickensian class room and given an old oak desk bearing the carved names of former occupants. The teacher arrived wearing his gown, at which we were told to stand and stop talking. Our form master was also the Latin teacher, and for a Latin lesson we had the additional task of standing to attention and intoning *Salve, magister*, when he entered. Even at lunch the grace was said in Latin, and one such grace was written by one of the school classics masters, with a pun on the name of our founder, Dr Favour. In fact, the reality fell short of the aspiration. It was difficult to instil these essentially public school virtues into a bunch of largely working class kids, but they never wavered in their efforts to do so. I recall one frustrated teacher forcibly reminding us we had better shape up as we were the elite of the town. Curiously it was not a thought which had occurred to many of us.

The teachers were an interesting assortment. They all had degrees of one sort or another, some had industrial experience and many had seen war service. The chemistry teacher had the catchphrase 'When I was in Abadan...'. We had nicknames for all of them, Biddy Taylor, Nuffa Morris, Doco Fleet, Whisky Haigh, Archie Littlefair, Com Maclay, Honkie Peace and so on. Doctor Fleet, one of the geography teachers, had been a pilot in the First World War and his nerves were all gone, not surprising when you consider that 14,000 British airmen lost their lives in the war, they were not issued with parachutes and life expectancy for a new pilot was just seventeen days. I discovered much later that his doctorate had been awarded for work on erosion surfaces in the Grampian Highlands of Scotland, which almost qualified him as a geologist. Mr Hallowes, the maths teacher, I discovered was a prominent member of the Halifax Antiquarian Society, and it struck me as odd that a maths teacher could have such unrelated interests. I should mention, in passing, that there were no electronic calculators in those days. In maths we used log tables or slide rules. Mr Birchall, the PE teacher, had competed as a pommel-horse gymnast in the 1936 Olympics, which reminds me of an anecdote. At those Olympics in Berlin the greatest athlete was a black American, Jesse Owens who won four gold medals. The story was circulated that Hitler refused to shake hands with him, but on being interviewed later he said, 'no it wasn't Hitler, it was President Roosevelt'.

Academically I was not very good. I found Latin a struggle, and I was mediocre in other subjects, but I enjoyed history, English and geography and I learnt the basics of the scientific method through the chemistry lessons where we had to record objectives, method, observations and conclusions – basic, but fundamental nonetheless.

I made some good friends at school, Bryan Hartley, who I had been my close friend at junior school, Michael Balmforth, who later usurped Bryan's position as my 'best' friend, Daryl Nash, and John Wiggen, who joined the school in the third year having moved to Halifax from another part of the country. We formed a little coterie with shared and common interests both at school and in our leisure time. Michael lived in a big house with an extensive garden. On our cross country runs his garden provided a very handy short cut, which improved our timings no end.

My first overseas school trip, at the age of fifteen, was to Switzerland, supervised by a couple of teachers. On that occasion, we were told *not* to wear our school uniform, and certainly not our caps. The whole experience was novel – by boat across the Channel, taking a very foreign-looking train with warnings about *ne cracher pas*, or *pericoloso sporgersi*, travelling through the night and waking up to see Lake Geneva and the Castle of Chillon, which we knew from our English lessons on Byron. And then by a little rack and pinion railway from Martigny to the alpine village where we were based. It was magical to see the high alps still with a covering of snow, and the alpine meadows full of flowers, and the honey-coloured cows with their cowbells. It was like an advertisement for milk chocolate. One day we took an excursion up to the Great St Bernard Pass at an altitude of almost 2,500 metres, famous for the hospice housing the St Bernard dogs which were trained to locate snowbound travellers. It was a St Bernard which found the body of the youth in Longfellow's poem *Excelsior*, and two hundred years ago one of them (called Barry), is said to have saved the lives of forty people. What remains of him is on display in the Natural History Museum in Bern. They never did carry little barrels of brandy attached to their collars; that is a myth which originated with the artist Edwin Landseer. Alas, modern traffic now passes through a tunnel, and the few remaining St Bernards are now purely decorative. Napoleon and his army passed this way in 1800, commemorated by David's highly romanticized painting of Napoleon on his horse Marengo, pointing the way to the top of the pass.

By the fifth year selection began, when we were assigned to one of three groups, arts, sciences or general. Not being great at languages, physics or chemistry I was put into the general stream. This was the low point in my school career and my mother became concerned that I was failing. I had my Damascene conversion, curiously enough, on the playing fields of Eton (where, according to the Duke of Wellington, the Battle of Waterloo was won). We had gone to visit my aunt Eva who was living with Dick in

Windsor at the time. We walked, my mother, my aunt and myself, over the bridge to the Eton playing fields, where they raised the topic of university. I said I really was not good enough to go to university, to which they replied in typical Yorkshire style, 'nonsense you can do anything if you set your mind to it'. It was a defining moment. I thought 'yes, damn it, I can'. I scraped enough 'O' levels to progress to the sixth form where teaching was on a much more personal level. I concentrated on history and geography for the two-year slog to 'A' levels, and I finally started to blossom. We had good teachers for both subjects, particularly for geography where Frank Haigh taught just two of us, myself and John Wiggen. He was a motivating teacher. By this time, I had become interested in geology, which he encouraged by giving me a book on the geology of West Yorkshire and by suggesting other books which I might read. So I added geology to my curriculum and became the first person in the school to take the geology exam. I also won a prize for my achievements in geography. I had left it very late, but things were finally beginning to look up.

Letter 9
My childhood friends Henry and Margaret left the street about the time I went to grammar school, and I saw little of them afterwards, although strangely Margaret married my best friend Michael, and Michael's sister married my sixth-form fellow geographer John Wiggen. My friendship with Bryan continued, particularly through cricket and train spotting. The cricket championship during the 1950s was frustratingly dominated by Surrey, but Yorkshire had Len Hutton, who held the record for the highest test score of 364, Fred Trueman, the demon fast bowler known as Fiery Fred, and Johnny Wardle, the left-arm finger-spinner, who had the best bowling average of any English Test spinner since the First World War.

I supplemented my meagre pocket money by delivering newspapers before going to school, by doing shopping for one of our neighbours, and by typing advertisements for a grocery store down the road on my dad's ancient typewriter. It was a laborious process involving the use of carbon paper, so only three or four copies could be done at a time. From the age of about fourteen I started reporting local cricket matches for a sports paper in Halifax called the *Green Final*. Each week they would send me a postcard telling me which match they wanted me to cover and I would head off to cricket grounds all around Halifax. I bought myself a scoring book and would sit with the scorers, where I obtained the names of the players and kept tabs on the scores. I wrote a 200 or 300 word report, took a bus to the newspaper office, and they would print it the same evening. But the best money was to be made at the Post Office. During the Christmas break they would take on casual staff to help them with the huge number of Christmas cards and packages, and along with most of my friends, I would sign-on as a delivery boy. They gave us a big canvas bag with bundles of letters for a particular area of town and send us off to deliver them. Occasionally they would ask us to sort outgoing letters into pigeon holes marked with the names of towns and cities. That was interesting, but we were very slow compared to the regular Post Office sorters who were incredibly fast. They paid us by the hour, and we would collect twenty or thirty pounds in the ten days before Christmas, which was a fortune by our standards.

In 1951 we went to London to visit my aunt Eva and uncle Dick and my mother took me to see the University boat race at Mortlake. It was a dull and blustery day and we stood on the bridge waiting for the boats to appear, but our wait was in vain. The Oxford boat sank in the choppy conditions and the race was abandoned. It was re-run the following Monday and Cambridge won by an enormous margin. Ever since then I have been a Cambridge supporter.

In the same year we went to the Festival of Britain, a brave attempt to demonstrate that after the war Britain was still a creative, dynamic country. It was timed to coincide with the centenary of the Great Exhibition held in 1851 which dazzled the world with its opulence and brilliance. The Festival was a more modest affair, assembled on derelict land on the South Bank near Waterloo Station. A new concert hall, the Royal Festival Hall, was constructed on the site, there was a Dome of Discovery, a futuristic cigar-shaped pylon called the Skylon, and the shot tower, a building preserved from the old industrial site, in which lead shot had been manufactured. In addition, there were exhibition halls dedicated to rural England, minerals, power, ships, transport, schools, health, sport and such like. I recall being very impressed; Britain, it seemed to me, was still in the forefront of invention in electronics, nuclear power, jet engines and cars. The feeling was not to last. Within a decade Britain had been left way behind by the United States, and Germany and Japan were fast catching up. My most treasured souvenirs of the festival were four or five specially minted silver crowns, which I think I still have somewhere.

But then, disaster. My aunt Eva and uncle Dick, who now had two children (my cousins John and Margaret), moved to an army base at

Pirbright about 20 miles from London, and later to Windsor. We visited them once or twice a year and in 1952 we arrived to find my aunt distraught. John had contracted polio. It was a shocking, devastating blow. A four-year old boy stricken down with an incurable disease which left him paralysed in both legs. My aunt and uncle were devastated. How could it happen? Where did it come from? Was there any treatment? The questions kept coming. My aunt tried everything, physiotherapy, aquatic exercises, homeopathy. She even went to a faith healer, but all without result. They resigned themselves to the sad fact of having a disabled son. In later life John achieved some remarkable successes, but at this stage the outlook was bleak.

We spent many of our vacations with Eva and Dick, and my cousins. For several years we hired railway caravans, located at scenic locations on the coast at small rural stations. They were converted railway carriages with bedrooms, sitting room and kitchen, surprisingly spacious and comfortable. One year we went to Seascale on the Cumbrian coast, very close to the secretive and sinister Windscale Nuclear Reactor where plutonium was processed, and from there I was able to roam around in the Lake District; the delightful Ravenglass and Eskdale miniature railway, the dark and sinister Wastwater, the deepest lake in England, and Scafell Pike, the highest mountain in the area at 3,200 feet which I tried to climb, but failed. But I was only about thirteen, and ill-equipped for fell walking. On the way back I was given a lift in a car and the driver put his hand on my knee, once or twice, which I thought was rather peculiar. I didn't realise the implications at the time, but fortunately he dropped me off without further incident. Another time we went to the village of Sandsend near Whitby, the town where Captain Cook was born. Whitby is a quaint, and at that time unspoilt, fishing port on the Yorkshire coast. It is famous for its Jurassic fossils which are exposed on the foreshore, and this I think was the spark which ignited my interest in geology. I dug out a large ammonite about eight inches across, which to me was a major find, and discovered that the town's coat of arms features three coiled ammonites. Whitby has a romantic ruined abbey on the cliff top and nearby a wonderful parish church full of old box pews and memorials to drowned sailors.

My mother had made friends by this time through her work at school. The son of one of these friends was engaged to be married, and we were invited to the wedding in the Peak District of Derbyshire, fifty miles or so away. We went off with our presents and smartest clothes and arrived at the house of the bridegroom the evening before the wedding, only to be told that the wedding was off. The bride had got cold feet and decided she couldn't go through with it. It was a bit of a bombshell. The groom later married someone else and became a well-known doctor.

I was just about tall enough to ride my dad's pre-war BSA bike, and I started going on cycling trips with my friends. The bike had been stored since he died, but was still in good condition. It had a carbide headlamp, which worked by slowly dripping water onto crushed calcium carbide to produce acetylene which was lit with a match to form a bright flame. Cyclists didn't wear crash helmets in those days and on one occasion I crashed into a boy who suddenly swerved in front of me and hurt my side and thigh, but I daren't tell my mother because I knew she would stop me from riding it.

Train spotting was all the rage at that time. It developed a bad reputation later on, being regarded as totally nerdish; little boys standing on the end of station platforms recording engine numbers and crossing them off in a little book. But this is to totally misunderstand the nature of the game. These were the final days of steam; month by month locomotives were being withdrawn and scrapped and within a few years all would be gone, along with 2,300 stations and 5,000 miles of track. We wanted to be witness to the end of an era. There is a poignant song by Flanders and Swann which captures the nostalgia of the times which contains the stanzas

> *No churns, no porter, no cat on a seat*
> *At Chorlton-cum-Hardy or Chester-le-Street*
> *We won't be meeting again*
> *On the slow train.*
>
> *No whitewashed pebbles, no up and no down*
> *From Formby, Four Crosses to Dunstable Town*
> *I won't be going again*
> *On the slow train.*
>
> *No one departs and no one arrives*
> *From Selby to Goole, from St Erth to St Ives*
> *They all passed out of our lives*
> *On the slow train.*

Because of my granddad's connection with the engine shed at Sowerby Bridge we could always talk our way in there, and we would often cycle further afield to an engine shed at Mirfield about ten miles away. In those days there were no health and safety regulations, and so long as you were not a nuisance they would let you wander around. But the best place to see exotic engines involved a trip to Leeds where we would take up position at a place called Nineveh, from which vantage point we could see trains entering and leaving both Leeds City and Leeds Central stations. Here were some serious engines. Many of them carried evocative names like *Meg Merrilies, Bonnie Dundee, Quicksilver, Auld Reekie*, and there were Jubilees with names redolent of the British Empire – *Quebec, Tasmania, Punjab, Bihar and Orissa, Tanganyika, Aden*. There were streamlined engines like the A4s, which included *Mallard* the fastest steam locomotive in the world, and big, powerful engines with blinkers (smoke deflectors) which handled the London expresses. Many of the trains themselves had names – *The White Rose, The Yorkshire Pullman, The Queen of Scots*, all guaranteed to be pulled by prestigious engines. The engines carried head codes and shed numbers so you could identify where they were going and where they were based. There were 'specials' in those days, too. Special trains (identifiable by their head codes) would often be put on for big sporting occasions, and for holiday excursions, but clubs (like railway enthusiasts for example) could also charter a special for their 'outings', often to obscure places or to branch lines which had not seen a passenger train in years. The royal family had a special train (I later met a chap who had been a steward on it), Churchill hired a special for a canvassing tour, and Professor Moriarty hired a special in *The Final Problem* in pursuit of Sherlock Holmes. In Germany in 1911 (admittedly a bit before my time) they put on specials to take hundreds of opera lovers from Berlin to Dresden to see the premiere of Richard Strauss's *Der Rosenkavalier*. Imagine doing that today.

British Railways at that time was organised on a regional basis, and the engines 'down south' were quite different from those in the north. We often visited my aunt and uncle in Windsor from where I was able to take a tiny diesel railcar, almost like a bus on rails, to Slough where I could see the Castles and Kings and Manors of what had been the Great Western Railway. These were not as visually impressive as the engines I was used to, but they were fast and efficient and had an enviable reputation. I once put a penny on the rail at Slough and after the train had passed it was squashed to the thickness of a post card. I also got to visit a station on the Southern Railway at Woking where you could see the strange but impressive streamlined locomotives of the West Country, Battle of Britain and Merchant Navy classes (known as Spam cans because of their light construction), which were used on the boat train services to Southampton and Portsmouth. By the time I was fourteen or fifteen I would take the train from Windsor to London and trawl around the London termini stations in search of unusual engines. You could buy a platform ticket for a penny and get a close-up view of the engines heading up the *Golden Arrow*, the *Flying Scotsman*, *Brighton Belle*, the *Irish Mail* or the *Cornish Riviera Express*.

Halifax Parish Church (now elevated to the status of a minster) is a rather splendid building, and for a year or two I attended services there. The present church dates from 1438, but contains fragments from earlier periods. The main features are a magnificent medieval font cover, carved Jacobean pews, ceiling panels bearing the heraldic arms of local families, and a few relict plain-glass leaded windows from the Puritan period, when stained glass was regarded as popish. With my friend John Wiggen, we somehow got involved in attending bible classes conducted by the Vicar of Halifax, a bluff, hearty chap called Archdeacon Eric Treacy. The meetings were not so much about the bible as discussions on religion in general, morality, Darwinism, humanism and a wide range of topics. They were always inconclusive, but at least they made you think. We decided that we should get some first-hand experience, so we started visiting churches of different denominations, Congregational, Unitarian, Baptist, Quaker, and even Catholic. Eric Treacy later became Bishop of Wakefield. He was a very keen photographer of trains, and published several books of splendid railway photographs. He died from a heart attack on Appleby station whilst on one of his photographic excursions.

Michael and I also developed an interest in archaeology. We had read that there were many places on the Pennine moors where prehistoric flint implements could be found so we went in search of them. But searching and finding are different things, and after several abortive expeditions we concluded that previous discoveries must have been purely a matter of chance. So we turned our attention to Roman roads. There are no Roman roads across the Pennines which can be followed in their entirety but there are several places where partial remains are visible. The roads led from York and Ilkley to Manchester and Chester, and there seem to have been several attempts to find the optimum route. Two Roman forts guarded the most prominent route, at Castleshaw on the Lancashire side, where the fort outlines can clearly be seen, and at Slack on the Yorkshire side where the remains are now hidden beneath a golf course and the M62 motorway. Further north at Blackstone Edge, on the very crest of the Pennines, there is an excellent paved road showing conspicuous wear and tear, with a deliberately cut channel down the centre. This was long regarded as another of the trans-Pennine Roman roads on the route from Manchester to Ilkley, but it could equally be an early turnpike road – or perhaps both. It may originally have been a Roman road, which was 'improved' and repaved in the eighteenth century. In medieval times Wakefield, 20 miles to the east of Halifax, was the principal town of the West Riding of Yorkshire, since it stood at the extreme limit of navigation on the River Calder. It was a centre of the wool and tanning industries, and considerable trade took place between the two towns, along a route called the Magna Via. This was a narrow, paved road constructed principally for pack horses, which is still well preserved on Beacon Hill to the east of Halifax. In fact it is one of the best preserved medieval roads in the country.

When I was fourteen or fifteen my mother and I were invited by the Commonwealth War Graves Commission to visit my dad's grave in Italy. He is buried in a war cemetery by the banks of the Arno, not far from Florence. The war cemeteries are built to a standard design with uniform white limestone headstones, and they are immaculately maintained, but to see my dad's headstone with his age of 28 and the inscription which my mother had chosen 'until we meet again' was unutterably sad, and brought back all the pain of his loss. We were taken to the place where he was killed, and we just stood there numb and tearful. I have returned several times since, and the reaction is always the same.

Letter 10
Osbert Sitwell's entry in *Who's Who* contains the ironic line 'educated during the holidays from Eton'. My entry would be 'educated at Heath Grammar School and by Mark Kirkbright'. Mark is the most remarkable person I ever met. He was a year ahead of me at school and I didn't get to know him until I was in the sixth form. He was an original thinker – and then some. The first time he came to our house his initial comment to my mother was 'this house would be easy to break into'. His subject was history, not just conventional, school-book history, but the interesting bits. He knew that William the Conqueror's son, William Rufus, had been shot with an arrow by one of his retainers in the New Forest in very murky circumstances which suggested murder. He knew that Charles II had nine mistresses by whom he had fathered sixteen children, and that George IV

had had a secret, unconstitutional marriage to a twice widowed, Roman Catholic commoner. He was an authority on Richard III and was persuasive and eloquent in his defence against the charge of plotting the murder of the Princes in the Tower. He was a mine of information, not just on British history, but more widely. I wrote an essay on Russia before the First World War, the 1905 revolution and the establishment of the Duma, but he was able to enlighten me that the *Potemkin* mutiny was caused by the sailors being served rotting meat crawling with maggots, and that Stolypin, the reforming Prime Minister, was assassinated by a Jewish activist in a theatre, much like Abraham Lincoln.

It was around this time that I became an admirer of Oliver Cromwell, and I suspect Mark had something to do with that. They say you can tell a lot about someone's political leanings by asking which side he would have supported during the English Civil War. I never had any doubt which side I would have been on. Cromwell was a quiet country squire until Charles I's dissolved parliament and instituted arbitrary personal rule, based on the disturbing belief that he was divinely appointed, so could do no wrong, and he could not understand why his subjects were so contrary. It was a curious position to take for the son of James I, who had been invited to accept the English throne only thirty years before. Cromwell became leader of the opposition, and when civil war erupted in 1642 he became one of its most successful military commanders. The Parliamentarians triumphed in the end, thanks largely to Cromwell's New Model Army, the king was arrested, tried, found guilty of treason and beheaded in 1649. The monarchy was then abolished and a republic proclaimed. All of Charles possessions, including a priceless collection of 1,500 paintings, plus the crown jewels, were auctioned off, except for items deemed 'offensively catholic' which were burnt or thrown into the river.

The wars cost the lives of 34,000 Parliamentarians and 50,000 Royalists, plus perhaps another 100,000 who died from wounds, disease, and famine. Cromwell was seen as the only man who could impose order on the many groups of extreme and vocal republicans, and he was appointed leader, with the magnificent title of 'His Highness by the Grace of God and Republic, Lord Protector of England, Scotland and Ireland' (he was offered the crown, but turned it down). He was initially a moderate, who constantly strove to bring the rival factions together, but ironically, he was forced to use the army to maintain control, which did not endear him to the public. Nevertheless he established the British Republic as a force to be reckoned with, and set up an efficient civil service (of which Samuel

Pepys was Secretary to the Admiralty), but within 20 months of Cromwell's death the republic collapsed, and the monarchy was restored. Cromwell was a despised figure for many years (Thomas Gray's *Elegy* contains the line 'Some Cromwell, guiltless of his country's blood'. But his reputation has gradually recovered, and his statue now stands outside the Houses of Parliament (rather incongruously facing a small head of King Charles I on the building opposite).

Mark was an avid chess player and he played for the Yorkshire junior team. He tried to teach me the rudiments and some of the tactics - the English opening, the Queen's gambit, the Sicilian Defence and the famous principle of *reculer pour mieux sauter*, but I found it impossible to think three or four moves ahead.

He had a Protean talent, which included an encyclopaedic knowledge of British criminal trials. He was a connoisseur of crime. We had fascinating discussions on Oscar Slater, a petty criminal who had come to Britain to escape military service in Germany. He was arrested for the murder of a Miss Gilchrist, a rich widow, at her house in Glasgow. A brooch was missing and Slater pawned a brooch the following day and left for America shortly afterwards. He was arrested and brought back, identified at an identity parade by Miss Gilchrist's maid and by a passer-by in the street. He was tried and found guilty, and spent nineteen years in prison. In fact, he had nothing to do with the murder. The brooch that he pawned was not Miss Gilchrist's, his trip to America had been planned long before, the police rigged the identity parade, and the passer-by revealed years later that she had not even seen Slater, but had identified him in the hope of receiving part of the reward money. Slater was eventually released and paid £6,000 compensation, less than £1 for each day that he spent in prison.

Then there was the Balham mystery, a case worthy of Agatha Christie. Florence Campbell had a colourful past. She came from a rich family and at 19 had married Alexander Ricardo, a Guards officer who turned out to be an abusive and drunken cad. She abandoned him, and had a brief affair with a doctor old enough to be her grandfather. She bought a large villa overlooking Clapham Common in London and hired a companion called Mrs Cox, who introduced her to an old acquaintance, Charles Bravo, a lawyer with a practice in London. He too had a shady past involving a mistress and a small child. Both agreed to break-off their former relationships and never mention them again, and with this understanding (and the fact that her former husband was now dead) they were married. Charles had considerable debts and he assumed that Florence would make over her fortune to him, but she refused. He believed it was his right to be master of the house, and became more and more authoritarian. He accused Florence of being a spendthrift and told her that she must get rid of her companion Mrs Cox, the gardener and her horses. Once again Florence refused. Twice she became pregnant and twice miscarried, and Charles decided eventually Mrs Cox had to go, she was too much of an influence on Florence. Not long afterwards Charles was violently sick on his way to work. The following month he was greatly shaken when his horse bolted, and later the same day he received an angry letter from his father accusing him of gambling foolishly on the stock market. That evening Charles and Florence had another row and she went to sleep in a spare room with Mrs Cox for protection. During the night Charles was heard shouting for water and was obviously in great distress. Doctors were called and he said he had taken laudanum for a tooth-ache but nothing else. His condition deteriorated and on the next day he made a will leaving everything to Florence. Mrs Cox took one of the doctors aside and said that Charles had told her in private 'I have taken some of that poison, don't tell Florence'. Despite the efforts of six doctors he died the following day. A post-mortem examination revealed that he had died from antimony poisoning with five times the lethal dose found in his stomach. An inquest was held and the coroner returned a verdict of suicide. However, it turned out that a groom who had been dismissed by Charles had bought antimony to treat horses, and he had made the ominous prediction that Charles would not live four months. Furthermore, Florence's former lover had supplied medicine to Florence, supposedly for insomnia, but which carried a poison label. Another inquest was held, which lasted for 32 days and became a sensation. The jury returned the verdict that Charles did not commit suicide, but was murdered by person or persons unknown. The case was never brought to trial and Florence and Mrs Cox were allowed to walk free. So, who did kill Charles Bravo - his wife?, Mrs Cox?, the doctor?, the groom? Did he drink the poison by mistake, or did he commit suicide after all? He didn't say, and now we shall never know, but Mark and I loved to speculate on the possibilities. To paraphrase Oliver Goldsmith:

> *When lovely woman stoops to folly*
> *And finds her husband in her way*
> *What charm can sooth her melancholy*
> *What art can turn him into clay?*

But for me there was one case that outshone all the others, the case of Madeleine Smith. To me she was *the* woman. Madeleine was the daughter of a much-respected Glasgow architect. She was attractive, vivacious and passionate. She had been sent to a school for young ladies in London and had returned to Glasgow at the age of eighteen. Not long after her return she was seen by a local warehouse clerk, Emile l'Angelier from Jersey who was immediately smitten by her. He contrived an introduction and Madeleine was flattered by his attention. He deliberately walked where he might meet her, and soon enough a clandestine romance developed. They met in secret and had the help of a go-between, but her father found out and put a stop to it. But Emile was determined. They continued to meet illicitly and exchanged letters arranging meetings and trysts. Her letters were very frank and explicit and in time she was addressing him as her husband. She would habitually end her letters with 'a kiss, a fond embrace'. They became intimate in the summer of 1856 and began to plan a wedding, and if necessary an elopement. But then William Minnoch, a wealthy merchant appeared on the scene and became a neighbour of the Smiths' at No. 7 Blythswood Square in Glasgow, and he began to take an interest in Madeleine. It slowly dawned on Madeleine that a future with Emile would be bleak. Her father would cut her off and she would be forced to live on the income of a junior clerk. Minnoch pressed his attention and her interest in Emile began to wane. She went to the theatre and to balls with Minnoch and after much procrastination she told L'Angelier that she was ending their relationship and asked him to return her letters and 'likeness'. Emile believed that, after what had passed, they were as good as man and wife, and he threatened to send all her letters to her father to prove it.

Madeleine pleaded with him not to disgrace her, and she begged to meet him before he did anything irrevocable. About this time Madeleine purchased several packets of arsenic on different dates from different chemists which she said she wanted to kill rats at their country house. Madeleine's bedroom was in the semi-basement of the house, with a barred window at ground level facing onto the street. She arranged for Emile to come to the window which he did on more than one occasion, and she pretended to acquiesce to his demands. She was in the habit of giving him a cup of cocoa through the bars 'to keep out the cold'. After one of these visits he became violently ill and needed the help of his landlady to reach his bed. He had several similar attacks and finally after spending a few days' vacation at Bridge of Allan he returned to Glasgow and arrived at his lodgings late at night in a pitiful state. He took to his bed and died in agony the following day. An autopsy revealed large amounts of arsenic in his organs. When L'Angelier's room was searched a package of highly embarrassing and compromising letters was found, all signed 'Mimi'. Mimi was quickly identified as Madeleine and she was arrested and sent for trial in Edinburgh. Madeleine's family retained the best lawyer that money could procure, and an epic courtroom battle ensued. Madeleine's demeanour impressed the jury. She was attractive, wore a different dress each day, and appeared modest and confident that she would be acquitted. At that time defendants were not permitted to give evidence, so the prosecuting counsel was not able to cross-examine her. The defence relied on the key point that there was no proof that L'Angelier had visited Madeleine after his return from Bridge of Allan, and after much deliberation the jury returned the uniquely Scottish verdict of Not Proven. Her defending council left court without congratulating her and when asked later, he said he would be happy to dance with Madeleine but would be wary of taking tea with her. She admitted to a reporter many years later when she was living in America that she had poisoned L'Angelier, and said that in similar circumstances she would do the same again. The letters that so scandalised Scottish society in 1857 are still preserved in the legal archives in Edinburgh, and when examined a few years ago by a crime writer, they gave off the smell of musk from the perfume with which Madeleine had doused them all those years before.

I was enthalled by these accounts and developed a fascination for dramatic court cases. Mark's knowledge was not limited to British cases but included American ones too, like Lizzie Borden.

> *Lizzie Borden took an axe*
> *And gave her mother forty whacks.*
> *When she saw what she had done*
> *She gave her father forty-one.*

He also introduced me to the now largely forgotten novels of Ernest Raymond, which were popular at the time, with novels like *We, the Accused* and *Gentle Greaves*, and to the plays of George Bernard Shaw, the Irish playwright, who delighted in parodying the double standards of society with plays such as *Major Barbara* and *Man and Superman*, and to the delicious aphorisms and *bon mots* of Oscar Wilde ('I can resist anything but temptation', 'Work is the curse of the drinking classes', and on fox hunting 'the unspeakable in pursuit of the uneatable'). It is interesting to note that after the scandal of Wilde's conviction for homosexuality in 1895 the name Oscar almost disappeared as a given name in England (as did Adolf in Germany after 1945). But Oscar had the last laugh: his tomb in Père Lachaise cemetery in Paris is covered with the imprint of hundreds of kisses in red lipstick left by his admirers.

Mark had lighter interests too, Gilbert and Sullivan with their very clever lyrics like the Major-General's song in *The Pirates of Penzance*, and *A Policeman's Lot* from *HMS Pinafore*, and Noel Coward with *Mad Dogs and Englishmen*, *The Stately Homes of England*, *Mrs Worthington* and his impish play *Blithe Spirit*. These were subjects they didn't teach at school, and I owe it to Mark for greatly expanding my horizons. I like Churchill's comment that 'My education was interrupted only by my schooling'. From Mark I learnt to be careful of accepting things at face value, and that even the most mundane situation is likely to conceal something of interest buried beneath. Mark, not surprisingly, won an open scholarship to Oxford, and got his name on the honours board. We arranged to meet a year or two later when I was in London, but he failed to show up, and I never heard from him again.

Letter 11
I cannot pinpoint precisely when my interest in geology began. It may have been on a trip we made to Whitby when I was thirteen or fourteen, or it may have been when I discovered some beautifully coloured stones used as gravel on an unmade road not far from our house (which turned out to be basic slag from a blast furnace), but certainly by the time I was fifteen I was hooked. I found a box of unlabelled rocks and minerals at school, which I sent off to the Geological Museum in London for identification, and later arranged them as a little display in the geography classroom. I bought myself *Geology for Beginners*, a rather antiquated text book, but I read it with a sense of discovery. I still have it, with its schoolboy annotations and underlining. I had found a subject which really turned me on. I also acquired *Principles of Physical Geology*, a more modern textbook, which contained photographs of the Grand Canyon, the Giant's Causeway, Fingal's Cave, the Great Barrier reef and many others which fired me with the desire to see these places.

> *There rolls the deep where grew the tree*
> *O earth, what changes hast thou seen!*
> *There where the long street roars hath been*
> *The stillness of the central sea.*

*The hills are shadows, and they flow
From form to form, and nothing stands;
They melt like mist, the solid lands,
Like clouds they shape themselves and go.*

Tennyson should have been a geologist. Mention of Tennyson puts me in mind of a story told by our English teacher at school about a boy who went to Oxford to sit an entrance exam in English and was asked to write an essay on the theme of 'the horns of Elfland faintly blowing'. I would have struggled a bit with that.

But then I bought *An Introduction to Palaeontology* by A. Morley Davies. Good old Morley Davies, what a debt I owe to him. I was familiar with dinosaurs and ammonites, belemnites and fossil shells, but this book got down to the nitty gritty. Every page contained a new technical term. I began to understand that palaeontology was a science with a language of its own, precise, specific and unambiguous, so that a description of a fossil in England could be understood in America or New Zealand or Russia. Another well-known text-book called *An Introduction to Stratigraphy* by Sir Dudley Stamp (known to generations of students as The Deadly Stamp), was written largely from memory while sailing between Britain and Burma or 'while waiting for one's bullock cart to arrive'. He died in Mexico claiming that he had finally visited every country in the world.

Mr Haigh, our geography teacher, gave me a book on *Geology and Scenery around Leeds and Bradford*, and I acquired another on the country around Halifax. With these I was able to get out into the countryside and see rocks with a new understanding. I went to the local park where there was a sandstone cliff full of pebbles which I discovered was Millstone Grit, the remains of a delta which once covered the whole of northern England, and to Shibden, where I had played as a child, where coal seams can be found in the streams. I was able to tell my grandparents that if they were to excavate beneath their house they would encounter the lowest four coal beds of the Coal Measures before reaching the Millstone Grit. The Millstone Grit contains fossils called goniatites which had been shown to evolve very rapidly and could therefore be used to determine the relative age of the various beds from one locality to another. Squashed impressions of these fossils could be found in the shales, but in the limestone beds they were preserved uncrushed. The problem was they were very difficult to extract. I discovered that if you placed a piece of the limestone containing the goniatite in a fire and heated it, it would crack, and if you were lucky the fossil would pop out. I knew that limestone would dissolve in hydrochloric acid and would leave behind any non-calcareous material. I tried this at school in the chemistry lab, but was reprimanded in no uncertain terms by a chemistry teacher for conducting unauthorised experiments, and he told me that I would never make a geologist as I didn't have the necessary background. So much for motivation. I resolved to prove him wrong. Darwin had it right when he said 'Geology is a capital science to begin, as it requires nothing but a little reading, thinking and hammering'. Geology used to be a pursuit of leisured gentlemen and the nobility (the first two presidents of the Yorkshire Geological Society were Earl Fitzwilliam and the Marquis of Ripon), whereas 'the father of English geology', William Smith, a mere land surveyor, was for years ostracised by the Geological Society of London, and was never elected a Fellow.

I found the naming of fossils intriguing. There are common names like the Dudley bug or the Devil's toe-nail, but fossils also have formal names like living plants and animals, using the Linnaean system, after Carl Linnaeus, a Swedish naturalist. This is called a binomial system because it consists of two names, the genus (a grouping of related species) followed by the species (members of which can interbreed). The concept of species in fossils is necessarily indistinct because we do not know for sure whether they could interbreed or not. It becomes a subjective judgement. So you have 'lumpers', who take a broad view of what constitutes a species and 'splitters', who use quite minor differences to separate one species from another. I was a definitely a lumper. Names, in Latin (I should have paid more attention to the Latin lessons), are assigned by the person who first described the fossil (or living organism) and can be either descriptive or named in honour of someone or something, so you have *Tyrannosaurus* (terrible lizard) *rex* (the king), or *Brontosaurus* (thunder lizard) *excelsus* (noble). This is the art of systematics which can become very arcane, with holotypes and lectotypes, paratypes and syntypes. The species may be subsequently reassigned to a different genus, or it may be found to be identical to a fossil already described under a different name. And then there are clades….it beats Sudoku or crossword puzzles any time.

Then I discovered geological maps, 'sheets of many colours'. These could be bought from the Geological Survey in London. In those days they sold one-inch maps, where one inch equals one mile, but much larger scale maps were available for areas of economic interest, which included the coal bearing district around Halifax. With these I was able to form a much clearer idea of how the surface rocks could be traced underground, using evidence from boreholes and mines, and I began to think in 3D which is the most basic requirement for a geologist. With the aid of these maps I was able to find quarries and stream sections and relate the outcrops to the map so that I knew what I was looking at (or for). I could also pin-point the location of long abandoned mines and bell-pits. A bell pit was a very primitive and dangerous method of mining coal. A pit would be opened up and the coal extracted until it became too dangerous to proceed further. The pit would then be abandoned and a new pit would be dug further along the outcrop.

Another fascinating aspect of geology is the relationship between the rocks and the landscape. We often used to go to the Yorkshire Dales which is limestone country, twenty or thirty miles north of Halifax. Here there are rivers which disappear into the ground, whereas a lake nearby does not. There is a great amphitheatre of rock which was obviously a spectacular waterfall at one time, but is now dry. There are collapsed caverns, and cave systems containing stalactites and stalagmites, spoil tips with fragments of lead ore and quartz crystals from some long-forgotten lead mine, and fossil corals and shells which once lived in warm, tropical seas. Many of the valleys are 'U' shaped, which is an indication that they were excavated by glaciers during the Ice Age, and near Ingleton there are large boulders of a quite different rock type which were transported from far away on the backs of glaciers, just as you see in the Alps today. In the same area, there is a road which has fine limestone cliffs on one side and low boggy ground without a trace of limestone on the other. What was going on? It was like a detective story. How could all this be explained? I decided this was the subject for me. My granny's attitude was 'stones? What kind of a job can you get wi' stones?', but I think my granddad was rather pleased.

And so I persuaded them to let me take the 'O' level geology exam (I was the first pupil in the school to do this), which I passed with flying colours. Unfortunately, I could not continue with it to 'A' level as there was no teacher available to teach it, so I had to make do with Geography.

Letter 12
At the end of my school career I received the offer of a place to study Geography at Durham University. Michael got a place at Trinity College

in Dublin, and Bryan a place at the London School of Economics. And so our paths diverged.

Durham is the third oldest university in England, after Oxford and Cambridge, and is modelled on them. It is a collegiate university with about eight or nine colleges of which two were for women (Mary's Fairies and Aidan's Maidens). I had secured a place at a college called St Cuthbert's Society. (St Cuthbert was a Northumbrian saint who was Prior of Lindisfarne Abbey in the seventh century and is buried in Durham Cathedral). Durham is a fascinating city with a long and impressive history. It is located on a ridge within an incised meander of the River Wear, much like Toledo in Spain. A castle guards the neck of the ridge and a magnificent cathedral dominates the high ground. Palace Green separates the two, and narrow cobbled streets wind around them. St Cuthbert's was located in one of these streets called South Bailey, and was little more than a large house with a garden behind. On one occasion a lecturer was reeling along the Bailey towards Cuthbert's when the Warden of a women's college called out 'Drunk again, Mr Holmes' to which he replied 'And so am I.' Cuthbert's had very limited in-house accommodation for undergraduates, and most of us, myself included, had to find 'digs' in or around the town.

The cathedral is a massive Norman construction, 'half house of God, half castle 'gainst the Scots', which aptly sums up Durham's role. Durham is a Palatine county which means that it was ruled by a Prince Bishop who was granted special authority and autonomy by the crown. The principal purpose was to provide defence against invasion by the Scots, and the Prince Bishops held enormous power, being able to hold their own parliaments, raise their own armies, appoint their own judges, administer their own laws, levy taxes and issue their own coinage. One Prince Bishop went so far as to say 'there are two kings in England, one in London wearing a crown, and one in Durham wearing a mitre'. This privileged position was drastically curtailed by Henry VIII, and during the Civil War, and came to an end with the union of England and Scotland. Durham Castle was formerly the residence of the Prince Bishops. When I was a student it was home to University College, but also the temporary residence for the Judge of Assize who would visit Durham twice a year to conduct the assizes at which the most serious criminal cases were heard. He would process in his scarlet robes and judge's wig with great ceremony accompanied by the barristers and court officials from the castle to his waiting car which would take him to the court at Durham Jail.

The jail had gallows permanently in place and hangings were carried out there until December 1957. I attended one or two trials, but I never saw the judge put on his black cap and intone the famous litany ending 'to be conveyed to a place of execution and there be hanged by the neck until you are dead, and may God have mercy on your soul'.

The bishopric of Durham was founded in 995 and the present cathedral dates from 1093. It is one of the finest Norman cathedrals in Europe and has been designated a World Heritage site. It contains the relics of St Cuthbert, St Oswald and the Venerable Bede (known to schoolboys as the Venomous Bead). It also possesses three copies of Magna Carta, which spelt out in unequivocal terms the ancient principle that the king's power (or, for that matter, that of the Prince Bishop) is not absolute. From 1093 to 1536 the cathedral formed part of a Benedictine priory, of which a large part has survived. St Cuthbert's shrine was one of the most sumptuous in England, but was destroyed on the orders of Henry VIII and is now marked by a simple black slab. During the Civil War the cathedral was used as a prison for 3,000 Scottish soldiers captured at the Battle of Dunbar who were kept in inhuman conditions with little food or water and no heat. In order to keep warm, they tore down the wooden memorials and fittings and burnt them. More than half of them died, and the survivors were shipped off as prisoner-of-war labourers to North America.

The situation and the architecture are glorious, and a leading architectural historian has compared the assemblage of castle, cathedral and monastery to those of Avignon and Prague. The thing that impresses most is the massiveness and solidity of the structure. None of the delicate tracery of later architectural styles is evident here. The tower is square and forbidding, more like a castle than a cathedral. I remember climbing the tower and being surprised to see the prisoners exercising in the prison yard at Durham Jail. Inside, the columns are massive and decorated with deeply cut zig-zags and reticulations which were cut in the quarry prior to assembly, and the arches between the columns are rounded Romanesque arches with characteristic zig-zag ornamentation. The impression is one of power and strength. As building progressed it was discovered that greater height could be obtained by the use of rib vaulting in which the arches are pointed. Rib vaulting was used for the roof of the nave, and Durham is the earliest example of its use in England. It had its problems, with some of the early arches collapsing, but they were rebuilt and strengthened, and have stood the test of time ever since. The north door has a sanctuary knocker, which allowed fugitives to rap on the door and claim sanctuary.

There were monks permanently on duty ready to open the door for them. They were granted 37 days either to prove their innocence or arrange their escape, a system which persisted from 890 to 1624.

But there is another side to Durham. It has a proud industrial past. Durham lies on a coalfield and as the Industrial Revolution took hold at the end of the eighteenth-century dozens of coal mines were developed in the area surrounding the city. Even on the surface coal seams can be seen in the banks of the incised meander of the River Wear at Durham. The coalfield extends from Durham into Northumberland and coal has been mined there since the thirteenth century. The location of the coalfield was uniquely favourable since it extended to the coast, allowing locally-mined coal to be easily exported in small ships called colliers, along the coast to London to supply the insatiable demand of the capital for 'sea coal'. (The only alternative fuel in London was charcoal or wood). On arrival in the Thames the coal was transferred to barges and transported up the Fleet River, where to this day there is an Old Seacoal Lane and a Newcastle Close. The expression 'carrying coals to Newcastle', meaning a pointless exercise, probably dates from this time. The coal was exported from the Tyne at Newcastle and the Wear at Sunderland, but both these rivers were shallow and inaccessible to the sea-going colliers. A system developed where small, shallow-draught vessels called keel boats transported the coal from staithes on the river bank to the colliers waiting at the river mouth. This was a major industry with over a thousand keel boats in operation by 1800, each with a four-man crew. The keel men were a tough bunch immortalised in the folk song *The Keel Row* which refers to the Sandgate area of Newcastle where they formed a closely-knit community, and to the blue bonnet which formed part of their traditional dress along with yellow trousers (not very practical, you might think). The keel-boat system continued until the rivers were deepened in the nineteenth century and the Tyne opened up above Newcastle by the construction of a swing bridge to replace the earlier stone bridge. As new coal pits were opened up further from the river waggonways were built, with wooden or iron rails, on which coal waggons were pulled by horses from the coal pits and the riverside staithes. This, quite literally, marked the birth of the railway age. There is a nice legend about a French ship wrecked on the Durham coast at Hartlepool during the Napoleonic wars of which the only survivor was a monkey. Never having seen a Frenchman or a monkey, the locals took the monkey to be a French spy, held a drumhead trial, and since the monkey would not speak, they found it guilty and hanged it.

Hartlepudlians are often referred to as 'monkey-hangers', and they even erected a commemorative statue of the monkey on the promenade.

Coal mining in the area expanded rapidly, and children or ponies were regularly used to pull the coal tubs from the coal face to the shaft. The Marquess of Londonderry, one of the main coal owners, said that a boy of twelve 'should be learning his trade not wasting his time with reading and writing'. At its peak between 1870 and 1914 the Durham-Northumberland coalfield employed a quarter of a million men working four hundred and fifty collieries, with an annual output of 50 million tons. It was by far the principal industry of the area, but it was a dangerous occupation, with over 150 deaths every single year in the period 1860 to 1914. Some of the mines extended far out beneath the North Sea, by as much as eight miles in the case of Easington Colliery. The worst disaster was at Hartley Pit near Blyth in 1862 when 204 miners died when the massive cast iron beam of a pumping engine broke and fell into the only shaft of the mine, cutting off the ventilation supply to the miners below. Even as late as 1951 83 miners were killed at Easington Colliery when a cutting machine struck a spark off a nodule of pyrite which caused a massive gas explosion. From 1911 canaries were used to detect the presence of gas, since they are particularly sensitive to toxic gases. The last ones were not officially 'retired' in 1986. A typical colliery would work five or six coal seams at different depths, and by law, following the Hartley pit disaster, had to have at least two shafts. The seams all carried evocative names like the Hutton, Busty, Plessey, Brockwell and Marshall Green, names perpetuated in a well-known folk song. The coal industry was nationalized in 1947 and thrived for another twenty years, but coal was regarded as dirty, polluting and old-fashioned and became uncompetitive against cheap foreign imports. Despite large remaining reserves the industry declined rapidly from about 1970, and in the Durham-Northumberland coalfield mining ended completely in 1994 when the last colliery closed at Wearmouth near Sunderland. Sunderland's football ground, the Stadium of Light, marks the spot.

Along with coal mining came heavy industry, particularly iron (and later steel) making at Consett and Middlesbrough, and shipbuilding at Newcastle and Sunderland. The first public railway in the world between Stockton and Darlington was opened in 1825, using a combination of steam locomotives, stationary engines and horses. Locomotive design evolved rapidly, and at a trial held in 1829 Stephenson's *Rocket*, manufactured in Newcastle, outclassed all the others and established a design which set the standard for the next fifty years. Stephenson's gauge of four-feet eight-and-a-half inches, adopted from the horse-drawn Durham waggonways, became the standard gauge for railways in Britain, Europe, North America and many other countries around the world. Durham lay at the centre of all these developments and when a railway was eventually built from London to Scotland, the route included a station at Durham.

Letter 13
My introduction to university life was daunting. Many of the students came from a public school background and from more affluent families than mine. They had a confidence and ease which I lacked, and it took a while to find my feet. We took our meals at college where the wearing of gowns was obligatory. Gowns were worn at lectures too; the only exception being in the science laboratories. The Master of Cuthbert's was a clergyman, and quite a number of the undergraduates were theology students. The college had a Junior Common Room used for recreation, which contained a television on which it was considered hip to watch a kids' show called *The Magic Roundabout*, and a gramophone with a pile of mostly classical and jazz records. This was my introduction to classical music. When I came home on vacation the first thing I did was to buy recordings of Beethoven's *Pastoral Symphony* and Dvorak's *New World Symphony*. This marked the start of a love affair with classical music. There was no pop music then, and by the time it appeared I was into other things.

I found lodgings with a widow called Mrs Hewitt in High Shincliffe, a village on the road to Stockton, which I shared with another student from the same college. Occasionally we would eat there, and I caused a scandal one day by eating my landlady's dinner, in the mistaken belief that she had left it for me.

The Geography Department was located in a modern building on the Science campus, and most of the lectures and practicals were held there. The department was presided over by Professor Fisher, a Human Geographer, who had obtained his doctorate from the Sorbonne for a study of the effect of hygiene on population distribution, or something along those lines. Geography, I perceived, was a wide-ranging subject. We had some interesting projects, for example determining the extent of Viking and Anglo-Saxon influence from a study of place names, or studying the crops grown on farms around Durham in relation to the soil type, which we measured with a pH meter. We even did some rudimentary plane-table surveying. There were field trips to the dales west of Durham to see examples of villages which had grown by ribbon development along old established highways, or nucleated settlements centred on a church, a village green or a colliery. The dull, grey villages often had optimistic names like Sunnyside or Mount Pleasant.

My closest friend at this time was a fellow geographer called John Parr, who in his spare time was a caver. He had some hair-raising tales about crawling through caves where the roof was only two feet high and emerging into caverns containing stalactites and bones left by animals, who knows, maybe hundreds of years before. Geomorphology was a fashionable science at the time, and we had long inconclusive discussions on the merits of the Davis versus Planck theories of slope development. I can hardly imagine that such esoteric topics are still taught. We went on a field trip together to Malham, an area I knew well, where the instructor was a geomorphologist, and despite torrential rain and being soaked to the skin, I learnt a great deal about karst, limestone pavements, clints, grykes, karren, dolines and poljes, which was all new to me. We speculated at length on the origin of the splendid amphitheatre of Malham Cove and the gorge at Gordale, and came to the conclusion that they were meltwater features from the last glaciation (about 12,000 years ago), and whilst it was obvious that the grykes had been formed by solution along joints and cracks in the limestone, we came to the view (not widely accepted at the time) that they formed beneath a soil cover, which served to accelerate the solution process by being permanently damp, and containing abundant calcophylic lichen. Interestingly, in the very wet winter of 2015 for a period of just two or three days, a waterfall developed at Malham Cove for the first time in living memory with a free fall of 230 feet (60 feet higher than Niagara) to the pool below.

As subsidiary subjects I chose Geology and Zoology. Much of the basic geology I was already familiar with, but I was able to get a feel of the Department, and I liked what I saw. Zoology was a subject new to me, and I found it fascinating (with weird names like the Crypts of Lieberkuhn and the Islets of Langerhans). Each week we would do a dissection: the optic nerves of a dogfish, the urinary-genital system of a rat, the circulatory system of a frog (or was it the other way round?). It was a revelation, and it led to me dissecting my landlady's goldfish which had died, when I made the discovery that it had swim bladder by which it regulated its buoyancy.

In the winter of that year Michael and I joined a group from Cambridge on a ski trip to Austria. I had never tried skiing before, and it was a novel experience. I don't think I learnt much, since, when I tried it again several years later I was pretty inept. During the Easter vacation I went to visit Michael in Dublin. Trinity College was established by the Tudor monarchy in the sixteenth century on the Oxford and Cambridge model, and specifically as a Protestant university. Catholics and Nonconformists were excluded until comparatively recent times, and for many years the Catholic Church forbade its members to enrol. It was well endowed, splendidly built and in a prime location, but to most Catholic Irishmen it was an unwelcome symbol of English domination. I spent a weekend there with Michael and his two room-mates. They introduced me to the songs of Tom Lehrer, a mathematician with a Harvard degree, and later a lecturer at MIT, who became an entertainer, writing and singing satirical and cynical songs characterised by black humour, such as *The Irish Ballad, The Old Dope Peddler, Poisoning Pigeons in the Park, We'll all go together when we go* and *Lobachevsky*. The lyrics of these songs are so clever and original that they gained cult status, and are still widely known today. These were the pop songs of my generation. When, much later in life, I wrote a book, I frivolously considered dedicating it to Lobachevsky. If you want to know why, you will have to listen to the song.

I can't resist retelling a tale which for me typifies Ireland and the Irish. A well-known Irish composer and entertainer called Percy French arranged to give a concert in Kilkee on the west coast of Ireland. He travelled with his *ensemble* from Dublin, which involved changing onto a branch line run by the West Clare Railway, which was noted for its inefficiency. The train stopped for hours because of a problem taking on water for the locomotive, and he arrived very late at the concert venue to find that everyone had gone home. He sued the railway company for loss of earnings and was awarded ten shillings. He wrote a humorous song about his experience called *Are you right there Michael* upon which the railway company sued him for libel. He arrived late for the court hearing and when the judge asked for an explanation, he replied 'I came on the West Clare Railway'. Without further ado, the judge dismissed the case.

I returned home for the summer vacation to find that our dog Major, my boyhood companion, was terminally ill. He had lost a lot of weight and ate almost nothing. I found him one day in the long grass by the scout hut and I had to carry him home. I took him to the vet, who said there was nothing he could do, he was dying. So the vet gave him a lethal injection, and I came away alone and desolate.

At the end of that year I managed to arrange a transfer to the Geology Department. I was accepted with no problem and so began a new chapter, which was to shape the rest of my life.

> *Two roads diverged in a wood, and I,*
> *I took the one less travelled by,*
> *And that has made all the difference.*

Letter 14

My digs with Mrs Hewitt at Shincliffe were not available in my second year, so I found new accommodation with a Mrs Lynn in Bowburn, another mining village along the Stockton road. She lived on an estate that had been built for the miners at the local colliery with her husband (who she always called Lynn) and her brother Bob. Both were retired miners from the pit at the end of the street. But she had a large spare room, the place was clean and she was very welcoming. The only problem was that they spoke with a thick Durham accent which it took me a while to master. Instead of saying yes, they would say, 'why, aye, man'. So I moved in, and spent the rest of my undergraduate years there. They treated me as one of the family.

I now had to find my feet in a new department. The Geology Department was much smaller than Geography, but it had a fine reputation. The professor was Kingsley Dunham, who had spent his formative years in the USA where he had worked on 'Mississippi type' ore deposits in which different mineral assemblages are found in 'halos' around the granite from which they originated. Similar ore deposits are present in Britain, and in Cornwall where granite outcrops at the surface, tin and copper are found, whereas in the northern Pennines, where no granite had yet been found, only lead and zinc are present. The implication was that at depth, adjacent to the assumed granite, tin and copper ores might be found. His predecessor was equally distinguished. Arthur Holmes (who had written the textbook which had so impressed me at school) had, more or less single-handedly, developed a method of radioactive age dating, based on the decay of radioactive minerals, which revolutionized the dating of rocks. It made it possible to determine with some precision the age of previously enigmatic rocks which had frustrated geologists for decades. It was one of the most significant geological breakthroughs of the twentieth century.

The teaching staff was small, but distinguished. There was Hoppy, Dr Hopkins, old now and close to retirement, who had established a method for the precise correlation of coals within the Coal Measures (on which Durham was located) from one area to another, based on fossil 'mussels'. Dr Johnson was an enthusiastic ('oh, jolly good show') and highly motivating stratigrapher of the Carboniferous Limestone. Dr House, who had been recruited even before he finished his PhD, was a renowned palaeontologist. Martin Bott was a young and enthusiastic geophysicist keen to try out new techniques and technologies, and Henry Emeleus was an expert on igneous petrology. Mr Phillips was a no-nonsense mineralogist who always gave the impression of being in a hurry to get back to his research on reflected light microscopy, rather than teach.

I enjoyed the courses and the practicals, apart from Optical Mineralogy and crystallography, and I was not greatly turned on by igneous petrology, (but I am not alone in having struggled with pinacoid faces, eutectic diagrams and stereo nets). We had some wonderful field trips – to a working face in a coal mine at Ferryhill, to see the Whin Sill on Hadrian's Wall, and the cannon-ball limestone at Marsden Bay. In stream beds and on the beach nearby you can find pebbles and cobbles of exotic igneous rocks like larvikite and rhomb porphyry which do not occur in England, but were transported by glaciers from the Oslo region of Norway during the last Ice Age. We had longer trips, one of which was to Devon to see the classic outcrops at Hope's Nose, Kent's Cavern, Meadfoot Bay and Bonhay Road in Exeter, and we visited the serpentine outcrops on the Lizard and the ruined engine houses of the derelict tin and copper mines. Dr House, who specialized in the Devonian System, was in his element. But by far the most spectacular trip was to Inchnadamph in Scotland, a mecca for geology students, where some of the oldest rocks in Britain are exposed. Not only that, but the Moine Thrust, the Torridonian, the Dalradian, the schists and gneisses, and the volcanic tuffs provided a wonderful insight into the geological evolution of Scotland. One of the guys later had a tee shirt printed with the logo 'Tuff schist'.

During my first year in the Geology Department my granddad died from pneumonia following a fall and a broken hip, and only a month later my granny died too. They were in their seventies so it was not totally unexpected, but to see the coffins being carried out of the house where I had spent so much of my youth was heart wrenching. My mother and aunt Eva had the sad task of disposing of their effects, which didn't amount to much, and most of them were given to the local chapel or to charity. We heard later that their elegant Victorian chiming clock had appeared on the sideboard of the chapel caretaker. My mother and I got my granddad's glazed mahogany bookcase which I put in my bedroom and crassly removed the doors and painted it white. I still had much to learn.

Shortly afterwards I decided to take my bike to Durham, and instead of putting it on the train I opted to ride it. It was ninety miles and I had never ridden anything approaching that distance before. The first thirty or forty miles were no problem, but then the next twenty was along the A1 into a strong headwind, and I arrived at my digs after a ride of four or five hours in a state of acute discomfort.

I still kept contact with Michael and in the summer vacation he invited me to accompany his family on a car trip to Bavaria. We took the car ferry to Hook of Holland and drove, mostly on the German autobahns, to Munich. I was very impressed by the autobahns, sweeping across the country and by-passing the towns and cities. The Nazi regime had left one useful legacy at least. We spent two weeks touring around King Ludwig's fantasy castles at Neuschwanstein, Herrenchiemsee and Linderhof, and Lake Starnberg in which Ludwig met a mysterious end. Ludwig was obsessed with German legends and became an avid patron of Wagner, who rather exploited poor old Ludwig. His castles, which bankrupted the Bavarian treasury, reflected his taste. Neuschwanstein is like something from a child's fairy tale book. Herrenchiemsee has a hall of mirrors larger than Versailles, and Linderhof has an underground grotto with a swan boat in which Ludwig was rowed whilst being serenaded with the music of Wagner.

Only seven of us advanced to the Honours group, and of those the one with whom I formed the closest bond was Richard Freeman. He was a precise and serious chap with wide ranging interests, in archaeology, churches and music. I rather fancy it was Richard who first introduced me to Beethoven. We went to Hadrian's Wall which was built 1,800 years ago to keep out the Picts (Scots) and is still remarkably intact. There are full scale Roman forts built to house an entire cohort of 800 legionaries, a temple dedicated to the Persian god Mithras, where the altar portrays the god with a perforated halo, which would have been illuminated by a lamp from behind, and the amazing Vindolanda tablets (admittedly not discovered until after our time), which are the oldest hand-written documents in Britain. They were written on thin sheets of wood in cursive Latin with carbon-based ink, and contain shopping lists, dinner invitations and contemptuous comments about *brittunculi*, the nasty little Brits. They owe their preservation to being buried in a bog where they were protected from oxidation.

I learnt to drive when I was in my second year, and bought myself a 1938 Rover 12. It was a sort of artichoke green colour and cost £100. It had leather seats and a device which enabled it to free-wheel when going downhill, which reduced petrol consumption. With that we were able to take longer trips. We visited churches, and discovered two exceptional examples. The church of Jarrow Abbey has a porch and tower dating from 674, which would have been familiar to the Venomous Bead, who was a monk there, and a tiny church of about the same date at Escomb near Bishop Auckland. Stones from Hadrian's Wall had been used in its construction, one of them bearing the imprint of the 6th Legion which built the wall, and which was based there for more than 300 years.

In Northumberland we went to the Cheviots Hills near the Scottish border (shades of my granddad's *Wilson's Tales of the Borders*), and to see the white native cattle at Chillingham, and to the unspoilt, windswept coast of Northumberland with the great sentinel of Bamburgh Castle on its rock overlooking the sea, and the rather splendid Howick Hall, belonging to the Earl Grey family. We checked out the rather sinister Seaton Delaval Hall, designed by Vanburgh, with its central block still a ruin, 200 years after a devastating fire, and Cragside the home of Lord Armstrong, the arms manufacturer (shades of *Major Barbara*), the first house in England to be lit by electricity. Northumberland is truly an under-appreciated county.

But I digress. Our studies continued. We became familiar with most of the common British fossils from the reference collections in the department. We learnt the stratigraphy of the British Isles from the Precambrian to the Recent. We studied sedimentology and petrology, geophysics and geochemistry, and in our second year Professor Dunham taught us the art of geological mapping. Then we were sent off to Hadrian's Wall where each of us was given an area to map. Mine was at Fozy Moss near Housesteads, and I stayed at an inn with the charming name of Twice Brewed. I didn't win the prize for the best map, but I got a *proxime accessit*, which was pretty close.

Both Richard and I had shown a keen interest in palaeontology, and Dr House took us to Redcar on the Yorkshire coast to examine a population of Jurassic fossil oysters called *Gryphaea*, a rather abnormal creature which had two very unequal valves. The idea was to collect examples from different beds to see whether we could detect any evolutionary trends, whether for instance the bottom shell became thicker and more curved in order to lift the gills higher above the muddy sea floor. He gave me a detailed map to bring along which showed all of the reefs on the foreshore in great detail so that we could plot our collecting points very precisely. We drove down there in the Departmental van, and as we arrived Dr House asked for the map. I had left it at home

Professor Dunham was a consultant for the Weardale Lead Company, a mining company which worked several mines in Weardale. The company was a major lead producer before the war, but at this time it mainly produced fluorspar for the steel industry. In our final year Professor Dunham asked Richard and I if we would be interested in doing some underground mapping in Stotsfield Burn Mine at Rookhope in Weardale. We jumped at the chance. Rookhope is a village on a tributary of the River Wear near Stanhope located close to a mineral vein called the Red Vein which had been worked in a number of mines along its length, one of which was at Stotsfield Burn. Rookhope was a place of special significance to W.H. Auden, the poet. His brother was a geologist and Auden visited the area several times. He was fascinated by the technology and by the physical remains, the abandoned mine buildings, the adits, smelting mills, hushes, washing floors, and the enormous water wheel which powered the pumps at Boltsburn Mine. His poems *The Old Lead Mine, Rookhope, New Year Letter, Paid on Both Sides* and *Amor Loci* all reflect his enthusiasm for the area, and they contain numerous reference to mines, localities and features. He also published a tourist itinerary of the northern Pennines, 'for travellers with six days to spare'. Just west of the village is the Rookhope Arch, part of a collapsed flue, over a mile and a half long which conveyed the hot poisonous fumes from the smelt mill at Rookhope to the open moorland. Lead and silver condensed onto the walls of the flue and were periodically scraped off and washed into settling ponds. The metals recovered paid for the cost of the flue within just a few years.

Metal mines are quite different from coal mines. The ore minerals are found in vertical veins which contain a mixture of quartz, calcite, barite and fluorspar, along with small amounts of lead and zinc. The lead occurs in veinlets and pods and in flats, and is easy to spot by its dull silver lustre. Stotsfield Burn mine operated from 1863 to 1882 and then reopened in 1929. Few records were preserved from the earlier working, and new areas had been opened up in more recent times. There were four main working levels, a horizontal adit at the surface, and levels at 15, 25 and 42 fathoms, linked by vertical shafts. The Red Vein mostly composed of quartz and fluorspar varies from six to seventeen feet in thickness, with lead ore frequently occurring in the middle. Where the vein widens and becomes economic, the minerals are extracted by stoping, that is drilling out the vein material above. This can then be carried away in tubs. Stopes can be opened up to 40 or 50 feet in height, and often require timber supports, or backfill to prevent collapse. There was the ever-present danger of roof falls in the stopes, and now and again the workers would break into the old nineteenth century workings (known as 'the old man'), which were often full of water and slurry, and could be a serious hazard. Our job was to map the vein thickness in the three levels and check the cross cuts for evidence of any subsidiary veins. The ultimate objective was to determine whether there was any evidence of improvement of vein thickness which would justify the expense of extending the levels further along the vein. We found accommodation in the village and worked in the mine for two or three weeks, producing some very nice maps and sections, along with a report on our findings, which were eventually deposited in the Durham County Archives.

Whilst we were there work began on the Rookhope Borehole. I mentioned earlier that Professor Dunham had predicted that beneath the orefield a large intrusion of granite would be present, which had given rise to the ore minerals. A gravity survey was conducted by Dr Bott from our department which indicated the likely presence of such a body, and Professor Dunham, who had lots of contacts in industry, managed to raise the funds to drill a borehole to test his theory. Sure enough, at a depth of 1,280 feet granite was encountered and Dunham's prognosis was vindicated – or so it seemed. Further work showed that the granite was a good deal older than the mineral veins, so further work was required. It was subsequently established that the mineral veins were generated by a later magmatic event which utilized the fractured granite as a conduit into the overlying strata. The hope of finding tin and copper in commercial quantities however was not realised (although the lead ore contains an average of 3.5 ounces of silver per ton of lead, and during the boom years over 5 million ounces of silver were recovered).

I was presented with my Bachelor's degree at the University Degree Congregation in the summer of 1961. Our group of seven produced one who would go on to become Director of the British Geological Survey, another who became Exploration Manager of the British National Oil Corporation and a third (me) who became Chief Geologist of Britoil. Science faculty graduates got to wear a rather exotic silk hood in Palatinate purple, edged with ermine fur, and my mother, aunt Eva and cousin Margaret came to the ceremony. Now I had to decide what to do next.

That summer I got a temporary job with the British Geological Survey in Leeds at a new regional office which they had opened whilst mapping in the area. I was assigned to work with two survey palaeontologists, Drs Ramsbottom and Calver, helping to unpack and catalogue some of the hundreds of boxes of reference material sent up from the BGS in London, and I ended up preparing specimens for Dr Ramsbottom to photograph for a paper he was writing on Carboniferous fossils (and received an acknowledgement in his paper).

We then went on a motoring tour around Ireland with Eva and Margaret. We kissed the Blarney Stone and visited Michael and his family who had a caravan at Barleycove, on the south-western tip of Ireland. We drove around the Ring of Kerry with its exotic fuchsia hedgerows, and then around the Dingle Peninsula to Gallarus Oratory, a mysterious structure which could be anything from early Christian to late eighteenth century. Margaret became increasingly disinterested and we assumed she was just being maungy, to use a Yorkshire expression. She wouldn't even get out of the car to walk up to the oratory. We had arranged accommodation in Tralee, where it became clear that Margaret was sick. We sent for a doctor who immediately took her off to hospital. She had a ruptured appendix and they operated immediately. The hospital was run by Catholic nuns who were very kind, but they said it was providential that she had been admitted in time because she would not have survived the night.

Letter 15
On the basis of my recent acquaintance with the British Geological Survey I applied to them for a job, but they were fully staffed at the time. I discussed the possibility of doing a research degree at Durham on goniatites. These fossils had been very successful in establishing a detailed subdivision of the Millstone Grit, and I wanted to extend the study downwards into the Carboniferous Limestone. The problem was

that goniatites are extremely rare in the limestone, and Professor Dunham felt (correctly) that I would not have sufficient material to constitute a viable research project. Dr Johnson came up with the idea of applying to do a Master's degree in micropalaeontology, for which several options were available. Micropalaeontology is the study of microscopic fossils, and they have particular value in being able to date rocks in which larger fossils are absent. I applied to join the course at University College London and was accepted.

My mother and my aunt came up with the ingenious suggestion that my cousin Margaret could move to Halifax and go to school there, and I would go and lodge with Eva and Dick in London. By then Dick had retired from the army and was working as caretaker of a Territorial Army centre in Balham. My mother was glad of the company of Margaret, and had good friends in Halifax, but I sometimes wondered whether she regretted encouraging me to go to university. She must have known that I was unlikely to return and work in Halifax. I suspect she would have preferred me to be a teacher and remain nearby. But that was not to be.

University College London is a famous college with a very high reputation. Its foundation was based on the principles of Jeremy Bentham, a humanist, political radical and social reformer. Oxford and Cambridge, the sole universities in England at that time, only accepted students who were communicants of the Church of England, and were able to pay the rather exhorbitant fees. UCL was established on the basis that it would accept students irrespective of race, religion or political belief (although it took another fifty years to accept women). The college was built in Gower Street, not far from the British Museum, and because of its humanist principles it became known as 'that godless institution in Gower Street'. Bentham was a polymath who designed the panopticon system for prisons, where the corridors and cells could all be observed from one central point, and he spent some time in Russia with his brother, who was chief factotum for Prince Potemkin, the viceroy of the southern lands under Catherine the Great. Bentham specified in his will that his body should not be buried, but should first be dissected and then preserved as an 'auto-icon'. His head and skeleton were preserved. The skeleton was padded with hay and dressed in his own clothes and placed in a glass-fronted cabinet. The attempt to mummify the head was not successful so a wax model was made to which his hair was attached, and his real head placed underneath the chair. The auto-icon is on permanent display in the college cloister and occasionally is wheeled in to meetings of the college governors, where he is listed as 'present but not voting'.

There were only four or five of us enrolled on the micropalaeo course and we were based in a laboratory around the back of the college, probably because of the noxious fumes we produced. The Micropalaeontology Lab formed part of the Geology Department, but was run more or less independently. It had been established by Dr Tom Barnard and over a number of years had gained a considerable reputation. There was a one-year Masters course, and facilities for three or four research students. I felt now to be among equals, and in retrospect that course proved to be a watershed.

> *When I was a child, I spake as a child, understood as a child, and thought as a child: but now that I am become a man, I have put away childish things. For now we see through a glass, darkly, but then face to face.*

I had finally come of age.

The lab had a teaching area surrounded by cubicles for the research students, a preparation room for processing samples, and offices for the secretary and for Dr Barnard. There was always the powerful smell of simmering hydrogen peroxide and other smelly chemicals (like hydrofluoric acid which was so corrosive that it would dissolve glass), which were used to extract microfossils from rock. The course extended over three terms and was conveniently split into three parts, foraminifera taught by Dr Barnard, ostracods by Dr Robinson and palynology by Dr Chaloner. The three of them had very different teaching styles. Dr Barnard was a cynical, rather sardonic man (he once made the comment about someone that 'he would spoil a shilling pair of gloves to pick sixpence out of the gutter'), but a very competent teacher. He had a bias towards the Jurassic and Cretaceous, on which he was an expert, and rather less enthusiasm for the Palaeozoic, but we all have our preferences. Eric Robinson on the other hand was what would now be called a polymath. He had a very broad range of interests ranging from ostracods to the geology of building stones and graveyards and, as I later discovered, to railways. He was open and friendly and treated us as fellow geologists rather than as students. But the one from whom we learnt the most was Dr Chaloner. He was a gifted teacher, precise, rigorous and exacting. He taught us how to write scientifically. We each did a project to identify an unknown sample and wrote up a report on our findings, which he promptly tore to shreds. He had a clever system of punched cards which could be used to identify individual species (such was the technology before computers), and he showed us how to write a professional report, how to compile synonymies, provide precise taxonomic descriptions, quote references and assemble a bibliography. It was an invaluable lesson.

We had some very interesting characters in the lab, and it became a sort of club. We didn't start at nine and finish at five. Often we would stay late into the evening, working, bull-shitting or playing darts. There was a very canny Scottish gal from Edinburgh, one of the PhD students, who was always one step ahead of everyone. Bach's Brandenburg concertos came into the conversation one day, and I naively assumed she would not be familiar with them, but she said 'did you know they were lost after his death and not rediscovered until 100 years later?' One of the guys had a brother who occasionally dropped by, who worked for a pharmaceutical company. His job was to check out various organic compounds for potential medicinal use, including material extracted from extinct or even fossilized animals and plants, which was vividly brought back to mind when *Jurassic Park* appeared some years later (which is based on a not too dissimilar concept). There was another research student noted for his taunting and mocking comments, although not done with any malicious intent. Tragically he was killed a few years later whilst swimming at Galveston in Texas. He was swept out to sea and drowned. Curiously we had another PhD student who died at a young age while doing field work in Morocco. He was asphyxiated when a gas heater malfunctioned in his hotel room. The survival rate amongst PhD students seemed to be rather low.

I formed a special friendship with John Church, a Londoner with a similar background and outlook to mine, and we have remained friends ever since, occasionally meeting up to reminisce about 'the good old days'. I also made friends with Raymond Wright, a very urbane and handsome Jamaican, who certainly knew his way around. He introduced me to some of London's more disreputable clubs, but there is more to education than is taught in school. There was another chap too, very earnest and serious, a singer of madrigals and a counter-tenor, I recall, who laid himself open to all kinds of ribaldry. Then we had a mature student from Iran in his late thirties, Ali Gollestaneh, who liked chocolate, so one day we gave him a sizeable slab of Ex-Lax, a powerful laxative which looks like chocolate.

We didn't see him for the rest of the day. There was a cheap and cheerful refectory at UCL where we would often go for dinner (no gowns or formality there). The sausages and mash were famous, and I recall having the same dish every day for a month – or was it a term?

I discovered that in Mill Hill to the north of London there was a specialist second-hand geological bookshop, and I would occasionally make the trip out there to check the stock. The owner would buy up geological books and maps from various sources, and I found all kinds of treasures, a copy of the first-ever Geological Survey Memoir, published in 1839, Forster's *Strata* (of the area around Rookhope where I had worked in the lead mine) a splendid work of 1809, with all kinds of weird rock names like plate, clunch, trap, hazel, girdles and galliard, which I later discovered he had shamelessly plagiarised from unacknowledged sources. I also found a book called the *Nummulosphere*, written by a guy who worked at the Natural History Museum in London, in which he claimed that *all* rocks, including granite, basalt, gneiss and schist, originally had an organic origin. He came up with the classic comment that 'the Nepunian scale-armour is absolutely impenetrable to the prongs of Plutonian pitchforks'. And that was written in 1913.

Tom Barnard organised a field trip to Normandy during the Easter vacation and three of us went along with him as field assistants, and to learn something about sampling and the geology of the Normandy coast, where the Allied landings had taken place during the war. There are still plenty of wartime relics there, particularly the Mulberry Harbour which was towed across the Channel, and which is so massive and broken up that it is not worth the effort to remove it. This is where my uncle Dick landed, with his tank in the summer of 1944. The German pill-boxes are still there, and it is not hard to imagine the thousands of troops, scores of ships, armour of all kinds, the planes and the noise of that far-off summer. But we were there on a different mission. We collected samples from the Middle and Upper Jurassic rocks on the coast near Bayeux (and went to see the famous tapestry), and had picnic lunches near the beach. Tom's comment on the French baguette was that 'Frenchmen must have jaws of steel to eat this stuff'. We came back over the newly opened Tancarville Bridge spanning the Seine estuary, past the medieval battlefields of Harfleur, Crecy and Agincourt, and along the *Côte d'Opale* to Calais.

Letter 16

My desire to study the palaeontology of the Carboniferous of northern England was still smouldering, and having completed the micropalaeo course, I saw a possible way of achieving my goal. Foraminifera, which we had studied under Tom Barnard, are single-celled microscopic animals most of which are marine, and most have calcareous shells, and they cover a very long time span from the Cambrian to the present day. The lab subscribed to the formidable *Catalogue of Foraminifera*, published in America and updated annually, which runs to thirty or forty volumes, with descriptions of over 40,000 species. I asked Dr Barnard if he would consider accepting me for a post-graduate study on Carboniferous forams from northern England. He was not much interested in things pre-Jurassic, but I pointed out that almost nothing had been published on the subject in Britain, whereas they had been extensively studied in Belgium and Russia and the United States. He consulted with Eric Robinson, who was a Carboniferous expert, and must have received the thumbs up. I suspect he felt that Eric could share the role of supervisor, or maybe the funding of the department was related to the number of students enrolled. In any case, he accepted my proposal, which then left the problem of funding.

The governments of the 50s and 60s had a much more enlightened attitude to education than at present. I had received free education at grammar school thanks to the 1944 Education Act, and free education for my bachelor's and master's degree courses courtesy of the Halifax Education Authority. I now applied to the Department of Scientific and Industrial Research for a grant to cover three years' research work. I outlined the scope of my planned research and its potential industrial applications, along with the endorsement of my supervisor, and on the strength of this submission, I was awarded a grant to cover fees, field expenses and living costs. They operated on the simple principle that with a research degree I was likely to obtain a well-paid job and therefore would contribute substantial taxes to the government, and that my research might have tangible benefits to industry. It was a mutually beneficial policy, which regrettably has long been consigned to history. Nowadays, research is totally commercialised. The government subsidy to universities has been greatly reduced, universities are therefore obliged to charge student fees, the government provides student loans, which have to be repaid, so it is ultimately the student who pays for his education – a sad reversal of role, in which commercial interests now rule supreme. Had these conditions prevailed at the time, I suspect I would not have gone ahead with my research.

We had now started taking a summer vacations in Scotland, usually with Eva and Dick, and before starting my research we toured most of the west coast from Loch Lomond to Glencoe, Arisaig, Eigg, Skye, Gairloch and the spooky island of Gruinard, which was used during the Second World War as a testing ground for anthrax as a biological warfare agent. It remained contaminated and closed to visitors for fifty years. We explored Suilven and Canisp, Stoer Point and Kylestrome Ferry, Badcall Bay and Scourie, and the north coast to Loch Eriboll and the Kyle of Tongue, and eventually to John O'Groats, just about the furthest point you can reach on mainland Scotland. Most of the roads at that time were single track with occasional passing places. This is a truly wild, desolate and remote part of the British Isles.

In preparation for my planned fieldwork I bought myself an ex-Army Land Rover from an Army Surplus auction. I think I paid about £150. It was a bit temperamental and on occasion I would have to crank the engine, or get someone to push to get it started, but I gradually tamed it, and it became my vehicle about town for the next three years.

My uncle was offered the job of commissionaire and caretaker of the Westinghouse Brake and Signal Company offices and workshops at Kings Cross, which was much more convenient for me, since UCL is just a short walk away. I was provided with one of the research cubicles in the Micropalaeo Lab, and set about planning my project. I elected to base my study mostly on the Askrigg Block, the area between Skipton and Stainmore, taking in four principal dales, Wharfedale, Nidderdale, Wensleydale and Swaledale which I knew well, and restricted my attention to the Yoredale Series, a sequence of ten repeated cycles of limestones, shales, siltstones, sandstones and coals about 1,000 feet thick, and about 330 million years old. The limestones and the basal shales were deposited in a marine environment whilst the rest are brackish or non-marine. The main purpose of my research was to see if one cycle could be distinguished from another simply on the basis of the foraminifera they contained. My collecting therefore concentrated on the limestones and calcareous shales. I collected about 200 samples that first year, all documented with location, and the cycle from which they had been collected. I got to know the Dales very well. I checked out most of the stream sections and quarries, and wandered high on the hills among the remains of long abandoned lead mines, where the only sound was the *krk, krk* of the grouse and the *pee-wit* call of the curlew. Until relatively recent times there were shepherds up there who still used a Brythonic Celtic system for counting sheep. This was a vigesimal system (based on the number twenty), which started over again when the number twenty was reached. The numbers one, two and three are yan, tan, tethera, by which the system was known. The Land Rover turned out to be the perfect vehicle for the job. I was able to drive over rough, rocky roads known as drove roads or green roads, previously used for driving sheep to market, and up some of the gentler stream sections. One winter I went with Michael to Semerwater, a small lake in the Dales. The lake was completely frozen over. I was able to drive the Land Rover onto the ice and take pictures of Michael standing nonchalantly beside it. The photographs are impressive, but the lake was actually only about six inches deep at that spot.

There is a famous walk in Yorkshire called the Three Peaks, which involves a cumulative ascent of 5,200 feet and a walk of 24 miles. It is usually done in summer. That first Christmas Michael and John and I decided to do it in winter, but using the Land Rover to drive between the peaks, since the days were short and the snow quite thick. It was an exhilarating experience, and it became a regular Christmas trip for the three of us for several years.

The following summer John married Michael's sister Hilary, and whilst they were away on honeymoon, Michael's father invited my mother and I to accompany the rest of the family on a motoring trip to Norway. We sailed from Newcastle to Bergen, visited Edvard Grieg's house at *Troldhaugen*, and drove on unpaved gravel roads (this was before the Norwegians became rich from North Sea oil) through the stunning fjord country, all the way to Trondheim, and then to a little place called Hell. So we were able to say we had been to hell and back.

Back in the lab I processed quite a number of shale samples and was able to extract some forams from them, but they were almost impossible to identify without making a thin section to see the internal structure. This was a laborious and unrewarding exercise, so I decided to concentrate on the limestones, where the forams are as abundant as raisins in a fruitcake. It is impossible to extract them from the limestones, but by making thin sections of the rock – so thin that they become transparent, the fossils are revealed in section – as if they had been sliced open. The sections are of course random, but the forams were so abundant that it was usually possible to find one specimen out of maybe nine or ten which showed sufficient features to be identifiable. This is where the *Catalogue of*

Foraminifera came in useful since almost all of Carboniferous foraminifera are identified on the basis of their internal structure. A great many species had been defined by Russian palaeontologists (mostly female, with splendid names like Rauser-Chernousova and Miklucho-Maclay) working on the Russian Platform in the 1940s and 50s, and I contacted some of them to request copies of their papers. They usually obliged, and on one occasion addressed the envelope to Sir Donald Hallett. If you begin a letter with Dear Sir, I guess it makes sense. I decided that I needed to learn a little basic Russian in order to be able to scan their publications, so I enrolled at a local institute and got to the stage where I could read the fossil descriptions which are always written in very terse, scientific language, so it was largely a matter of simple transliteration, but I gave up when I learned that there are fourteen different prefixes to express 'to go'.

We got to know some of the research students in the main Geology Department, and one day we had a visit from Henry Meyer, who was researching fluid inclusions in diamonds. He had been supplementing his income by teaching evening classes at Westminster Technical College, but had been offered a place in the United States so he asked if John Church and I would like to take over his teaching role. We were doubtful because the course was on geology for civil engineering students, but he said it was a doddle and he gave us an outline of the syllabus and his lecture notes. We accepted and each of us began teaching one class a week to students who were not very different in age from us. We gave them exercises on geological maps, and lectures on rock properties, landslides, earthquakes, aquifers, superficial deposits and so on. We also took them on field trips to see the engineering problems of the Sevenoaks by-pass which had to be re-routed because of ground instability, and to see the London Clay (through which the London Underground railway is cut), the structure of the Weald, and other places of engineering interest. We tried to arrange as many field trips as possible because they paid by the hour, and not only that but they covered our expenses too.

My cousin David, Rhoda and Leslie's son, had finished school and had joined a printing company as a compositor. He was now twenty-one, his career was advancing and he was engaged to be married. He had a bubble car, which were all the rage at the time, and one morning as he was driving to work he hit an icy patch, lost control and crashed into a concrete lamp post. He was killed instantly. Poor Rhoda. First she had lost her brother, my dad, at the age of 28. Now she had lost her only son at the age of 21. She must have felt that there was a curse on the family. Rhoda had a very sensitive nature and although she never claimed to be a clairvoyant she did have some strange stories. I mentioned in an earlier letter her 'vision' at the time my father was killed. Now she had another. Leslie said that on the day before David was killed Rhoda said to him 'what is that dark shadow in the corner?' Leslie said he couldn't see anything, but Rhoda took it as an ill omen. Later she claimed it was the angel of death come to warn her of imminent tragedy.

Letter 17
That summer, at the end of my second year, Michael and I decided to take a trip behind the Iron Curtain in my Land Rover. Eastern Europe at this time was still under rigid communist control, and it seemed a rather daring thing to do. We loaded up with supplies and camping gear, obtained visas and set off. We drove through France, up the Meuse valley, through the Ardennes, around Frankfurt and along the three-lane, busy, well-surfaced autobahns to the small town of Hof in Thuringia (a region full of associations with Bach), and shortly afterwards arrived at the East German border. Here we were told we would have to change the equivalent of £30 a day into East German marks 'to cover the cost of accommodation' (credit cards had yet to be invented). We crossed the border into a different world. There the autobahns were exactly as built in the 1930s, surfaced with concrete, two lanes wide (but the bridges had three lanes in anticipation of traffic which never came), and it was an event to see another vehicle. We drove past dark, dystopian industrial cities like Plauen, Zwickau and Karl Marx Stadt and eventually arrived at Dresden.

Before the war Dresden was known as 'the Florence of the Elbe'. It was a jewel of Baroque architecture - palaces and churches, elegant terraces and promenades. Towards the end of the war the Nazis were no longer able to effectively defend their cities against bombing. This was retribution time, and the allies conducted massive bombing raids on cities like Hamburg and Cologne in the hope of coercing the Germans to surrender. Dresden was a major communications hub, through which Nazi troops were streaming to escape the Russian advance, and it also had many factories supporting the war effort, so it was regarded as a legitimate target. The city was bombed for two nights and two days by British and American planes, resulting in the almost total destruction of the historic heart, and the creation of a fire storm which sucked people bodily into the maelstrom. The Germans quoted astronomical figures for the dead, but subsequent studies have shown that the death toll was about 25,000. We arrived in Dresden with my ex-Army Land Rover with British plates twenty years after the bombing. The place was still a ruin. It looked as though the war had recently ended. The rubble had been removed from the streets and the houses made habitable, but no effort had been made to restore the churches and palaces and public buildings, and there were large areas which were still derelict. We received some very unfriendly looks from people, and decided it might be best not to hang around.

We took the road to Prague on one of the routes that the Nazis must have taken when they invaded Czechoslovakia in 1939. Much of the road was paved with stone setts, unchanged since the war, and it was an eerie feeling to reflect that only 25 years before this route had echoed to the sound of German tanks and infantry. We drove up the slope of the Erzgebirge to a small frontier post at Zinnwald and tried to change our surplus Ostmarks back into pounds. They told us that they would transfer them to our bank accounts back in Britain, which we thought was a bit of a joke. We were most surprised when the transfer turned up a couple of weeks later, calculated at exactly the same rate as when we entered the country. On our way to Prague we stopped at Lidice, the scene of a horrific reprisal by the Nazis following the assassination in 1942 of Reinhard Heydrich, the Nazi 'Protector' of Bohemia and Moravia. The village was surrounded and the entire male population was executed, and the women and children taken off to concentration camps. In all about 340 people died. In Prague we visited Wenceslas Square, the Charles Bridge, the old town and the Jewish quarter, and we went to see the church where the Heydrich assassins had holed up, been betrayed and then killed.

Czechoslovakia at the time of our visit was under the control of the hard-line communist Antonin Novotny, who was exceedingly unpopular. He was replaced four years later, during the so-called Prague Spring, by a much more liberal regime headed by Alexander Dubcek who attempted to establish 'socialism with a human face', but the Soviets took fright and sent in the tanks, and the Prague Spring became the Prague Fall.

From Prague, we drove to Brno and over the border into Austria and on to Vienna. We spent a couple of days exploring the historic centre and were stuck that such a small country should have such a magnificent capital. What we didn't appreciate at the time was that Vienna had been the capital of an empire which stretched from Bohemia to Bosnia, and from Tyrol to Transylvania. Austria today is less than one seventh of its pre-

First World War size. In fact, after Russia, Austria-Hungary was the largest country in Europe. We went to Schönbrunn, the Imperial Summer Palace where Franz Joseph spent much of his life. The Habsburgs had a sad end. Franz Joseph's wife, the beautiful Elisabeth (Sisi), was profoundly unhappy with the formality of court life, and spent as much time away from it as possible. Her son Rudolf, the heir to the Imperial throne, a playboy and womaniser, died in an apparent suicide pact with one of his mistresses at Mayerling, a hunting lodge in the Vienna woods, and Elisabeth herself was killed by an Italian anarchist in Geneva. The next heir to the throne, Franz Ferdinand (who had been totally excluded from affairs of state because of a morganatic marriage) was assassinated in Sarajevo in 1914, triggering the First World War, in which Austria, for all its military pretensions fared badly. There was a quip at the time that the prowess of the Austrian army was in inverse proportion to the lavishness of its uniforms. Franz Joseph died in 1916, a broken man.

From Vienna we took the road to Budapest, a city which still bore the scars of the 1956 uprising. Even ten years later the castle on the hill was partly shrouded in scaffolding, and shell damage was evident on many buildings. The uprising was sparked by students in Budapest who marched on the Parliament building. One of the students was shot dead by the State Security Police, and the uprising rapidly spread. The protesters organised militia groups and freed political prisoners and set up workers' councils. Under this onslaught the government caved in, the State Security Police was disbanded and a new government, led by Imre Nagy, declared their intention to withdraw from the Warsaw Pact. The Soviets initially indicated a willingness to negotiate the withdrawal of Soviet troops, but quickly changed their mind and sent in 6,000 tanks to suppress the revolt. In the fighting which followed 2,500 Hungarians and 700 Russian soldiers were killed, and 200,000 Hungarians fled the country. Nagy was arrested and after a secret trial was executed. He was buried face down with his hands and feet bound by barbed wire, 'as a lesson to other Socialist leaders'.

Budapest is a city split by the River Danube. Buda is on the west bank and contains Buda Castle, the Citadella and the Presidential Palace. Pest is on the east bank and contains the Parliament building, the Heroes' Square and the Old Town. The two are joined by the famous Chain Bridge (and by several others). So far, we had been able to get by with English, schoolboy French and Michael's German, but in Hungary we found we could not understand a word. We tried the time-honoured principle that 'if they don't understand, just speak louder', but that didn't work either. We spent a day or so exploring the city and ended up at a nice restaurant with a terrace overlooking the city, and for once we found someone who spoke English. We asked for something typically Hungarian. The waiter said 'ah, yes, I bring the rosbif'. The other thing I remember about Budapest was driving into an underpass to find a tram heading towards us. I have never hit reverse with such speed. Stupid foreigners!

We then headed for Lake Balaton through the unpronounceable town of Szekesfehervar and lush, fertile country studded with vineyards. Balaton is the largest fresh-water lake in Central Europe, nearly 80 kilometres long, but rarely more than three metres deep, so the water is comfortably warm. It was a resort of the aristocracy in pre-war days, but by the time we visited was popular with the proletariat who were accommodated in workers' state-owned 'sanitoria' at subsidised prices.

From Balaton we crossed into Yugoslavia which had a more relaxed feel than Hungary. It was very rural, and I took a photograph of a farm cart pulled by oxen crossing a river, reminiscent of a Constable painting. The old part of Zagreb is dominated by the tall towers of the Cathedral. Many years later we took a visitor from Croatia to see Big Ben in London, which he said was nice, but not as tall as Zagreb Cathedral. We continued through the folded and crumpled limestones of the Dinaric Alps to Rijeka and Trieste, a region which has seen more changes of ownership than almost anywhere else in Europe.

Rijeka or Fiume in Italian, (both names mean 'river') was under Habsburg control from the 15th century and was Hungary's only sea-port. It was under French control during the Napoleonic era, and later became part of the Austro-Hungarian Empire, when there was fierce competition between the ports of 'Hungarian' Rijeka and 'Austrian' Trieste. The city even had its own language called Fiumano, which is a derivative of the Venetian dialect. After the dissolution of the Austro-Hungarian Empire following the First World War, the city was claimed by both Italy and the new Kingdom of the Serbs, Croats and Slovenes, but while discussions were in progress, the city was seized by Italian nationalist irregulars led by the poet Gabriele d'Annunzio, a sort of modern-day Byron. Under the Treaty of Rapallo Rijeka was declared a Free City, at which D'Annunzio declared war on Italy. It was the stuff of comic opera. When Mussolini took power, Rijeka was seized by the Fascists, but during the Second World War a bitter guerrilla war developed between the Italian and Slav inhabitants, and the Axis forces were not finally dislodged until May 1945, by which time most of the city was in ruins. The remains of the fortifications can still be seen surrounding the city. After the war the city was ceded to Yugoslavia and the Italian population was bullied and coerced into leaving, with Yugoslavs taking over the abandoned properties.

Trieste had an ironically similar history, but in reverse. Under the Habsburgs it was declared a Free City in 1719 and remained so until 1891. It was in the Viennese coffee houses of Trieste in the early 1900s that *cappuchino* coffee first became popular. The city was annexed by Italy in 1920 and the Slav population was given the choice of leaving or of being Italianized. Life was made miserable for those who remained by Mussolini's Black Shirt troopers. During the war Trieste's Jews and Slovenes were interned in Italian concentration camps where over 7,000 perished. Retribution came in 1945 when the city was occupied by Yugoslav forces and thousands of Italians, Fascists, Catholic priests, German collaborators and anti-Communist Slovenes disappeared without trace. Following an agreement in 1947 Trieste again became a Free State, and then a few years later reverted to Italy.

The two cities appeared to be tranquil when we visited, but memories are long and there is little love lost between the Italian and Slavic people in this troubled corner of Europe. The skyline of Trieste is dominated by the Victory Lighthouse, built to commemorate the Italian 'victories' during the First World War. It was built to be just a little higher than the Victory Column in Berlin.

We pushed on to Venice where we spent a day or two soaking up the atmosphere of this unique city, and even shelled out a pocketful of lire for coffee at Florian's in Piazza San Marco (but our resources didn't extend as far as Harry's Bar). We took the road to Padua and Vicenza, Verona and Brescia and near Milan we made a detour to Monza, the Italian Grand Prix motor circuit. To this day I cannot explain it, but somehow we drove onto the racing circuit, no one was there, no one stopped us, and we drove around the course in my battered old Land Rover, regretting that it wasn't a Maserati.

We continued beside Lake Maggiore and crossed the Alps by the Simplon Pass and headed into Switzerland and then to France. We spent a day in Paris and visited Versailles, which rather eclipsed Windsor in splendour (but we never had a Sun King, and their revolution was still to come). As we drove towards the coast we stopped to obtain a few francs for our final expenses in France and changed £20 at a small-town bank. The cashier made a mistake and gave us the franc equivalent of £200. When we realised the error we went back, but the bank had closed. On arriving back in England, I mailed the excess back to the bank, and received a fulsome letter of thanks from the cashier who said they would have deducted it from his pay. I don't believe I would do the same today, but it gave me a warm feeling at the time. It had been a unique and memorable trip. The old Land Rover had done us proud. We had driven 4,500 kilometres, with not even a puncture. *Impeccable*, as the French would say.

Letter 18
It will be clear from my earlier letters that to date I had very little contact with girls. I had been to a boys' school, an all-male college in Durham, most of my friends were male, and geology at that time was a predominantly male discipline. About this time Raymond, my Jamaican friend, introduced me to a Norwegian girl called Torill who was working as an *au pair* for a family in Epsom. The three of us went in my Land Rover to a motor racing event at Snetterton in Norfolk, and had a great time, marred only by heavy rain, and we all got wet through. We found refuge in a pub with an open fire where we are able to chill out and dry off. After that I met her quite often. We would go to cinemas or exhibitions or concerts, and I grew very fond of her. Unfortunately, her time as an *au pair* was almost up and she returned to Norway. I had also begun to visit the Royal Festival Hall, a concert hall built as the one lasting legacy of the Festival of Britain. It had a left-wing ethos, it was open to everyone and had a restaurant called the People's Palace. I went to concerts with legendary conductors like Herbert von Karajan, Otto Klemperer and Pierre Monteux, who were approaching the end of their careers. In particular, I loved the way Klemperer conducted Beethoven. He was slow, deliberate, solemn and majestic and he could raise an orchestra to a level of intensity beyond the reach of most conductors. There was a British conductor called Stanley Pope who also made a big impression. When conducting Berlioz'a *La Damnation de Faust*, his violent gestures coupled with his jet-black hair, waxed moustache and goatee beard made him look like some latter-day Mephistofeles. When he visited South Africa one time a concert poster advertised him as Pope Stanley, which guaranteed a full house every night.

I can't resist two other musical stories of the time. Sir Malcolm Sargent was a fashionable and dashing British conductor (known as Flash Harry) who was chief conductor at the Promenade Concerts, held each year at the Royal Albert Hall. He lived across the road in Albert Hall Mansions, forty or fifty yards from the hall, but he would insist on travelling there in his chauffeur-driven Rolls Royce in order to make a grand entrance. Sir Thomas Beecham was a conductor noted for his wit and sarcastic comments. He once described the dapper Herbert von Karajan as 'a kind of musical Malcolm Sargent'. He was a Lancastrian and said 'In my county, we're all a bit vulgar, but there is a certain heartiness – a sort of *bonhomie* about our vulgarity – which tides you over many a rough spot. But in Yorkshire, they're so damn-set-in-their-ways that there's no doing anything with them'. But the story I like best occurred at a rehearsal where the lady cellist was having some difficulty with the score. He turned to her and said 'My dear, you have between your legs an instrument capable of giving pleasure to a great many people, and all you can do is scratch it'.

But I digress. There was a rapid turnover in the lab as Masters' students came and went, and researchers finished their doctorates. John Church left to take up a job in Wales. Raymond Wright returned to Jamaica, another guy left to become a forensic geologist with the Metropolitan Police, Marj and Rob and Henry finished their PhDs, and Ali and I started the third year of our research projects. One day Ali mentioned that he was looking for a piano for his daughter, and it happened that my aunt Eva had one that she rarely used. I offered the use of my Land Rover, and with the help of three or four strong lads from the lab we loaded the piano onto the back, with one on each side to hold it steady. Ali lived in Chiswick in the south-west of London, and Kings Cross is in the north, so we drove, open-topped, through the centre of London, with one of the guys playing the piano as we cruised down Park Lane.

I spent time each summer collecting samples, from which the lab technicians made thin-sections for me to study, and by this time my sample coverage was fairly complete. I was fortunate that two absolutely key books had been recently published, one American and one Russian. They were both exhaustive, detailed and up-to-date. They allowed me to ensure that my attribution of a specimen to a particular genus or species was accurate and correctly named, which saved me from the embarrassment of using names which, for one reason or another, had become obsolete.

I made contact with Raphael Conil, a Belgian geologist based at Louvain University, who invited me to visit him and see the Belgian Lower Carboniferous rocks. He had worked and published extensively on the Belgian foram faunas, and there was nothing in Britain to compare to his publications. Part of the trip was done by motor bike which was a bit of a white-knuckle experience. It was common practice in Belgium to mark the formations with paint, so you would arrive at a quarry or road cutting and find a painted notation such as V3β, indicating a unit which he had identified by its foraminiferal fauna. I gave him a reciprocal tour in England of the Avon Gorge and the Yorkshire Dales. He spent a night or two with us at Westbury Terrace and would charm my mother with expressions like *'enchanté, madame'*. Raphael was a Walloon, a French-speaking Belgian, but Louvain is within a predominantly Flemish-speaking area. He told me that when I wrote to him I should address the envelope using Flemish names, because if I addressed it in French it would not be delivered.

It was about this time that the lab secretary left and a new girl appeared on the scene. She was a vivacious redhead called Wendy, gregarious and friendly and bubbly. She had been at secretarial college in London and was living with three other girls in Golders Green. I invited her out, and found that we got along well together. We went to see *My Fair Lady*, and to shows and concerts, and drives in the country. I visited her flat and met her flatmates (who were mostly rather superior upper class gals) and I introduced her to Eva and Dick and then to my mother (who were definitely not upper class). Wendy was a Catholic, which was regarded with some suspicion by my family, but it didn't seem like a major impediment. The following summer we took a trip together with the trusty old Land Rover to Paris, Switzerland, Florence and Pisa, and the affair became serious.

My cousin Margaret had spent almost four years with my mother in Halifax and now needed to return to London to complete her education, so I had to find alternative accommodation. Wendy introduced me to a chap who was in a similar situation, and we found three other guys and together we rented a house in Holly Park Gardens in Finchley, a rather posh area of north London. We were a very mixed bunch, one was a trainee hotel manager and chef, which was useful when it came to Sunday lunch, another was a crazy German guy called Eberhart C. Klein whose recreation was firing an air pistol in the garden, and who always ended up eating his bowl of breakfast cereals in the Land Rover on the way to college. I drew the lucky straw and got the main lounge as my bed sitting room.

In 1961 the giant oil company Esso, which had been marketing oil and petrol (gasoline) in Britain for decades, obtained a number of onshore drilling concessions, and hired Tom Barnard, our head of department, as consultant. The concessions were mostly in Surrey and Sussex to the south of London, and in 1963 they started drilling a programme of exploratory wells in this area. Tom passed some of the more routine work to me and John Parmenter, one of that year's Masters students. We got to know some of the Esso geologists and we visited their exploration office in London, which they were in the process of staffing up. We did several small projects for them, seeking out reference material, plotting well locations on maps, and changing well tops from feet to metres (or was it the other way round?). They must have found our contribution useful because they offered to hire us once we had obtained our degrees.

My thesis was finally finished in January of 1966 and proved to be quite a substantial compilation of 442 pages and 60 plates. I was rather pleased with it. This was all new data, 140 species were documented, most of them never recorded in Britain before, and fourteen of the species were new, so I had the privilege of naming them. I gave them names like *Aphralysia garwoodi* in honour of a former professor at UCL who specialised in Carboniferous studies, and *Tetrataxis brigantiae* in honour of a north of England tribe which resisted the Roman occupation of Britain and *Polyderma reitlingeri*, in honour of one of my contacts in Russia (the one who had addressed me as Sir Donald). Professor Garwood, a Carboniferous expert, had at his own expense, built up a reference collection of specimens at UCL for the use of students. After his death in 1949 his successors regarded the Carboniferous collections as excessive, so they were unceremoniously thrown out, and suffered the indignity of being used as hard-core in the foundations of a new extension around the back.

I submitted my thesis which was read by an internal and an external examiner (my supervisor Tom Barnard, and Professor Alan Wood of the University of Wales) and I then had to appear before the examiners to 'defend' my thesis. Fortunately, all went well, and my thesis was 'approved for publication'. I had my PhD., and could now call myself 'doctor'. I received my degree at Senate House, the administrative headquarters of London University, and got to wear a crimson gown with

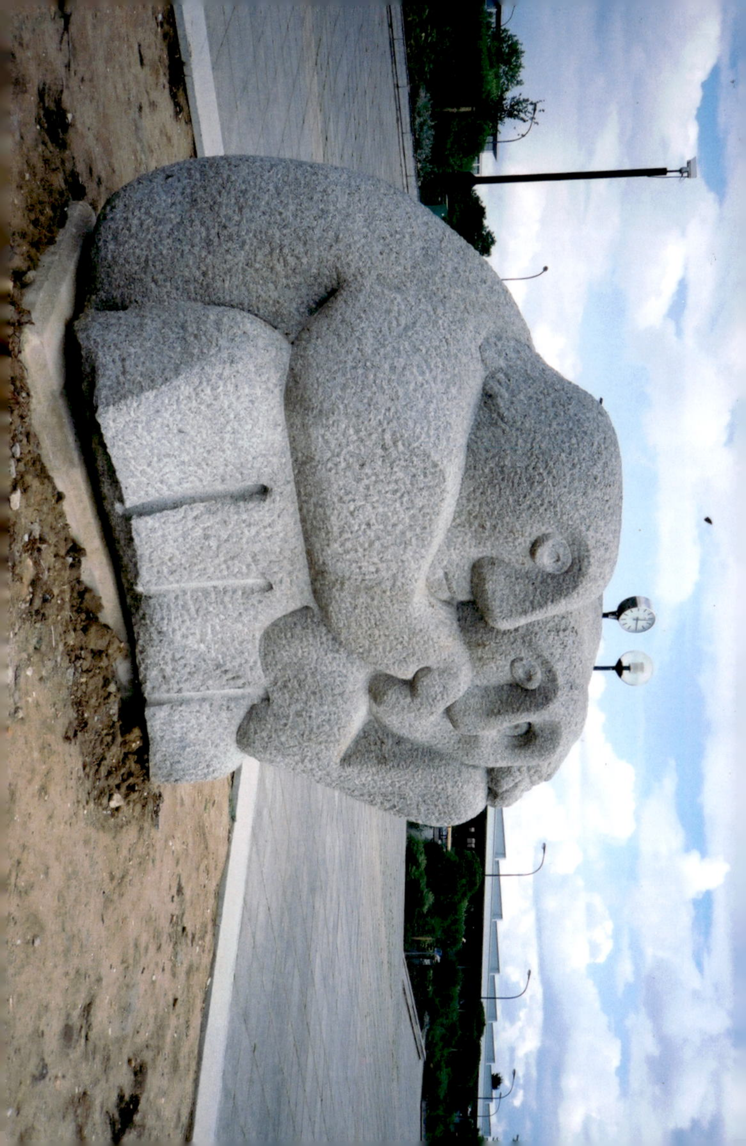

gold edging (indicating the Science Faculty) and a black velvet Tudor bonnet. My mother and Eva and Wendy all came to the ceremony. I reflected that my fifteen years of study, grammar school, university, master's and doctorate, had all been funded by government grants. It was an enlightened policy for which I was immensely grateful. I celebrated by getting rid of the faithful Land Rover and buying Ali's Mercedes.

Letter 19
Shortly after I received my degree one of Wendy's flatmates got married. A couple of months later we visited them in their new home and were very struck by how happy they were. On the way back I proposed to Wendy and she accepted. We had known each other for over a year, we had grown very fond of each other, and it seemed the natural thing to do. When I left UCL, I went to thank Tom Barnard, not only for supervising my research but also for providing me with a job and a wife. When Wendy and I discussed a date for the wedding she said she would have to obtain a dispensation from the priest to marry a non-Catholic, and that I would need to receive instruction on the fundamentals of Catholicism. Perhaps that should have sounded warning bells, but if it did I didn't hear them.

The decade from 1960 to 1970 became known as the swinging sixties. Up until that time British society had been very straight-laced. But the sixties saw the appearance of the mini-skirt, pop music (*Beatles, Rolling Stones*), provocative films (*Look Back in Anger, Room at the Top*), satirical television (*The Frost Report, That was the Week That Was*), sexual liberation and political activism like the 'ban the bomb' marches. Within just a few years London was transformed from a dowdy, post-war city to a vibrant centre of style and fashion and youth culture unmatched in Europe.

We felt to be part of this scene, especially Wendy who, being younger, found it easier than me to adjust to the new-found freedom. She was vivacious and outgoing and made friends easily, whereas most of my friends were work-related or neighbours. My mother and Wendy got along well together, I think both were able to appreciate each other's merits, but I did not have a comfortable relationship with Wendy's parents. Her father John was an accountant who had climbed the ladder from fairly humble origins to a partnership in the firm, and although not rich, had achieved considerable success. As was common in his profession, he had accomplished this by cultivating his clients, playing golf and sharing cocktails with them, and he seemed to think that I should follow a similar path. He apparently could not see that my situation was different. Norah, Wendy's mother, was a brisk and efficient woman who worked as his secretary, but she had little patience for those who were not as brisk and efficient as herself, she had a sharp tongue and did not suffer fools gladly. We did not get off to a good start. Wendy's parents came to visit my mother at Westbury Terrace and were very disapproving of our very ordinary terrace house, our Yorkshire accent, and our obvious working-class background – which is odd since one generation earlier they had been in exactly the same situation. They sat in the front room bristling disapproval. It was a very fraught visit. My mother's comment afterwards was 'what did he do in the war?' On another occasion, in winter, I drove Wendy to Scunthorpe in the Mercedes. I had the misfortune to slide on an icy road just before picking her up and made a bit of a mess of the front wing, so we arrived at her parent's house in my left-hand drive, damaged Mercedes, and I overheard John saying to Norah, 'he's bought a pup'. I was seriously pissed off; particularly since his car at the time was a Riley. John was also very straight-laced. I recall he disapproved of Eartha Kitt, who sang zany songs in a deep, sexy voice, but he forbade Wendy from having any of her recordings in their house. As the saying goes, you can choose your friends, but not your relations.

John Parmenter and I joined Esso at the same time and we started work at their office in Charles II Street off the Haymarket in London. My starting salary was £1,500 per year. There was no formal training; you were expected to learn on the job. The Exploration Manager was an oldish guy named Harold Stoneman (reputed to be descended from native American stock). He cultivated his image of a corporate boss by wearing mohair suits and smoking expensive cigars. He had a puffy face with eyes so narrow that they were mere slits. The rest of the team – all Americans, included Jerry Graham, Leroy J. Perry and Larry Vinson, middle-aged and experienced, and Pete Blau, Don Brehm and John Stone, young and ambitious, and keen to make their mark. Leroy was a dapper little man with a Habsburg jaw, and one of the most cultivated Americans I have ever met. He had a wide range of interests in music, painting, theatre and literature which was very rare in most of the Americans I had come across.

During the lunch break we would sometimes play darts, where I discovered just how competitive Americans can be. Otherwise we went for lunch to a pseudo-Victorian pub called the *Cockney Pride* with a live pianist (wearing a derby, coloured waistcoat and sleeve garters), or to the *Stockpot* bistro in Panton Street, or on special occasions to *Stone's Chop House*, a men-only dining establishment founded in the 18th century. We scoured London for geological information on southern England - the Geological Survey, the Geological Society, libraries, universities, even second-hand bookshops. John and I were given the task of compiling a map of the Palaeozoic Floor of the south of England from the data we had collected and from well and borehole information. This gave a picture of the structure of southern England and an indication of the areas where oil and gas might be expected. The presence of small gas accumulations was already known, for instance at Heathfield in Sussex, where natural gas had been used to light the railway station from 1896 to 1963. The map revealed a number of structures, some of which had been drilled in earlier decades. One of the undrilled features was near the wonderfully-named village of Farleigh Wallop in Hampshire and I recall checking available borehole evidence to see what might be present there, and writing a brief memo highlighting the structure. It was very rudimentary, but it marked my first attempt at a prospect description. Interestingly the structure was later drilled by BP, but the results have not been disclosed.

Esso management was also interested to check out the oil potential of Northern England and in view of my earlier experience of mapping in Northumberland, I was asked to escort Larry Vinson to the area to collect samples for laboratory analysis. Larry was an easy-going, laid-back Texan who I felt would have been more at home in Amarillo or Abilene, except for the fact that he lived in some style in a fancy house in Walton on Thames. We called in at Halifax on the way north and I introduced him to my mother. I thought what an incongruous situation to have a Texas oilman sitting in our little front room. But he was a perfect gentleman and treated my mother with great courtesy. In the event Esso did not pursue this line of enquiry. They had their eyes on a much greater prize: the North Sea.

At this stage I need to debunk a popular myth. Oil and gas are not found in great underground lakes or caverns, but simply in the pore spaces between grains of sand or limestone, or sometimes in fractures in the rock. Both oil and gas constantly strive to migrate upwards towards areas of lower pressure, and they continue to do so until they either reach the surface (where they dissipate or form oil or gas seeps), or until they are trapped against an impermeable barrier. Petroleum geologists spend their lives trying to find these elusive oil and gas traps. It becomes a game of hide and seek. Where is the source rock? Is it mature? When did the oil begin to be expelled and migrate? Which direction did it go? Are there any structures in which the migrating oil might be trapped? Is there a suitable reservoir rock in which they might be contained? This is why the study of ancient sediments, geochemistry and fluid flow are so important. It is a source of enormous satisfaction to grapple with these questions, which just occasionally come up trumps.

Esso's first well was drilled at Bolney in West Sussex in 1963, on trend with the Heathfield gas discovery and about 20 miles further west, but the well was dry. (The location where the well was drilled is now a thriving vineyard). The following year a second wildcat well was drilled near Gatwick Airport called Collendean Farm, targeting a porous Jurassic limestone called the Corallian ('wildcat' wells are named for Wildcat Hollow in Pennsylvania, where many trial wells were drilled in the pioneering days of oil exploration). The limestone proved to be well developed, but water-bearing. It is likely that the well was drilled outside closure, because in 2014 the same structure was drilled again at Horse Hill a short distance to the east, and was announced as a significant oil discovery. In 1965 Esso drilled a third wildcat to test the Corallian at Bletchingley, 6 miles to the north-east of Collendean Farm, and this time they were successful. The well flowed at a rate of 3.6 million cubic feet a day, which was a quite a decent flow rate. This was pounced upon by the press, who in their usual gonzo style, inflated the discovery out of all proportion. 'Gas discovery in London's back yard', 'Environmental concerns over gas discovery in area of outstanding natural beauty', 'Oil major plans to develop Surrey gasfield'. Why do they always have to exaggerate? Curiously history repeated itself in 2014 with the Horse Hill discovery mentioned above, which was even more outrageous, with claims that the well indicated oil reserves of 50 to 100 *billion* barrels – a ludicrous and totally misleading claim. What they actually meant, but did not explain, was that the well indicated that the oil *source* rock could contain this amount - but *in the entire basin*, and virtually none of this would be recoverable by conventional methods because the rock was neither porous nor permeable. The figure was meaningless.

Letter 20

I submitted my PhD thesis at the beginning of 1966, and was then free to become involved in Esso's drilling programme. From our examination of other wells in the area we had built up a picture of the Corallian limestone forming a narrow rim around the southern flank of the London Platform.

Northwards it pinched out; southwards it passed into poor quality sandstones. We wanted to check what happened to it further west so we drilled two stratigraphic test wells, Strat A1 near Worplesdon in Surrey, and Strat B1 at Stratfield Mortimer near Reading. Both wells showed that the Corallian became non-prospective to the west; the entire reservoir section became much thinner and passed into a non-porous mudstone.

This was my first experience of a drilling operation, and it was a quite an eye-opener. There is a very clear pecking order on a rig which is under the control of the toolpusher, followed by the driller and then the labourers, colourfully called derrick men, roughnecks and roustabouts. The geologist fits somewhere in the middle. He can request certain operations, but it is for the toolpusher to decide whether such operations are safe.

A well is drilled by rotating cones of hardened steel on the end of drill pipe. The string of drill pipe is turned by a rotary table and mud is pumped into the pipe, through the drill bit and back up the outside. The density of the mud is carefully controlled to ensure that it collects the drill cuttings which are carried up to the surface where they are examined by the geologist. Every thirty feet a new stand of drill pipe has to be added, and when the bit is worn out the whole lot has to be pulled out, stacked in the derrick (by the derrick man on his monkey board – the oil industry has a colourful vocabulary), while a new bit is fitted, and then lowered back into the hole. Periodically the well is lined with steel casing to prevent the hole from collapsing. There is a mud engineer who monitors the mud properties, keeps a plot of the rate of penetration and keeps an eye on the gas detectors. The job of the geologist is to check the cuttings coming out of the hole and to evaluate the significance of any gas or oil shows. If he encounters a zone of interest he can request a core, in which case the drill string is pulled out, a core barrel is attached and lowered back in the hole. A round-trip at Bletchingley might take three to four hours, but in a deep well it can take 12 hours or more. Drillers are understandably reluctant to cut cores as it is essentially dead time for them, so they are not called for lightly. Frequently there are problems. If the bit penetrates a highly porous or vugular rock the drilling mud may disappear into the formation which results in a loss of mud circulation. Cones from the drill bit may be left in the hole and require a fishing operation, or the drill string may become stuck in the hole, which can be a major headache. There are various remedial measures, but all of them are uncertain and can be very time consuming. Stuck drill pipe is not good news. Very occasionally a well may penetrate a highly pressured zone and result in a blowout, but modern rigs are equipped with blowout preventers which (theoretically) can be closed to kill the well. If oil or gas is encountered it is tested using special tools which collect samples and measure the pressure, temperature and rate of flow. Before each string of casing is lowered into place the well is logged with a series of electrical, acoustic and radioactive tools which produce continuous wiggle-trace logs through the rocks which have been penetrated. From these operations, the geologist will normally end up with a box of cuttings from the well (each sample covering 10 or 20 feet), cores (of any zones of interest), a mud log showing all gas and oil shows, a set of wiggle-trace logs, and the results of any tests which have been run. The logs record properties such as spontaneous potential, electrical resistivity, acoustic velocity, and such like which, in those days, had to be interpreted using complex charts and cross-plots. I recall, in the days before calculators, we were given a fancy log-log-decitrig slide rule, the latest state-of-the-art calculating gadget. It is the job of the wellsite geologist to interpret all these data, draw up a well completion log showing all features of interest, and write the geological part of the well completion report.

The wellsite geologist from whom I was supposed to learn the rudiments was a long-service Esso geologist from Tennessee who had little interest in teaching a young geologist the tricks of the trade. Due to local restrictions, we had to shut down the rig at nights, which is most unusual given that rigs at that time cost $20,000 a day to operate. We stayed at a nearby guesthouse, where he would begin each morning with a shot of gin. He was not a highly communicative man, and I was left to pick up whatever I could from the mud loggers, electric loggers and drilling personnel.

Following the two stratigraphic wells we returned to appraise the Bletchingley discovery. I was sent down to Bletchingley to act as an assistant wellsite geologist on an appraisal well, unsurprisingly called Bletchingley No 2. This gave me my first experience of a successful well. The Corallian was well-developed and on test flowed 9.6 million cubic feet of gas per day which is a very respectable flow rate. It is quite a thrill to see the well flaring gas, and to feel that we had made a significant discovery. Two more wells were drilled, one of which was successful, but the other drilled through a fault and the reservoir was absent. We calculated that the Bletchingley gas field contained reserves of about 20 billion cubic feet which sounds a lot, but is actually quite small. It subsequently supplied gas to a nearby industrial estate.

Esso drilled two more exploratory wells, Tatsfield near Westerham in Kent, drilled partly to earn a 50% interest in a BP concession, which turned out to be dry, and Leigh near Dorking which was a bit of a disaster. The drill-pipe became stuck in the hole and while trying to pull it free the rig buckled. The operation was shut down until the rig was repaired and then an attempt was made to recover as much drill pipe as possible by backing-off above the point where it was stuck. In doing so a partial blow-out occurred in which mud, gas and water surged out of the hole, a nightmare for the engineer in charge, but fortunately he was able to shut the blow-out preventer before it went out of control. It would have been possible to kill the well and re-enter the hole and sidetrack the stuck pipe, but in view of all the problems Esso decided to abandon the well without reaching the objective.

In the summer of 1966 Wendy and I were married at her local Catholic church in Scunthorpe, an industrial town in Lincolnshire, much like Halifax, but with an economy based on steel manufacture. It was a large, formal wedding with close to a hundred guests, including all of my family and several of our Halifax friends. Michael was my best man and the bridesmaids were my cousin Margaret and Wendy's cousin Anne. The wedding went smoothly (although my speech was what the Germans would call *witzelsucht*, a bit banal), and afterwards we departed on honeymoon to an idyllic spot called Limone on the shores of Lake Garda in Italy. My school friend John Wiggen married Michael's sister Hilary, and Michael married Margaret, my childhood friend from Westbury Terrace, so that summer marked the start of a new chapter for all of us.

We rented an apartment in Kilburn, a not very salubrious suburb in north London, on a one-year lease, and in my spare time I wrote a summary of my PhD work for publication, as required by the terms of the degree award. I presented the results at an International Carboniferous Congress held in Sheffield, which was attended by Dr Conil, my Belgian colleague, and by a whole host of Carboniferous experts, including some of my Russian contacts. The text and illustrations were then published in the proceedings of the conference, quaintly known by their French title of *comptes rendus*. That rather abruptly marked the end of my career as a micropalaeontologist, but curiously when my mother was in her final illness many years later she told the nurses, 'my son is a micropalaeontologist'.

Our domestic bliss in Kilburn was interrupted by a number of overseas and wellsite assignments. We had collected quite a number of cores from the Bletchingley wells which were sent to Esso's European labs in Bordeaux for analysis. The gas was reservoired in a porous Jurassic limestone called the Corallian (because of its abundant fossil corals). The lab carried out tests on porosity and permeability, and as part of my training I was sent to Bordeaux to assist in a sedimentological study of the formation. I spent three or four weeks there under the tutelage of an Esso sedimentologist called Harold Beaver. Unlike the well-site geologist, Harold was a good teacher, and he showed me how cores were slabbed (cut down the middle to produce a section through the rock), how sedimentological features in the limestone could be identified and interpreted, and how a picture could be built up of the environment of the formation at the time of its deposition. It was an excellent training assignment.

In Bordeaux I was based in the Hotel Splendide, an old-fashioned hotel overlooking the river. When I checked in I thought it would be polite to use my schoolboy French, but after a while the receptionist said 'it might be quicker if we speak English'. The hotel had a nice restaurant with an impressive wine list covering, if I remember, about four columns of fairly small type, all but half of the last column were clarets, the classic wines of the Bordeaux region. I started at the top and tried a different one each evening. I had free time at the weekends and the use of a company car, so I toured the area looking for chateaux offering *degustation de vins*. The best wine is produced from vineyards on the higher valley slopes and on the river terraces along the Gironde estuary. The quality of the wines seems to be determined not so much by the geology which is quite varied, but by the soils and the climate. Many of the soils are low in humus, and the absorption of minerals by the roots is poor. The critical factor which creates such distinctive wines appears to be the rate at which the vines take up water. The type of soil, the temperature and the amount of sunshine allow abundant water uptake early in the growing cycle, but a reduced uptake during the late summer. This limits the volume of wine produced, but also gives the richness, colour and bouquet to these famous wines. Such are the intricacies of wine production. To my mind there is no red wine anywhere which comes close to the subtlety of a good chateau-bottled claret.

I went into the bar at the Splendide one evening and found a group of British sailors. They were from the crew of a Royal Navy vessel which was paying a courtesy visit to Bordeaux. We chatted for a while and they invited me on board the following day. I duly showed up and was given a conducted tour of the ship. She was a Leander class anti-submarine frigate and carried a Seacat surface to air missile (which was kept under wraps), and a Westland helicopter. We ended the tour in the wardroom where I was given the traditional Royal Navy tot of rum, which turned out to be a tumbler full. Maybe they gave me two.

On one of my free days I went to see the largest sand dune in Europe at the Bay of Archachon on the Atlantic coast. The Dune du Pilat is three kilometres long, fifty metres wide and over 100 metres high. It is on the move, heading inland at a rate of three to four metres a year and has already overwhelmed part of the coastal pine forest, houses, roads and part of the Atlantic Wall defences, built by the Nazis during the war. The region from Bordeaux to the Pyrenees forms the ancient province of Gascony which was part of the Angevin Empire under the rule of 'English' kings from 1152 to 1453. It is home to *foie gras*, Armagnac brandy and claret. What a pity we lost it.

Letter 21
In 1960 Percy Kent, who went on to become Assistant General Manager of BP Exploration, reputedly offered to drink all the oil found in the North Sea. It was a spectacularly unwise claim, but it did not prevent him from being awarded a knighthood for his services to industry. In 1964 the UK government passed the UK Continental Shelf Act which opened up the UK sector of the North Sea for oil and gas exploration and Esso entered into a 50:50 joint venture with Shell. The partnership was awarded several blocks in the southern North Sea. Early the following year they discovered the giant Leman gas field, and Esso quickly reorganised its UK activities. They closed down the onshore exploration office in London and set up a much larger office in Walton on Thames designed to handle both North Sea operations and a new group set up to examine the prospectivity of the West African continental margin. John Parmenter and I were assigned to West Africa, with promotion from Junior Geologist to Geologist I. Esso operated a salary policy based on country of recruitment, so although we received a raise, our UK salaries were far below those of the Americans.

When the lease on our London flat came to an end Wendy and I bought a house in Camberley, a small town about 15 miles west of Walton. Camberley is dominated by the Sandhurst Royal Military Academy, the

British equivalent of West Point. We moved into an estate of modern houses nearby called The Glebe, indicative that it had once formed part of the benefice of the local church. This marked the real start of our married life – a place of our own, a promising job, promotion. Life was looking good.

The East Atlantic Study Group, as it was called, was a different set up to the onshore exploration office. It formed part of Esso Exploration Inc, a prestigious specialist organisation set up to scour the world for new opportunities. The Exploration Manager in Walton was a legendary figure. Dave Kingston belonged to a group of pioneering new-venture geologists who emerged in the 1950s, known as the Rover Boys, who were sent to conduct reconnaissance in virgin areas. They were dispatched to places like Libya or Mauretania or Senegal to evaluate the oil and gas potential of these areas. They disappeared for months on end, looking at surface outcrops, collecting samples, talking to officials and examining archives. They would eventually reappear and submit their assessment, and such was their status that they reported directly to the Exploration Vice-President in New York. It is an extinct breed. In those days, most African countries were still colonies, and travel and access were relatively safe. It is a different story today. The group was staffed by a number of high-flying American geologists and geophysicists, and we were privileged to work with them. The Chief Geologist was Walt Ziegler a jovial, rather cynical character, one of three brothers who were all involved in oil exploration. He had a slight German accent, and a mannerism of ending each sentence with a rising inflection, which made it sound like a question. We developed a good relationship, and it was Walter who first gave me the opportunity to travel overseas, expand my horizons, and become involved in some prestigious projects.

Esso Exploration had established the East Atlantic Study Group to evaluate the continental shelf of West Africa. To this end they had acquired exploration permits offshore Morocco, Senegal, Portuguese Guinea and Cote d'Ivoire, and small operations offices had been set up in Agadir, Dakar and Abidjan (but not Bissau, which was regarded as unsafe as at that time it came under frequent attack by anti-government rebels). In the summer of 1968 John and I were sent to Agadir as junior well-site geologists. Esso Norway had a plane on charter so we flew to Stavanger and then on the company plane to Agadir. My first visit to Africa made a deep impression. They say that once you have been bitten by the Africa bug you will keep returning, and so it was to prove. Agadir has since become a popular holiday resort, but at this time it was still struggling with the aftermath of a major earthquake which devastated the city in 1960, killed a third of the population and destroying an enormous number of buildings. Large areas of the city were still derelict. The kasbah on the hill overlooking Agadir had been flattened, and was simply abandoned and left uninhabited. There were shocking stories of hands and fingers being cut off bodies buried in the rubble, whether dead or alive, for their jewellery, and of a fisherman dragged far out to sea and then thrown back onshore by the earthquake tsunami. There was only one hotel, the Hotel Salaam, a simple single-storey building with a garden sloping to the sea. What struck me most was the exotic vegetation, the tough, coarse grass, the pink oleanders, the agaves, yuccas, and the carpets of purple-flowered ice plants. The office building was one of the few to have survived the earthquake. We were in the office one day when there was a slight tremor and the lights began to flicker, which was decidedly disconcerting given what had happened previously, but fortunately it was not serious. Agadir is built astride a major fault (like San Francisco) which forms the southern edge of the High Atlas Mountains, and it was movement on this fault which caused the earthquake.

Esso's Moroccan concessions were located offshore southern Morocco and a programme of three or four exploration wells had been planned, using the American drill ship *Glomar Grand Isle*. They assembled a small team of Operations Manager, Drilling Manager, several experienced American well-site geologists, John and myself (still rather low in the pecking order), and logistics guys, expediters and office staff, about 20 people in total.

Offshore drilling operations are inherently more complex and dangerous than onshore operations. The drill ship is dominated by the derrick which is located amidships above the drill floor and 'moon pool', a large rectangular pool open to the sea, through which equipment is lowered. The ship is held firmly in place by numerous anchors set radially around the ship. The drill floor is connected to the blow-out preventer on the sea floor by a telescopic riser designed to compensate for ocean swells, waves and tides. At the location of MO-1 (Morocco Offshore No. 1) the water depth was about 100 metres. Drill ships are extremely expensive to lease, so they operate 24/7, and any 'down-time' is a major cause of concern. Esso personnel worked a one-week tour (pronounced tower by the Americans) and then rotated with a second crew. The company chartered a Beechcraft and a Piper Cherokee which were used for the weekly crew-change, flying from Agadir to the tiny settlement of Tantan, about 300 kilometres to the south from where a helicopter took us out to the drill-ship, located about 100 kilometres offshore, near the edge of the continental shelf.

Tantan had been bombed by Spanish planes during the 'forgotten war' of 1957/8 when Moroccan forces tried to seize the Cape Juby strip and Ifni from the Spanish. The enclave of Ifni has a very curious history. In the fifteenth century the Spanish established a fortress on a lagoon near Cape Juby called Santa Cruz de la Mar Pequeña, which became a port for shipping slaves to the sugar plantations of the Canary Islands. They were expelled by the Berbers fifty years later, and the place was abandoned. But in the 1860s, during the Scramble for Africa, Spain decided to lay claim to the old trading station. Unfortunately they couldn't find it, and by mistake they acquired the Ifni enclave 500 kilometres to the north. Spain finally obtained the Cape Juby strip (which included Tantan, Tarfaya and Santa Cruz) in 1916. Santa Cruz was appropriately renamed Puerto Cansado (the sad port). Both Ifni and the Cape Juby strip were reincorporated into Morocco in the 1960s and 70s.

On the drill-ship the geologist had a cabin of his own, but when the rig was drilling he worked in the mud-logging unit checking samples as they came out of the hole. Daily reports were radioed to the office in Agadir in code, so that anyone intercepting the call would not be able to discover what was happening on the rig. For recreation, we fished for red snapper. The cooks would throw garbage overboard which attracted dozens of red snapper which we could catch with great ease. We would give them to the cooks who would prepare them for dinner - a perfect example of recycling. The first well had to be abandoned due to junk in the hole, but the replacement well was drilled to the target horizon at about 2,750 metres (coincidentally the same age as the Bletchingley gas reservoir) where it encountered traces of dead oil (oil which has been oxidised). The reservoir quality was good and the well was considered encouraging. At the end of the well we returned to the UK, drew up a completion log and assembled the geological part of the well completion report. It was invaluable experience which taught us a great deal about life at the sharp end of oil exploration.

Back home our married life prospered. We furnished our house, mostly from Heal's the London equivalent, I suppose, of Macy's; we went to shows, and to long defunct restaurants like *L'Escargot Bienvenu* and

Schmidt's, and others which still survive like *The Gay Hussar* and *Bentley's*. We explored the countryside, socialised with the neighbours and before long Wendy was pregnant. One of our trips was to the Cotswolds, an area of idyllic English villages, rolling countryside and old-world tranquillity and we spent a night at the small town of Burford. After dinner, we took a stroll and came to a little country church and churchyard. The first gravestone we saw was inscribed to the memory of Susannah someone or other, and we thought what a charming, evocative, feminine name. If the baby was a girl, we decided, that's what we would call her. Sure enough, it was a girl, your aunt, and as we had decided in that Cotswold cemetery, Susannah is the name we gave her.

Just a few weeks after Susannah was born I was offered the opportunity to relocate to Agadir with Wendy and Susannah for the next phase of drilling. Esso provided us with a new villa on the very edge of town, and a company car. We found a maid, Aicha (known in Morocco as *'une bonne'*), to help with the housework and with Susannah, and a gardener called Mubarak, and Wendy soon discovered other expats (expatriates, not ex-patriots) living in the town, mostly working for charities such as Save the Children and Oxfam.

The pacification of Morocco by Spain and France in the period 1905 to 1934 was neither quick nor easy. The guerrilla campaign against the Spanish in the Rif mountains led by Abd el Krim was particularly troublesome and was not suppressed until 1926. French progress against less well organised opposition was piecemeal. The area north of the Atlas mountains was occupied before the First World War, but the mountainous area further south was not secured until the 1930s, and even then depended on the cooperation of local chiefs. When we arrived, Morocco was still quite feudal. We were told that in some villages a girl would be carefully nurtured and when she reached the age of thirteen or fourteen she would be offered to the king. Morocco at that time was an absolute monarchy with Hassan II as king. He had succeeded his father seven years before and his position was decidedly precarious. Ostensibly a constitutional monarch, Hassan had dissolved parliament and ruled alone, and his record on human rights was not good. There was opposition from socialists and from the Istiqlal Party, and relations with neighbouring countries were strained. There was a brief military conflict with Algeria in 1963, which left a corrosive legacy, and with Mauritania after Moroccan forces seized three-quarters of Spanish Sahara following the withdrawal of the Spanish in 1975. He was nevertheless a resourceful man and survived two attempts to overthrow him. Once in his palace when troops tried to arrest him he asked to visit the bathroom, from where he was able to telephone loyal troops who came to his rescue, and another occasion when his plane was fired on by rebel air force planes, he seized the radio in the cockpit and shouted 'stop firing you fools, the king is dead'.

Letter 22
In 1969 Esso embarked on a much more ambitious exploration programme. A new structure was drilled 100 kilometres to the south-west, in a different geological setting. This was labelled the MO-2 well with a similar target to the earlier well, but this time the well was located on a salt swell and closer to a presumed source 'kitchen' – an area where the source rock is mature and has generated oil. The new location necessitated the relocation of the helicopter shore base from Tantan to Tarfaya, a village on the coast of what used to be Spanish Sahara (when it was called Villa Bens). There is a massive fort just offshore, crumbling and in ruins, built as a trading post by the British in the 1880s (trade took them to the most unlikely places). Onshore were the remains of what we took to be a former foreign legion fort half buried in the sand, which looked like something out of *Beau Geste*. The Cape Juby strip which includes Tarfaya was ceded to Spain in 1912 and was reincorporated into Morocco in 1958. In the 1920s the French established an airfield there which was used as a staging post on the route to French West Africa and beyond. The station manager was Antoine de Saint-Exupéry, a larger than life figure, author, aristocrat, aviator and poet. In 1935 he attempted to win a prize of 150,000 francs by establishing a new record for the flight from Paris to Saigon, but he crashed in the Sahara Desert. Both he and his mechanic survived the crash and languished for four days before being found by Bedouin tribesmen, adding further to his celebrity status. During World War II, although well into his forties and in poor health, he joined the Free French Air Force, and in July 1944 he flew a mission from Corsica in an unarmed Lockheed P-38 to reconnoitre German troop movements in the Rhone Valley. He did not return. In 1998 a fisherman recovered a bracelet bearing Saint-Exupéry's name from the seabed south of Marseilles, and later the scattered remains of his plane were found.

But I digress. From Tarfaya we flew by helicopter (chopper to the Americans) the 40 kilometres to the drill-ship. As soon as the well reached the Upper Jurassic objective oil shows began to appear in the mud. Cores were cut and the well was tested. The test flowed almost 2,400 barrels of oil per day, which sounds great - the ultimate reward for all the preparation, effort and expense that Esso had invested in the project, but there was a problem, the oil was thick and heavy, and obviously biodegraded. Somehow during its long gestation, it had become oxidised, which had reduced it to the consistency of tar. The oil flowed only when it was hot, and the cost of treating it was prohibitive. So, back to the drawing board; we would need to come up with some answers and explanations. We returned to the structure with our MO-8 well in 1970 and this time drilled deeper to test the Middle Jurassic which proved to contain oil which was not biodegraded although the quality of the reservoir was not great, and the oil would not flow. The discovery was named the Cap Juby Field and was subsequently drilled by Mobil who confirmed the deeper oil pool but considered it to be sub-commercial.

Morocco is a fascinating country, and we got to explore some of the more exotic locations. Fifty or sixty kilometres up the Souss valley from Agadir is the ancient walled town of Taroudant entirely surrounded by a huge battlemented wall set with bastions and gates, which dates from the sixteenth century. Inside the walls there was (and probably still is) a souk selling leatherwork, antique guns, daggers, metalwork and such like, and an old hotel called the *Palais Salam*, with a pool and a lovely cool shaded garden. My mother and Wendy's mother came to visit and so did my cousins John and Margaret. We went to the cascades at Imouzzer, which are spectacular during the wet season, but dry up during the summer months, and drove down an idyllic valley of pink rocks, pink houses, pink oleander flowers and date palms to the oasis of Tafraout, and on another occasion, south to see the camel market at Goulimine. But one of the most memorable trips was over the Tizi-n-Test pass at an altitude of 2,090 metres in the High Atlas, to Marrakesh, where we stayed at the Mamounia Hotel. Churchill had painted there in the 1930s, and he took President Roosevelt in 1943 (during the Casablanca Conference) to see the sunset over the snow-capped Atlas mountains. We saw the snake charmers and story-tellers on the Djamaa el Fna, restaurants with rather flabby belly dancers, and the souk with its quarters of goldsmiths, metal workers, tanners and dyers. The city is built at an elevation of over 500 metres where the air is crystal clear, and the light more intense than anywhere I know. Little wonder that it appealed to painters.

To celebrate our offshore discovery the local *caid* invited Esso employees to a *mechoui* at his *uzba*, twenty kilometres or so along the Souss valley. This was a traditional open-air banquet, with retainers in white robes and turbans and daggers in their belts. The feast was held in the courtyard, lit with flaming torches, and the guests sat on carpets around brass tables, with ten guests at each table. Whole sheep were roasted on spits on the watchtowers around the courtyard, and there was one sheep for each table. The lamb was served with couscous, Moroccan salad and *pastilla*, a dish made with chicken, almonds, pistachios, chopped dates and spices within fine layers of filo pastry. It was a night to remember, but who got the sheep's eyes I do not recall.

Perhaps nothing better illustrates Morocco's recent feudalism than the story of the El Glaoui clan. They were war lords in the High Atlas who controlled one of the principal caravan routes heading south from Marrakesh. Their main base was a fortress at Telouet, 2,000 metres up in the mountains. In 1893 the Sultan was traveling over the mountains with his entourage when they were caught in a snowstorm. They were helped and given shelter by the El Glaoui clan and in return the Sultan made the clan chief *caid* of the surrounding provinces, and presented them with a fully operational Krupp cannon. The family grew in influence and acquired the posts of Grand Vizier, and Pasha of Marrakesh, and became virtual king-makers in the treacherous world of corrupt and inept Sultans and pretenders, and shifting allegiances. The French gradually established control over Morocco, under the mantle of protecting French interests, with the active support of the Glaoui clan. Thami El Glaoui, the clan leader after 1918, became a virtual viceroy of southern Morocco on behalf of the French, and assumed the soubriquet of Lord of the Atlas. The family amassed great wealth and lived like feudal lords, with slaves and concubines, and numerous fortresses. They were instrumental in exiling Sultan Mohammed V first to Corsica then to Madagascar, but nemesis struck in 1956 when Mohammed returned to Morocco, negotiated independence from France and became king. Retribution was swift, Thami el Glaoui died in Marrakesh, the family was stripped of its titles, properties and wealth, and forced into exile.

We went to see one of their castles in Ouarzazate, the Taourirt Kasbah, a maze of almost 300 irregularly shaped rooms, narrow passage-ways, harem, courtyards, ksours, dungeons and kitchens. Some of the rooms were decorated with elaborate plasterwork, but all were empty, neglected and crumbling. The guide who showed us around was a former black slave who ironically was now master of the place. I made the trip with three of our friends from Agadir, since Wendy and Susannah were away visiting family at the time.

From Ouarzazate we continued down the valley of the Draa, the longest river in Morocco which originates in the High Atlas, flows through a valley of *palmeries* and olive groves before eventually petering out in the desert (except in the wet season when it flows all the way to the sea). Until 1958 the lower Draa valley formed the border with Spanish Sahara. The countryside became progressively more arid, and at the end of the road is a dusty little town called Zagora. Beyond there is nothing but desert, and at the end of the main street there is a crudely painted sign showing a camel caravan and an arrow pointing to 'Tombouctou, 52 jours'.

A second drill ship was acquired and in 1969 we drilled four more exploration wildcat wells each one on a different structure in the hope of finding the ideal combination of trap, reservoir and the presence of non-biodegraded oil. On the rig, there are considerable periods, during tripping to change the bit or to run casing or test the well when the geologist is free to catch up with his paperwork or update his daily log. It was during one of these lulls, whilst working on the MO-5 well, that we heard over the ship's radio that 'the eagle had landed'. It was 20th June 1969. Apollo 11 had landed on the Moon. I was on a crew change flight on the Piper Cherokee one Sunday from Tarfaya to Agadir when the charter company assigned a young French woman as pilot because the regular pilot was unavailable. We were not encouraged to see her reading the instruction manual before take-off, and even less so to learn, when we were past the point of no return, that the control tower at Agadir was unmanned on Sunday afternoon because there were no scheduled commercial flights. She was forced to fly without ground assistance, and land 'by the seat of her pants' - not a procedure to be recommended. On another occasion flying back to Heathrow in winter on a Royal Air Maroc Caravelle the plane landed on a very wet runway, started to aquaplane and began to slide sideways. It was clearly out of control, and ended up on the grass. All the lights went out and there was a good deal of confusion. The emergency fire trucks were on the scene in minutes, but fortunately no one was injured and the plane was not seriously damaged, but not a pleasant experience.

I worked on a couple of the new wells and witnessed what a dangerous environment it could be. We always had divers on board in case of problems on the sea floor or with the riser. They were all professionals, most of whom had been trained by the navy. Most of the time there was nothing for them to do, but when the rig arrived or left a location they went down to attach or detach the riser from the blow-out preventer, and occasionally there were leaks which needed to be fixed. The water depth at the drilling locations was mostly between 90 to 120 metres, which requires several hours in a decompression chamber after the dive to avoid 'the bends'. The divers were very well paid, but it was a dangerous job and most of them retired from diving by the age of thirty. On one occasion, at the beginning of a well, a diver went down to latch the riser to the blow-out preventer, and as he guided the riser into position the ship rose on a swell and as it fell, the riser came down and chopped off his fingers. They got him up and into decompression and patched him up, but that was the end of his diving career. The drill floor too can be a very dangerous place with heavy chains being tugged by winches to operate the tongs. There are always a number of roughnecks and roustabouts on the drill floor and many of them carry tools in their pockets. One day a panel of the metal drill floor was lifted to carry out repairs underneath, and one of the roughnecks inadvertently stepped backwards where the panel had been and fell through the hole into the moon pool. His pockets were full of spanners and tools; he disappeared from view and was never seen again.

At the end of that drilling season my assignment in Agadir came to an end and I was transferred back to the UK. Esso continued their Moroccan operations until 1975 and drilled four more wells, none of which was commercially successful. All of the wells, except the final one, had oil shows, but none contained all three essential ingredients of source, reservoir and trap, and they decided to call it a day. Other companies have subsequently explored in the area with an equal lack of success, and the great hope of oil production from the southern Moroccan offshore remains unfulfilled. Early in 1970 Esso shipped all their reports and documentation on a flight back to the UK, accompanied by the office secretary. The plane crashed approaching Casablanca, spreading Esso's files and reports across the countryside and killing 61 of the passengers including Mercedes, the secretary, who had a young son waiting for her at home.

I was then sent on a brief assignment to Barcelona where Esso had a regional office in a villa in a residential part of city. The manager was a French Canadian, Pierre Coty, who had a roving eye, and hired secretaries more for their looks than for their secretarial skills. At lunchtime, they would lie by the pool and look very decorative. The house had a conference room in the centre of the building without windows and with steel walls – a relic, I was told, of the Spanish Civil War.

Letter 23
Change was in the air. Wendy was pregnant with our second child, the Moroccan venture was winding down, John Parmenter had been transferred to the North Sea Group, and Dave Kingston, the flamboyant manager of the East Atlantic Study Group had been replaced by a much more business-oriented manager called Henry Schaetti. Henry was an 'organization man' who took pains to obtain a consensus before proceeding. He was dedicated and thorough, and introduced tight security, insisting that all our maps and reports should be locked in a vault overnight, and instituted a system where articles sent by company mail were placed in two envelopes, the inner one marked 'Proprietary, addressee only'.

Concurrent with the operations in Morocco the company was engaged in evaluating the offshore areas of Senegal and Portuguese Guinea. Seismic data had been acquired and drilling began in Senegal in late 1967. Two wells were drilled north of the Cap Vert peninsula and three to the south. In addition, French companies had drilled many wells onshore Senegal and had made some small discoveries, and the information from these wells was useful in interpreting the offshore geology, so I was sent to Dakar to collect information on these wells.

Senegal was a world apart from Morocco. It is situated in tropical Africa and the climate is very hot and humid with lush vegetation. The population is diverse, but is dominated by the Wolof people who are very black, very tall and very slim. The country obtained its independence from France in 1960 and at the time of my visit was a fairly stable, mildly socialist, pan-African state. Dakar was a staging post for flights to the south and across the Atlantic, and there was a constant stream of airline personnel passing through the hotel where I was based. Esso had an office in Dakar and my contact was a French geologist who was resident there. Senegal had been a major source of slaves in the seventeenth and eighteenth centuries. They were captured in the interior by coastal tribes, brought to the coast and sold to European slave traders. We visited the island of Gorée where the slaves were held in the *Maison des Esclaves*, with its 'door of no return', and its underground vaults, complete with shackles and all the accoutrements of slavery. The slave trade was active between about 1670 and 1810, and whilst the number of slaves who passed through Gorée is disputed, the place had a very chilling feel to it.

The vegetation is completely different from Morocco and includes the bizarre-shaped baobab tree with a very wide, partially hollow trunk and a canopy of branches at the top resembling the bristles on a chimney-sweep's brush. It used to be the custom to place dead bodies in the hollow trunk to be eaten by the vultures which sit ominously in the branches. I got into conversation with a group of local youths, most of whom were wearing rather elaborate bracelets. These were *gri gri*, fetishic amulets which they believed brought good luck and protected them from harm. I offered to buy one, but no amount of money would persuade them to sell.

One evening my French colleague offered to take me for a drink in the port area of Dakar. I imagined a typical French bar down by the sea, but the waterfront was dark and ill-lit with sinister looking buildings, a perfect setting for a *film-noir* movie. It got worse. The bar looked very disreputable, dimly lit, smoky, with secluded couches around the edge. It suddenly dawned on me that this was a bordello, frequented by sailors fresh off one of the numerous ships in the port, and several of them were already enjoying their shore leave. I can't imagine my French friend was a *habitué*, but then again, who knows? Those French are funny guys, but it scared the shit out of me.

Senegal turned out to be another Morocco as far as Esso were concerned. I was not involved in the drilling operations, but I did work on preparing some of the prospect reports and in contributing to some of the completion reports. The company drilled five wells which were frustratingly unsuccessful. We were trying to locate various Cretaceous reservoirs on the continental margin, but we discovered that a major fault was present between the onshore and offshore and the prognosis for the first well was seriously in error. The other wells were drilled either too far proximal or too far distal of the shelf edge to be viable, and the final well found the objective, but on the wrong side of a fault. Exploration drilling is a high-risk business.

I never got to visit Portuguese Guinea, which may be no bad thing. At that time the country was still a Portuguese colony governed from Lisbon. It had a grindingly poor economy and one of the lowest standards of living in Africa. There is a commonly held belief that Portugal treated its colonies better than most other European countries, by integrating more with the local population both by marriage and by commerce, but a movement for independence which began in 1956 had steadily gained momentum, until eventually the insurgents controlled almost the entire country. Six years after our involvement Portugal was forced to grant independence, and relinquished sovereignty of a colonial possession which it had held for exactly 500 years.

We had a concession which covered the entire offshore area of the country in an area called the Casamance salt basin. This region contains beds of salt, and when the overburden pressure on salt reaches a certain point, it becomes plastic and begins to flow. It finds the line of least resistance and rises up fault planes and areas of weakness, and will eventually pierce the overlying strata, almost like lava, forming domes and walls of salt. The arched-up rocks above the domes form potential traps for migrating oil, and those were the principal target of our exploration effort in the area. Three of the five wells tested salt dome structures but all three were dry holes. One of the other wells tested a small amount of oil but the reservoir quality was poor, and shortly afterwards Esso relinquished the concession. Esso's campaign in West Africa had not been a success. The company had spent over $100 million and had not made a single commercial discovery. The main problem in all three countries was a lack of a good quality, mature source rock capable of generating large volumes of oil.

Our second child was born in September 1970 and I attended the birth at the local clinic near Camberley. Some men find it an inspiring experience, but to see your wife in pain and all the blood and fluids is not for the squeamish, and personally I would not recommend it. But maybe that says more about me than the event itself. It was a boy and we called him Simon. We had him home for a few days, but it became clear there was a problem and we had to return to the hospital. Tests showed that he was suffering from hydrocephalus, a build-up of fluid in and around the brain, which causes the brain to swell. He was transferred to the Atkinson-Morley Hospital in Wimbledon, which was one of the most advanced brain surgery centres in the country. They inserted a cerebral shunt to drain fluid from the brain, but there were further complications and his condition did not improve. He did not seem to be in any discomfort and would give us a lovely smile, but they kept him in hospital. He survived for four months, but his condition proved to be incurable and he died the following January. It was a very emotional event when he was buried in a little plot in the churchyard close to where we lived in Camberley, and it was infinitely sad to see two-year old Susannah putting a posy of primroses on the grave of her dead baby brother.

I spent the winter of 1970/71 writing a number of 'post mortem' reports on our exploration efforts in Morocco, Senegal and Portuguese Guinea, following which I was assigned to a small group charged with assessing the potential of a number of offshore basins in Namibia, South Africa and Mozambique. It involved searching published accounts, obtaining information on earlier discoveries and dry holes, checking the activity of other companies, and working with our geophysicists who were busy interpreting regional seismic data which the company had acquired. The geophysicist with whom I was working was not very cooperative and appeared to resent my participation in what he seemed to regard as his project. By this time my salary had risen to about £3,500 a year, still far below that of the American personnel, which rankled, when one day I saw an advertisement for an oil exploration geologist for a job in Algeria. The company which placed the ad was a British firm called CJB, an offshoot of the famous John Brown shipyard in Glasgow, which had branched into civil engineering projects, particularly those associated with the oil industry. They had undertaken several projects for the Algerian oil company Sonatrach, and had been commissioned to hire staff for the company. The salary offered for the geological post was three times what I was making with Esso, tax free and with free housing. With our recent experiences, we decided that it was time to move on. I applied for the job and got it, and was hired on a two-year contract.

I had gained invaluable experience with Esso for which I was extremely grateful, and which provided a firm foundation for my future career. When I submitted my resignation, although I was paid for the 30-day notice period, they told me to clear my desk immediately and leave the same day. They do not like to have people hanging around who have resigned from the company. Such an arrangement is called 'gardening leave'. In 1973 Esso was rebranded as Exxon, followed by a major upheaval which the staff referred to as the great double cross. Esso Exploration Inc was wound-up in 1983 having completed its world-wide review, and its conclusions were uncharacteristically pessimistic. They felt that all the great oil-bearing basins of the world had already been discovered and all that remained was 'mopping up'. 'Mopping up' however can mean many things, and the application of increasingly sophisticated methods of oil extraction was to give the industry a much longer life than Esso had anticipated.

Letter 24

In May 1970 we sold our house in Camberley, and the car, packed our things and left for a new life in Algeria. Having worked in Morocco we had no qualms about another overseas assignment, albeit in a different country.

Unlike Morocco, Algeria had been under French rule for more than 130 years. The conflict between the Barbary regencies of North Africa (Algiers, Tunis and Tripoli) and Europe had a long history. Nominally under the control of the Ottomans they were in fact largely independent, and their principal source of income was derived from piracy and slavery and ransom. From the early 1500s Barbary corsairs ranged far and wide, seizing ships and landing at secluded spots as far distant as Ireland and Cornwall (not to mention Spain and France and Italy) to seize prisoners. This was big business. It is not widely known that hundreds of thousands of Christians – perhaps as many as a million - were sold into slavery in North Africa between 1530 and 1800. Samuel Pepys mentions a conversation in 1661 with two seamen who had been captured and enslaved in Algiers and later released for a ransom. By 1800 the European powers were belatedly beginning to accept that their own involvement in slavery was immoral and unjustifiable, and that it was also time to put an end to the Barbary corsairs once and for all. Improved technology meant that by this time the corsair ships were no match for European and American ships. The United States fought a campaign against Tripoli from 1801 to 1805 (hence the reference to the 'shores of Tripoli' in the Marines' Hymn), and in 1816 the British navy launched a punitive raid on Algiers and destroyed a large part of the corsair fleet and freed 3,000 Christian slaves. In 1827 an argument between the Dey of Algiers and the French consul led to a blockade of the port, and three years later the French seized the opportunity to send an invasion force which occupied Algiers and quickly established control over the coastal areas. After several years of resistance led by the Emir Abdel Kader, Algeria was declared an integral part of France in 1848.

Hundreds of thousands of French citizens had established themselves in Algeria, as professionals, farmers, wine producers and traders, and many had been there for generations. They (of course) acquired the most productive land, owned the best houses, controlled the economy and administered the country. The northern part of Algeria was incorporated into metropolitan France as three *départements* – Oran, Alger and Constantine, but the vast, sparsely populated interior was administered as a 'protectorate'. The French inhabitants were known as *colons* or *pieds-noirs*, and included well known figures like Albert Camus, Jacques Derrida, Louis Franchet d'Espère and Yves St Laurent. The ethnic mix of French Algeria was roughly one million French colons, eight million Berbers, and two million Arabs, Turks and others. The Berbers were the indigenous inhabitants of Algeria, descended from the Phoenicians, Romans and Byzantines, and the Turks were remnants of the Ottoman occupation.

The movement for separation of Algeria from France began on VE day in 1945 when Moslem protestors in Sétif organised a march demanding independence. The march turned violent and a massacre occurred in which 100 pieds noirs were murdered and over 1,000 Muslims killed by French troops in retaliation. Following the French defeat in Indo-China, the independence movement gained momentum and the FLN was formed. It issued a proclamation calling for armed revolt against French rule and the establishment of an independent Algeria. A bloody civil war ensued which claimed the lives of 150,000 FLN fighters, 70,000 Muslim civilians, 25,000 French soldiers, 3,000 French civilians and 40,000 *harkis* (Moslem auxiliaries who fought with the French). At the height of the war there were more than half a million French troops in the country. Most of the FLN military commanders were killed during the war (and their names are now perpetuated as street names in Algiers). The war caused the collapse of the French Fourth Republic, and the establishment of de Gaulle as President with almost dictatorial powers, and he was strongly backed by the French forces in Algeria. On a visit to Algiers he made the cryptic statement *'Je vous ai compris'*, but just whom he had understood was left unsaid. By 1959 public opinion convinced de Gaulle that independence was inevitable, which incensed the *pieds-noirs* and the French military commanders in Algeria (who ironically by this time had more or less established control over the country), and a new phase of the war began in which the generals, supported by many French settlers, formed the OAS whose objective was to keep Algeria French. The uprising was eventually quelled at the cost of a further 2,000 lives and two attempts on the life of de Gaulle (one of which was portrayed in the movie *The Day of the Jackal*). The war ended with the signing of the Evian Agreement of 1962, which contained provision for *pieds-noirs* to retain their civil rights, but such was the bitterness and suspicion between the communities that the 95% of the *colons* chose to leave. They simply abandoned their properties, left their cars on the quayside and took the first available ship to France.

By the time we arrived, the government of Algeria was in the hands of Houari Boumédiène, a secretive and rather sinister figure of Berber descent who had been a colonel in the FLN. He was chairman of the so-called Revolutionary Council which controlled a one-party Socialist state, euphemistically named *La République Algérienne Démocratique et Populaire*. Almost all of the major enterprises had been nationalised, so there were dozens of company names beginning with Sona- or SN. One of these was the state oil company, snappily called the *Société Nationale pour la Recherche, la Production, le Transport, la Transformation, et la Commercialisation des Hydrocarbures*, happily shortened to Sonatrach, which was established in 1963.

Algeria was a totally different kettle of fish to Morocco. The mood was generally more sombre, some things were hard to obtain – there was a six-month waiting list for a new car (which I was able to circumvent by buying one from a guy who was shipping one in, but in the meantime had decided to quit), there was a great deal of bureaucracy, and there were frequent road blocks where the police would check papers. On the other hand, there were still many relics of French life – hotels, nice French restaurants, bakeries, delicatessen, shops, and wine was still freely available. The climate was good and there were some delightful places (*stations balnéaires*, as the French charmingly called them) along the coast. The language in Algeria was a strange mixture of French and Arabic plus many Berber words (particularly in place-names). Classical Arabic speakers, I am told, regarded it as rather droll.

The regime was strongly Socialist, and almost all of the major enterprises had been taken into state control. This included agriculture, and most of the former French-owned estates had been converted into collective farms. The oil and gas industry, which had begun under the French in the 1950s, was by far the most important money-earner in Algeria. Following independence, the state oil company, Sonatrach, was set up to participate in the exploration and production of oil and gas. Some French companies were reluctant to accept the involvement of Sonatrach in their activities and opted to pull out. In retaliation the French embargoed the importation of Algerian wine, which they had long used as a cheap source of *vin ordinaire*. Wine production in any case was regarded by the Algerians as unethical for a Muslim country, and many old-established vineyards were torn up *(arraché)* and replaced by artichokes (which seemed a shame, but I guess they had different priorities). The departure of many French oil-field specialists forced the Algerians to recruit experts from other countries – notably from the Soviet Union and Romania, and through contractors like CJB, which is how I came to be hired. I was employed by CJB, but on permanent secondment to Sonatrach. The CJB office, which I had to visit from time to time, was in Birmandreis on the hill overlooking Algiers. To get there I had to take the colourfully named *Route du ravin de la femme sauvage*.

We had the option of free housing at a Sonatrach complex at Boumerdès, on the coast 50 kilometres east of Algiers, or of finding (and paying for) our own accommodation. We opted for the former, and were given villa A-74, a furnished prefabricated bungalow near the beach, which was quite spacious and comfortable, and surrounded by other similar houses all occupied by ex-pats. In 1961 Boumerdès (then known as Rocher Noir) had been chosen as a refuge for the *Gouvernement général français* during the final stages of the war (and in case of the need for a quick getaway). Since independence it had become *'Cité Sonatrach'* where a Petroleum Institute had been established along with laboratories and a research centre, and homes for expat employees of the company, so there was a mixed population of western specialists, teachers at the institute, Algerian students and Eastern Europeans, mostly Russians, who were housed in a separate part of the town.

And so, we started a new chapter. The company provided a bus to transport personnel to the Sonatrach offices in Algiers that left at 6.30 in order to begin work at 7.30 am. On arrival in Algiers we habitually went to a little café in Place Agha where they served coffee and freshly-baked croissants straight from the oven. The Sonatrach exploration office was on the seventh floor of *Immeuble Mauretania* downtown Algiers, with a view over the port. The company operated a French-style policy of having a *planton* stationed on each floor to monitor the comings and goings of staff. To go out during working hours you had to obtain a *bon de sortie* which you showed to the *planton* on the way out. At lunch time, we normally went to a bistro, of which there were many, but our regular was the *Brasserie des Facultés* facing the university on the Rue Didouche Mourad, where they served an excellent cassoulet. Almost all of the streets had been given new names following independence. The only survivor that I could see was Rue Victor Hugo, which I think still exists. There is another French-sounding name in Place Maurice Audin, but thereby hangs a tale. The square was named after independence in honour of a young mathematics assistant at the University of Algiers. He was born in Algiers, the son of French settlers and he had become a

Communist, and therefore a man of suspect loyalties as far as the French military were concerned. He was arrested at home by French paratroopers in June 1957 and was never seen again. The French claimed that he had escaped whilst being transferred from one place to another, but they never made any effort to search for him. It became a modern Dreyfus case. His wife hired lawyers and demanded a judicial enquiry, letters were published in *Le Monde*, demonstrations were held, and questions asked in the National Assembly. A commission of enquiry was set up, but all to no avail. To this day the fate of Maurice Audin remains a mystery.

I was assigned to Sonatrach's Exploration Department, presided over by a brisk little man called Ahmed Said (who had previously headed up a totally unrelated national enterprise), and to District 4, the Triassic Basin, under the management of Mohand Ait Hamouda. The Chief Geologist of this group was a Palestinian, who was capable and proficient, but not too fond of the English, and I reported to the *Chef de Secteur* of the Eastern Area, an experienced Russian geologist called Alexei Markov. It was a very international organization. Since I had worked on Jurassic carbonates with Esso, my first assignment was a regional study of the Jurassic carbonates of eastern Algeria, a rather neglected topic since almost all the oil in the basin was present in rocks older than the Jurassic. This study allowed me to become familiar with the wells which had previously been drilled and the discoveries which had been made, but there was very little potential in the Jurassic, so I wrote my report and moved on to work on the Triassic which was already producing oil from several fields and had potential for much more.

Wendy made friends with other expat wives in Boumerdès, particularly with a young Indian woman called Nayeer, who had a daughter about the same age as Susannah. Susannah was pale skinned with ash-blond hair, Nipa was dark-skinned with black hair, and the contrast was very remarkable. They spent a lot of time together, at each other's houses and on the beach. In 1971 Wendy became pregnant again, and in view of the problems with Simon, she went back to England for the birth. Your dad, who we called Jonathan (maintaining the theme of biblical names), was born in Scunthorpe in February 1972. She returned with Jonathan when he was just a few weeks old, and he spent the first four years of his life in Algeria. When Susannah was four we enrolled her in the local *garderie* which was run by French teachers (who opted for teaching rather than National Service), and she soon picked up the language. By the age of five she was speaking far better French than either Wendy or me. My mother came to visit us on several occasions, and she got to know many of the expat families, particularly Wayne and Billie, George and Circe and the Qazis. Wayne was an American geologist and Billie was a former model who had been Miss California in her youth. They had a house in Grapevine Texas, and my mother went to visit them there several years later. George and Circe had fled Hungary at the time of the Hungarian Uprising in 1956 and had emigrated to Canada. George was friendly with everyone and a great source of information. If you wanted the latest gossip, George was the guy to ask. The Qazis were a family from Pakistan who came to Algeria a year or two after us. Naeem Qazi was an ambitious geologist who later in life became a director of the Pakistan Geological Survey. My mother went to stay with them in Islamabad for several weeks, which must have been quite an experience, but my mother was a great one for travelling to exotic places. Our other neighbours included people from Turkey, Palestine, Egypt, India, Canada, United States, Tunisia, Yugoslavia and Peru. It was a veritable united nations.

We found a maid to do the cleaning, and help with the children and I remember one time going to the regional police station to renew my work permit. I mentioned that we had just changed our maid, and the policeman said 'yes, we know'. They kept very close tabs on everyone, both expats and locals. Expats heading home on leave had to present the authorities with a *quitus* to prove that their taxes had been paid and debts discharged. It was a very bureaucratic society.

Letter 25
Algeria is a beautiful and very diverse country, and we used the car, my little green Fiat 850, which I had bought when we first arrived, to explore the countryside at weekends and public holidays. There is a rich and fertile coastal plain around Algiers called the Mitijda, with impressive former French estates, orange groves, orchards, fig trees and vineyards. Further east the Kabylie mountains form a very distinct region, occupied by Berbers who have lived there since before the time of the Romans. They have their own customs and language and have fought off numerous attacks over the centuries. For protection they established hilltop villages like Beni Yenni, a group of tiny hamlets situated precariously on a ridge, where there is a thriving *artisanat* industry producing silver filigree, coral and enamel brooches, pendants, bracelets and such like. Further south is the Djurdjura range of limestone peaks, resembling the Italian Dolomites. The mountains are covered in forests of cedar and cork-oak (parts of which were destroyed by napalm bombing during the war of

independence), culminating in the ski-resort of Tikjda (*station estivale et hivernale,* in the flowery language of the French), at an altitude of 1,600 metres.

South of Algiers the Mitijda plain extends as far as Blida, beyond which the road rises rapidly through cedar forests to the Col de Chréa, at a similar height to Tikjda, another winter ski resort. From the terrace of the church there are sweeping views to the coast 35 kilometres away, and it is said that you can ski at Chréa in the morning, and swim in the Mediterranean in the afternoon. Nearby another road climbs through the gorge of the river Chiffa where there is a small hotel beside a series of waterfalls on the *ruisseau des singes,* where monkeys come down to the road at meal time to see what they can scavenge (or steal) from the visitors.

The coastline too had some idyllic spots. Le Figuier, close to Boumerdès, we called Rocky Bay, and it was a perfect place for a swim and picnic. There was an attractive circuit up into the heart of the Kabylie mountains at Tizi Ouzou, then on a twisting, winding road over the mountains, through meadows of wild flowers reminiscent of Switzerland, to Tigzirt, and from there a rough and rutted unpaved road along the coast to Dellys, where there was an old hotel with the curious name of *Hotel Grande au Large*. The small ports along the coast have changed hands many times with evidence of occupation by Punic, Phoenician, Roman, Byzantine, Arab, Turkish, Spanish and French conquerors.

To the west of Boumerdès there were a number of seaside places – Cap Matifou, La Pérouse, Alger Plage, Les Ondines, but our favourite spot was Bordj el Kiffan, which we always referred to by its French name of Fort de l'Eau, where there was a street of sidewalk café's each one with an outside barbecue, grilling sardines, squid, prawns, *rougets, merlan* and other fish (the Algerian answer to fish and chips), and locally produced Kronenbourg and 33 lager. There were no plates or knives and forks, but plenty of serviettes and bowls of water with lemon to clean your hands. If you fancied a more up-market experience there was the *Poker d'As* which was a rather fancy French restaurant serving more 'refined' dishes like *soupe de poisson, loup de mer, mérou, dorade, St Pierre* and lobster.

The name Algiers comes from the Arabic *Al Jazair* meaning the islands, from the four islands which formerly stood just offshore but which have now been incorporated into the city. The city is particularly impressive when seen from the sea, with tier upon tier of white buildings rising majestically towards the wooded heights of El Biar, Hydra and Birmandreis. In the foreground are the port and waterfront buildings, on the right the prominent outline of *Notre Dame d'Afrique*, a nineteenth century Byzantine Revival basilica (the twin of a similar building in Marseilles), then the tightly-knit quarter of Bab el Oued, a stronghold of working class *pieds noirs* during the war. In the centre are the labyrinthine alleys and *escaliers* of the totally Moslem casbah, and further to the left the business district with the enormous Moorish *Grande Poste* and the tall Mauretania Building, where I was based, and finally the industrial areas of Hussein Dey and El Harach.

In those days it was safe enough to walk around Algiers. Even some of the mosques were accessible. We briefly visited the casbah, but the Algiers casbah is very different from that of, say, Marrakesh. This was not a place to linger. There were no tourist shops, no 'come and see my brother's shop', no begging. This was a place of blank walls, closed doors, hidden courtyards. The women wore a *hijab*, either an *amira* or a *niqaab*, and the men a *djellaba* or *burnous*. Foreigners were not welcome. The rest of the city was quite different. The buildings were European in style, there were fashionable shopping streets, coffee shops, department stores, impressive museums, fancy hotels like the Aletti, the St George (favoured by Edith Piaf, Churchill, Eisenhower and Simone de Beauvoir), and a monster new hotel called the Aurassi up on the hill. In many respects Algiers was deliberately developed to rival Marseilles across the Mediterranean. Near the St George Hotel is the *Palais du Peuple*, the former *Palais d'Eté* and residence of the French Governor-General. It was here that Admiral Darlan, the head of the Vichy government in Algeria, was assassinated on Christmas Eve 1942.

West of Algiers there is a string of pleasant seaside villages – firstly La Madrague with lots of sea-food restaurants overlooking the sea, and the vine-shaded pergolas of *Villa d'Este*, and then Club des Pins, an exclusive complex of villas among the pine trees with a beautiful beach, largely occupied at that time by business executives, but now an ultra-secure, closely guarded private sanctuary for the President, senior government ministers and *apparatchiks*, and for a while after 2011 home to members of the Gaddafi clan, following the overthrow of Colonel Gaddafi in Libya. Adjacent to Club des Pins is the vast *Palais des Nations*, a grandiose congress centre only a year or two old when we were there, where many of us from Sonatrach attended an OAPEC (Organisation of African

Petroleum Exporting Countries) conference in 1973 (largely, I suspect, simply to make up the numbers).

Further along the coast are the resorts of Moretti Plage and Sidi Ferruch, an attractive yachting harbour with the El Riadh Hotel, which was considered rather plush at the time. Sidi Ferruch is on a headland and is interesting historically since this is where the French landed in 1830 to begin their occupation of Algeria, and where the Allies landed as part of Operation Torch in November 1942 to seize Algiers from the Vichy government.

Beyond Sidi Ferruch, there was another government-sponsored vacation complex at Zeralda and further still the strange mausoleum known as *Tombeau de la Chrétienne*. This is a large circular structure built of stone on a hill overlooking the sea, capped by a stepped cone. It has four false doors facing the cardinal points and a concealed underground entrance beneath the eastern false door. Inside is a labyrinth of vaulted passages, chambers and galleries with niches for funerary urns, protected by sliding stone doors (long since broken). It is believed that the mausoleum was built for the Berber king of Mauretania, Juba II (who had been brought up in Rome and totally Romanized) and his wife Cleopatra Selene (daughter of Mark Antony and Cleopatra) in about 3 BC. Juba's capital was nearby at Caesarea, the modern Cherchell. The tomb was robbed in antiquity and apart from some bas-reliefs nothing remains inside. An eighteenth-century pasha used artillery in a fruitless attempt to open-up the interior, and in the nineteenth century it was used for target practice by the French navy (maybe in emulation of the Turks and the Parthenon).

The final location within easy reach of Algiers was Tipaza, originally a small Punic trading port converted by the Romans into a military base in about 50 AD. There are extensive remains of a Christian basilica, baths, theatre, amphitheatre and some vestiges of the ancient port. Not far from Tipaza we discovered the house of an artist from the French expedition sent to examine the prehistoric cave frescoes of the Hoggar mountains in southern Algeria, from whom we bought a couple of nice reproductions, which decorated our house for a number of years.

There was a large Russian contingent working for Sonatrach. A Russian Basin Studies Group had been in operation for some time charged with conducting a comprehensive review of the oil and gas potential of the entire country, staffed exclusively by Russian geologists, geophysicists, petroleum engineers and petrophysicists, and in addition Russian specialists had been assigned to each of the exploration districts (so they knew exactly what was going on throughout the Exploration department). It was interesting to observe the way they worked. I had had previous contact with Russian geologists during my PhD, but I was now working alongside them on a day to day basis. They were a mixed bunch. Some were extremely competent and sharp, using some novel techniques in log interpretation and petroleum engineering, but others were very slow and ponderous, hampered by the fact that many of them knew very little French or English, in which all the reports were written. The Romanians on the other hand were mostly fluent in French since Romanian (like French) is a Romance language derived ultimately from Latin. The Russians were friendly enough and were curious about prices in the west, the cost of a car, houses and salaries, but politics, by some unspoken agreement, were not seriously discussed.

The Russian Basin Studies Group completed their four-year project in 1971, and published a report in English. It was a very strange compilation. The technical content was comprehensive and competent and contained much valuable information, but the translation from Russian into English was diabolical, and the organisation of the material was so inept and confusing as to be almost incomprehensible. Many maps were produced without title, symbols key, field names or explanatory text. To find this information you had to cross reference the figure with the appendix where, if you were lucky, the information might be found. If you were unlucky it might say, 'symbols as in figure 16'. It was an extremely frustrating book.

Our work was mainly concerned with analysing the results of previous wells and using this information to produce maps of the various reservoir units to identify favourable areas. The principal target in the area was a Triassic sandstone reservoir capped by a formation of salt (which forms an effective seal). The challenge was to identify structures in areas where the sandstone was well developed, where the seal was present and where there was a migration fairway between the oil source rock and the trap; too far south and the seal was missing, too far off the fairway and no oil was present. We identified one such fairway to the south-west of a large oil and gas field called El Borma which straddles the Algerian-Tunisian frontier, and Sonatrach drilled two or three wells in this area. One of them discovered a small oilfield which was sub-commercial, and another had oil shows. However, about this time Sonatrach made some significant discoveries further west and the company switched its attention to those areas. The decision was premature. In later years, the eastern area was to produce some very large and spectacular discoveries.

My contact with some of the Russians developed to the stage where one of their team leaders asked me if I would give English lessons to his eight-year old son. I admit to a certain fascination with Russia and Russians, so I agreed, and started giving weekly lessons to his son Lonya. His parents were grateful and hospitable, and after the lesson they always offered me tea, or a shot of something stronger. When my mother visited Russia some years later they entertained her at their flat in Moscow with Lonya as interpreter. Around this time I came across a couple of Russian forenames, Danera and Melor, which sounded distinctly unusual to my untrained ear. I later discovered them to be hortatorical, representing on the one hand 'daughter of the New Era', and on the other 'Marx, Engels, Lenin, October Revolution'! It put me in mind of the English Puritan preacher and Fifth Monarchist Praise God Barbon, and his brother Jesus-Christ-came-into-the-world-to-save-Barbon. I wonder what his mother called him?

Letter 26
We settled into a routine. Susannah and your dad (who by now was toddling) enjoyed the beach and played with neighbouring children in Boumerdès. We bought supplies from the company store or from the nearby town of Boudouaou (*L'Alma* in French colonial days) where the fish and vegetable market (including a *charcutier de cheval*) was held in a former *Marché du Tabac*, and we would go off to the beach at Figuier or the roadside barbecues at Fort de l'Eau, or head off into the mountains with a picnic. Life was not so bad.

Later we became much more adventurous and travelled to more distant places. During one of the *eid's* (Moslem religious holidays), we drove to Oran and saw the harbour of Mers el Kebir where the British destroyed the French fleet in 1940 to prevent it falling into German hands (at the shocking cost of over 1,200 French sailors' lives). Oran was under Spanish rule for 200 years and Ottoman for a further 100 years until it fell to the French in 1831. In 1962 as the FLN entered Oran to celebrate the end of the war, they were fired upon, which provoked a savage response and the massacre of hundreds of *pieds-noirs*. At one time Oran had a larger proportion of Europeans than any other city in Algeria, but after the massacre the population of the city fell by half in just two months as the *colons* fled to France. We took a lovely corniche road out of Oran heading east which, although crumbling and deserted, had magnificent views at every turn. It became clear why the road was so quiet when we came to a bridge over a gorge that was completely destroyed and we had to turn back. It seemed to me a tragedy that some of the most glorious coast roads had been allowed to fall into ruin, but there were obviously other priorities at the time.

On another occasion, we drove east to Bejaia where we stayed at a lovely Arabic-style hotel overlooking the Mediterranean, and along the Corniche Kabylie, a fine stretch of coastline with cliffs, headlands, promontories, a beautiful beach at Tichy, and the *grotte merveilleuse* near Jijel. Just to the south are the Kherrata gorges, a deep defile between limestone cliffs where the road twists and turns for eight or nine kilometres, known in Berber as the 'valley of the dead'. Further south are the Roman ruins of Djémila set on a rather bleak plateau at an elevation of almost 900 metres. The city has a forum built during the time of Septimius Severus (a Roman emperor born at Leptis Magna in Libya, who died in York), a Christian quarter, an exceptional triumphal arch and a splendid theatre which had to be built outside the city because of the lie of the land. The ruins are exceptionally well preserved, some of the buildings still retaining their original height. The economy of the place was based on cereals, livestock and olives. It was a Christian city as early as 256 AD and following the departure of the Romans it became a Byzantine bishopric until the arrival of the Moslems in the seventh century. The Romans called it Cuicul, but the Arabs called it El Djémila – the beautiful one.

Constantine, the principal city of eastern Algeria, is spectacularly located on both sides of the deep gorge of the river Rhumel, cut through massive Cretaceous limestones, a setting reminiscent of Ronda in Spain. The gorge is crossed by several bridges. It was called Cirta by the Romans, but the name was changed in about 330 AD in honour of Constantine the Great, and the name has been retained ever since. At an altitude of 640 metres the climate is cool compared with Algiers. We stayed at a hotel called the Panoramic where Susannah was bitten by bedbugs, not quite what you expect in a four-star hotel. A few kilometres to the east is Hammam Meskoutine, or Baths of the Damned in Arabic. By the roadside is a group of twenty-foot high conical tufa cones which, legend has it, represent guests at an incestuous wedding, who were turned to stone by a disapproving god, which gives the place its name. But the main attraction are the hot springs and the petrified waterfall of travertine over which the steaming water cascades. There was an attractive spa surrounded by a park with citrus trees and olive groves and a courtyard shaded by a huge turpentine tree. The waters are radioactive and were used to treat arthritis and rheumatism.

South of Constantine is an area of *chotts*, salt lakes which are dry for most of the year, but flood during the wet season, beyond which are the Aurès Mountains. Like the Kabylie Mountains, the Aurès are home to a Berber tribe called the Chaoui, who fought off successive waves of invaders, and it was there that the war of independence against the French began. The locals still move their livestock between winter and summer pastures, a custom known as *achaba*, and spend the winter in tents with their flocks.

On the northern slope of the Aurès are the remains of the third and best preserved of the great Roman cities of Algeria. Timgad is remarkable in retaining the precise outline of the grid-iron pattern of streets and houses, as a result of being 'lost' beneath a covering of sand, from the Arab invasion until 1881, when it was rediscovered. The city was built by Roman soldiers in about 100 AD and adopted Christianity in the 3rd Century. It was a centre of the Donatist schism and there are the remains of a large Donatist monastery and cathedral. Satellite images of Timgad show the layout of the city in extraordinary detail.

There is one final area of Algeria that we visited several times which is the most spectacular of all. To get there involves a drive of 400 kilometres through the Tellian Atlas, across the *Hauts Plateaux*, and finally over the Saharan Atlas to Laghouat or Biskra, where the true Sahara begins. On one of our trips we were forced to stop by a swarm of locusts. It is an astonishing sight to see them approach like a cloud of smoke and then to be enveloped by millions of buzzing insects, blotting out the sun, and becoming enmeshed in the car radiator and in the engine. It put me in mind of the bible story of the ten plagues visited on the Egyptians by a vengeful god. There is an excellent circuit from Laghouat to Ghardaia, Ouargla, Touggourt, El Oued and back to Biskra, a distance of 860 kilometres, so it requires three or four days, but this is the real Sahara, a land of dunes, buttes, camels, sandstorms, salt flats - and oilfields, for it is here that most of Algeria's largest oil and gas fields are located. The Hassi Messaoud oilfield is near Ouargla, and the Hassi R'Mel gas field is not far from Ghardaia. Hassi R'Mel incidentally is one of very few fields around the world to contain economic amounts of helium. The French actually built a railway in colonial days to transport oil from Touggourt to the coast at Skikda, and I can recall seeing, south of Biskra, a wrecked and overturned train beside the track, like a scene from *Lawrence of Arabia*, which I guessed dated from the 1950s. These days, oil and gas are transported by pipeline, which you can see in the distance heading towards the coast 500 kilometres away. It was a curious feeling to drive over a featureless landscape and to know that beneath your feet was the largest gas field in Africa. It reminded me of my youth in Halifax when I knew that there were coal seams present beneath my grandparents' house. Everything comes back to geology!

Following independence, the Algerian government made a valiant effort to attract tourists to the country, and built some very attractive hotels, particularly on the desert circuit. Unfortunately, the service did not match the architecture. In a hotel restaurant, you might wait an hour between placing an order and being served, or you might find that the hot tap didn't work, or the electrical socket was broken. The town of Laghouat marks the beginning of the desert and is a convenient place to stretch your legs, but it is a modern town built in colonial days and has little of interest. Ghardaia on the other hand is the chief town of the M'Zab valley, a unique area of fortified Berber villages with mosques which double as watchtowers, *palmeries*, and markets specializing in hand-made carpets. From there it is 165 kilometres to Ouargla, past drilling rigs and desert tracks leading to producing oilfields. Ouargla is an oasis on an old caravan route along which gold and slaves used to pass from Sudan to the coast. Now its economy is based on over half a million date palms which almost completely surround the town, producing the famous *deglat-nour* dates. The town is a strange mix of old and new. The mosques and the circular market of the old town are juxtaposed against the new town of grid-iron streets and boulevards, but with other Berber villages dotted around the *palmeries*.

From Ouargla the road heads north to Touggout. Contrary to the common belief, the Sahara desert (a tautological expression since *sahara* means desert in Arabic) is not all rolling sand dunes and mirages. There are extensive areas of stony desert, flat-topped mesas and buttes and dry *wadis*. We were lucky enough to see a camel caravan travelling the old road south from Touggourt. There were about thirty camels laden with packs, white-robed men, colourfully dressed women and many children. We tried to approach them, but they also had some rather unfriendly dogs so we desisted. It brought to mind a couplet from one of my school books:

> *I teach them from a bloodless book*
> *To scan a bloodless chart*
> *And pray one day their eyes will look*
> *To find the throbbing heart.*

I felt that finally we were doing just that. Touggourt, which translates as gateway, is noted for its immense *palmerie*, containing over one and a half million date palms which stretch for almost 30 kilometres along the Oued Irharrhar. Touggourt has an old town of several distinct quarters. The houses are built of mud-brick and many of the alleys are covered to provide protection from the sun. Unfortunately, as is common in many such places, people prefer to live in modern houses or apartments, and many of the old houses had been abandoned and were falling into ruin. The traditional way of life is ending, and Touggourt was becoming a town of characterless modern blocks and dusty, wind-blown streets.

The highlight of the trip comes on the stretch from Touggourt to El Oued where the road crosses part of the *Grand Erg Oriental*. This is picture-book Sahara, with sand dunes rising and falling as far as the eye can see. There is something very romantic about vast unspoilt sweeps of sand without a footprint, and the sensuous curves and undulations of the dunes. In this area, the dunes are constantly on the move, blown by the prevailing wind, and they often spread over the roads which have to be regularly cleared by large earth-moving vehicles.

El Oued is by far the best of the five desert towns. It is an oasis in a hollow within the great sand-sea. The town has fortunately not succumbed to the trend of modernization and remains an authentic desert oasis. The view outwards from anywhere within the town is of huge dunes completely encircling the oasis. It is particularly evocative in the late evening as the setting sun casts mysterious shadows over the landscape, and the *muezzin* calls the faithful to prayer. El Oued is known as the city of a thousand domes, and this is no exaggeration, domes are the dominant feature of the place. Houses have domes, the markets have domes, the mosques and public buildings have domes; there are domes everywhere. In the market you could buy desert roses *(roses de sable),* clusters of radiating flattened crystals of gypsum formed by the evaporation of saline waters, and other curiosities of the desert – fragments of meteorites, fulgurites (formed by the fusion of sand grains from a lightning strike), and prehistoric flint implements. The markets were suffused with the smell of spices; cumin, cloves, turmeric, cinnamon, nutmeg and ginger. El

Oued is the chief town of a region called the Souf (Berber for valley) where the water table is very close to the surface, and it is an odd sight to see the entire area dotted with small lakes and lagoons in the middle of the Sahara desert. From El Oued it is a long drive back to the coast, but you arrive home marvelling at the diversity of this rich and varied country, and feeling privileged to have been able to experience it.

Letter 27
During our summer vacation of 1972 Wendy's dad arranged a couple of visits for me to the steel works in Scunthorpe, then owned by British Steel, where a new plant was being commissioned to handle iron ore imported from overseas, and to Conoco's oil refinery on the Humber estuary. They were both fascinating trips and I was grateful to him for arranging them. We were finally beginning to understand one another. My cousin John, who worked as a diamond sorter for the Diamond Corporation in Hatton Garden in London, had become engaged to Julia, a nice girl from Tottenham in north London, and we went to their wedding. In the fashion of the time John had hair down to his shoulders, I had a van Dyck beard, Wendy had a striped navy and white coat, my mother and Eva had frothy pink hats, and my cousin Margaret and the other girls wore miniskirts.

Back in Algiers I was given a new assignment, to conduct a regional study of the oil and gas potential of the Triassic sands in the central part of the Triassic Basin. The basin is divided into two parts by a broad ridge called the Messaoud Ridge on which there are some major oilfields including Hassi Messaoud, the largest oilfield in Algeria, but which produce from much older rocks than the Triassic. The study, based on examining dozens of well records, showed that the Triassic reservoir was either completely absent or very poor quality on the ridge, but improved progressively to the north, east and west where several discoveries had been made. The ridge itself could be written off as far as the Triassic was concerned. In October, a large delegation of Sonatrach directors and specialists went to Russia for discussions at the Ministry of Petroleum. They had a day or two in Leningrad (as it was then) before flying to Moscow. One of my Algerian colleagues on the trip was told to give up his seat to a more senior delegate, and so took the train instead. When he arrived in Moscow he found that the plane had crashed approaching Sheremetyevo airport in Moscow and everyone on board had been killed.

The following year Wendy's parents, John and Norah (who had by this stage built themselves a nice house in a village north of Scunthorpe), came to visit us in Algeria, and we took them on the tourist circuit to the beaches, to Djurdjura with the springtime flowers in bloom, and to the desert at Bou Saada, Touggourt and El Oued. I completed my Triassic regional project, and my contract with CJB came to an end in June of 1973. I was offered a further two-year contract and since we were now accustomed to life in Algeria, I decided to accept. It was a fateful decision. That summer I returned to England with Susannah by car whilst Wendy flew back with Jonathan. We took the car ferry from Algiers to Marseilles and had a nice cabin for the crossing. The ship seemed surprisingly uncrowded but when the passengers disembarked at Marseilles hundreds of Algerians appeared from the lower decks and were herded into a special customs shed. There was little love lost between the French and the Algerians. We drove up the Rhone valley, and stopped at Avignon to see the Papal Palace and did a little *danse* on the famous *pont*, and then across France to Calais and home.

I had an extended leave before starting the new contract and we went to Greece and explored Athens, Corinth and the volcanic island of Aegina, ending with a magical sunset at Cape Sounion where according to legend the father of Theseus threw himself into the sea when he saw his son's ship come into view under a black sail. Maybe it was an omen. We visited Halifax and Scunthorpe, and that was the last time that I saw Wendy's father.

We had been married for seven years, and I remember quite distinctly visiting a country pub called the Maypole near Halifax with Wendy at which we reviewed our life together. Generally, we felt content. We had two lovely children, I had an interesting job and a good income, and we had a varied set of friends. There were some negative elements – this was our fifth year living overseas in a third-world country, far from home, there was still something of a disconnect in my relationship with Wendy's parents, and we had differing views on religion and politics, but these seemed like minor matters. We decided quite deliberately that we needed to be more out-going, more independent, more adventurous. It turned out to be a fateful decision.

Back in Algeria I started my new contract. Many of the Russians had reached the end of their contracts and had returned to Moscow, including Markov, *my chef de secteur*, and Aristov and his son Lonya. I was offered

Markov's job and so became responsible for the geological interpretation of a sector of the Triassic Basin. The days are long gone when individuals selected wildcat drilling locations. Nowadays it is a team effort involving geophysicists, geologists, petrophysicists, and petroleum, reservoir and drilling engineers. Data is collected, analysed and worked up to determine whether a prospect merits the expense of perhaps five million dollars to drill. If a prospect appears viable a drilling recommendation is prepared, incorporating all the relevant information and a prognosis, which is then submitted to management for approval. The team closely monitors the well during drilling, and analyses the results on completion. A well completion report is then prepared reviewing what was found and explaining why it was a success or failure. A normal success rate (of finding some oil or gas, but not necessarily in commercial quantities) in this sort of area was about one well in five. My job was to coordinate the geological part of these activities. By this stage it had become a chore to ride the bus to work every day, so quite often three or four of us, Wayne, Frank, Nick and I, would drive to work. In the evening, we would buy a bottle of wine to share, which made for a convivial drive back home.

Susannah was now attending the French school and Jonathan was walking and talking and beginning to show his character. A new crop of Russians appeared on the scene, some assigned to our department, some to the Russian Basin Studies Group, and some to the Hydrocarbon College in Boumèrdes. Aristov had recommended me as an English teacher to some of the newcomers, so I found myself continuing to teach English to one of the new Russian children called Genia. Among the new arrivals were a number of secretary/translators, including a slim, attractive young woman with long flowing hair, totally different from the tough, rather dowdy image of Russian women at that time. Partying was part of the way of life in the expat community and so, in the spirit of our 'Maypole' resolution, I invited one or two of the Russians, including the slim-line secretary to one of our parties. In those days, the activities of the Russians were very closely monitored. Fraternization with westerners was not encouraged, but since we were all mixed up at work anyway, they tended to be fairly relaxed, so long as things were kept within limits. The girl showed up, accompanied (chaperoned perhaps) by another female, and I found that her name was Irina, that she was 23, married and from Moscow. She didn't speak English, but we were able to communicate in French. I was fascinated by her. There was an air of the exotic about her, and mystery and allure. She had moved into the villa formerly occupied by Aristov, along with another woman, which was just across the street from my young student. I called in there after one of the lessons on some trivial excuse, and that marked the biggest turning point of my life.

I do not propose to dwell on the events of the next few years. They were not edifying, and it was a traumatic and upsetting period for all concerned. I fell in love with Irina, and out of love with Wendy. The shocking thing is how quickly it happened. I had feelings for Irina which I had never had before. Maybe it was kama or the sexual imperative, or the fact that I had little experience of women, or that our relationship was clandestine. I don't know. But I developed an obsession with her, and became blind to all reason. Wendy was aware of the situation of course and she did everything she could to stop it, persuasion, threats, tit for tat, but to no avail. Opportunities for me to meet Irina were rare, but we managed a few trips, not always alone, sometimes with Susannah or with Irina's roommate. The Russian authorities knew that something was going on, and probably had a good idea what it was, but they had other concerns (which I will elaborate in another letter), so they left us alone. At home tensions rose, and I am ashamed of some of the scenes which ensued, but I was adamant. I would not let go. Irina reciprocated my affection and we began to think the unthinkable. Nemesis often follows in the footsteps of hubris, and so it was. Once on the slippery slope there is no way back. Wendy and I rapidly lost all respect for each other, and without respect and affection there is nothing left – except the children. I love my children dearly. I had watched them grow and develop. We had shared all the usual pleasures of happy childhood, bath-time, bedtime stories, trips to the sea. How could I think of abandoning them? Was that the price I would have to pay? It was a decision which has dogged my footsteps ever since.

Some love too little, some too long,
Some sell, and others buy;
Some do the deed with many tears,
And some without a sigh.

Not surprisingly Wendy took them back to England for extended periods and during one of these trips she called me to say that her father had died suddenly whilst on holiday in Majorca. I went back for the funeral of course. I collected his car which had been left at the airport and drove it back to their new house in Burton, which he had had little time to enjoy. It was a sad affair and I was sorry that I could not show more sympathy to Wendy and her mother, but not surprisingly my reception was deservedly frosty.

In October of 1974 Irina's contract came to an end and she had to return to Moscow, and Wendy probably felt, even at this stage, that our marriage could be saved. But it was too late, too much damage had been inflicted by both sides. We had passed the point of no return. I have to say plainly that it was my intransigence which led to our divorce. I had chosen this path. Now I would have to live with it.

Letter 28
In spite of our domestic problems I still had to work. Due to our limited success in the eastern part of the Triassic Basin we now switched our attention further west to the area between Touggourt and Ghardaia, and in the period from 1970 to 1976 we made discoveries at Takhoukht, Guellala NE, N'Goussa, Oulougga, Bou Khezana and Oued el Meraa. Then we discovered a second trend, further west which yielded discoveries at Draa Temra, Kef el Argoub and Haniet el Mokta, and near Ghardaia we found three more oilfields at Ait Kheir, Makouda and Djorf. These discoveries were relatively small but formed a compact group, and some of them were subsequently developed and put on production. We were rather pleased by our success. Of course we drilled a number of dry holes too, some of which I remember - Bou Goufa, El Meharis, Bou Chaffra, Bou Settach, Orbata, Daiet et Tarfa. But no one is interested in dry holes.

The expats at Sonatrach contained some interesting characters. There was a Pakistani woman geophysicist, of aristocratic appearance who used to wander around the offices quite freely and inquire what people were up to and what they were working on, and never gave the impression of doing a stroke of work herself. She appeared to have a privileged position with a line of communication directly to top management. It seemed like a very fishy situation to us lesser mortals. Just what was her status? We speculated of course, but never found out for sure. In our exploration district we had an American guy with a gammy leg, which he claimed was the result of a wartime flying injury (presumably in the Korean war), but others said he had crashed a plane at a flying club whilst being the worse for drink.

The Russians too had their stories. One of them, who was employed as a translator, spoke perfect English. He had no accent and only very occasionally would use a word not in frequent use by native English speakers. He had never previously been out of Russia, and had learnt his English at a special school in Moscow. He knew the lyrics to all the Beatles songs by heart and I remember thinking that it was a pity that he was not around earlier when they made such a hash of the translation of the Russian Regional Study report. A year or two later I arrived at Moscow airport, and who should be supervising Passport Control but the Beatles-lyrics Russian. Was it just coincidence?

Another of them, who also spoke very good English, was the son of a drilling engineer who had been sent to California in the 1930s to check out American drilling technology. He returned to Russia and recommended a number of innovations, which he had observed in the US, and was rewarded with a fancy apartment in Moscow. During the Second World War the Russians were desperate to find new oil deposits in case the Germans were able to reach the oilfields of the Caucasus (which they almost did). They drilled in the Ural Mountains where winter temperatures can reach -40 degrees Celsius, and the drilling crews would be given a shot of vodka every hour just to keep their blood flowing. He said that during the war his family was located near Perm in the Urals, and one day in the depths of winter an old woman came to their door begging for food. They took her in and fed her, and it transpired that she was a former aristocrat who had been deprived of her house and denied a job by the Soviet regime, and was destitute. She was an educated woman, fluent in English and French, so my friend's family took her in and gave her a place to stay in return for her giving English lessons to my friend and his sisters. He said that she arranged amateur theatricals for the children, and that they dressed up and acted scenes from Shakespeare, in English, in the Ural Mountains, in the middle of World War II.

All the Russians were very circumspect in their dealings with western expats – except one. There was a guy called Alex Nilov, who worked as an instructor at the Boumerdès Hydrocarbon Institute. He was young and personable, but a bit suspect because he didn't play by the rules. He was a relatively new arrival and he lost no time in befriending western expats. He would show up at parties, and was welcomed because he was very sociable and spoke good English. He knew that I had some Russian connections and he would occasionally show up at our house, out of the blue, for a drink or to bull-shit. I don't recall him being especially critical of the Soviet system, but he was more admiring of the west than most of the others. He got to know many of the expats in our neighbourhood through the party network, including a colleague and good friend of mine called Frank. Frank had lived in the States for many years, had married an American, and had acquired US citizenship. He had a daughter, who was about twenty-three or four. She had a young baby, but I think was not married. She brought the baby and came to live with Frank and his wife for several months. Alex fell for her in a big way. He was keen to marry her, and he said that she was equally willing. He told us he would consider defecting in order to be with her, but we thought he was being unduly melodramatic and over-emotional. In the end Frank sent his daughter back to the States with the baby, and shortly afterwards Nilov returned to Russia. Frank himself resigned not long afterwards and rejoined his family in the States. The dramatic *denouement* came five years later, but that is for another letter.

In June of 1974 I was invited to be an external examiner for the final year geology students at the Petroleum Institute in Boumerdès (where Nilov worked but in a different department). It was an interesting experience. Most of the teachers were Russian and they all had a vested interest in demonstrating that they were efficient, competent and up to date. The students presented a short review of their project work and the four internal examiners, and myself as the one external examiner, rated them according to content, relevance and presentation skills. One of them was very poor and I rated him as unsatisfactory, but the chairwoman pointed out that we could not actually fail anyone, they all had to pass, albeit some better than others. I thought how typically Russian, but I was to discover that the same attitude prevails in other places too, and I was to encounter it in an even more blatant form later in my career.

The following month Kamal, my Palestinian boss, left to take up a post in Kuwait, and I was appointed to take his place as Chief Geologist of District IV, to the chagrin of one expat in particular who thought he should have got the job. This brought me into daily contact with exploration management by attending daily meetings, and an involvement with partners like Total and Hispanoil who were operating a number of concessions with Sonatrach. At that time all the senior posts in Sonatrach were occupied by Algerians, but almost all of the technical work was done by expats who were also supposed to provide on-the-job training for young Algerian graduates from the Hydrocarbon Institute or from universities overseas. It was a very mixed bunch. My new boss was the Exploration Manager of District IV, an Algerian Berber called Mohand Ait Hamuda, who prided himself on his appearance, and would walk down the corridor like a ship under sail. He ran the department with a fairly light touch and was willing to listen and discuss matters with me and the Chief Geophysicist, and we developed quite a good working relationship. Most of the expats were competent, but came from a diversity of different backgrounds – American, Indian, Egyptian, Tunisian, British, Yugoslav, Russian, and it was no easy matter to weld them into an efficient team. There was none of the gung-ho enterprise and competitiveness that I had experienced with Esso. At Sonatrach it was more a matter of sorting wheat from chaff. A few of the Algerian juniors showed potential and one in particular went on to achieve considerable success, as author of a number of important publications and he showed up much later in Libya as a member of the National Oil Corporation. In my new role I got to meet some of the petroleum engineers who were based miles away in an office at Hydra on top of the hill. It was an odd system since geologists and engineers need to work closely together, but because of the physical separation our contact with them was minimal.

My cousin Margaret, who had lived in Halifax whilst I was at UCL, had become a tour guide with Thomson Holidays, and had led tours in many Mediterranean countries, but especially to Greece. Margaret and my mother had lived with a Greek family in Athens for several weeks in the Spring, and she now spoke Greek quite well. She had met and become engaged to a Greek called George Tsangarakis, and they were married at Caxton Hall in London in December of 1974. George came from Crete, but following their marriage they came to England and set up house in London.

My trusty Fiat 850 had provided good service for the best part of five years, but had had some rough usage and was now giving repeated problems so I decided to replace it with a Renault 5, a model introduced a couple of years earlier, so I ordered one from the Renault factory in Paris and flew there with the children to collect it, and then drove it home.

My mother knew what was going on between Wendy and I – she could hardly fail to notice, and she was clearly upset to see our marriage heading for the rocks, but her comment was 'You are my son, I will support you whatever you do'. I sometimes wonder what I would have done if she had said 'I'll never speak to you again if you do this', but of course she couldn't say that; she was a widow and I was her only child. So I ploughed ahead. I was able to keep contact with Irina through a go-between who, at considerable personal risk in those benighted times, shuttled letters between us. The method was never discovered and I am forever grateful to our faithful go-between for keeping our lines of communication open. I went to visit Irina in Moscow for her 25th birthday and we made plans for the future. Wendy and the children stayed in

Scunthorpe and I drove back to Marseilles and put the car on the ferry to Algiers. The car carried temporary French plates, which I was able to use in Algeria for six months without paying import tax, which would have been prohibitive.

And so things wound down. Wendy and the children came for brief visits, but spent much time in the UK. Ironically I was offered a new contract at an increased salary, and we had more drilling success in that last year than in any previous year, but it was obvious that 1976 would see the end of my time in Algeria, a career which had started out so well and ended in so much uncertainty. We started divorce proceedings at the end of 1975, as did Irina, and so the die was cast.

Letter 29
In 1963 Harold Wilson made his famous 'white heat of technology' speech in which he claimed that Britain's ruling elite was dominated by fuddy-duddies and elderly old-Etonians, who were almost wilfully ignorant and contemptuous of science and technology in an increasingly technological world. Wilson became Prime Minister in 1964 and inherited an economy that was in dire straits. With perfect timing the first gas discovery in the North Sea was made in September 1965 and the first oil discovery in December 1969. It could not have come at a more opportune time. North Sea oil became the saviour of the British economy. Since then dozens of oil and gas fields had been discovered. The first gas was produced in May 1967, and the first oil was brought ashore in June 1975. A new Labour government was elected in 1974, again with Wilson as Prime Minister. The following year he appointed Tony Benn, a left-wing Labour radical, Secretary of State for Energy. Benn felt that Britain was not getting a fair deal from North Sea oil and gas production. Much of it was being produced by foreign companies and shipped overseas, a process over which Britain had virtually no control. He decided that Britain needed the security of a guaranteed supply of North Sea oil in time of crisis, so in 1975 he set up the British National Oil Corporation to address this situation. There was nothing unusual or sinister in this arrangement; most major oil producing nations had national oil companies: Mexico (1938), Iran (1951), Brazil (1953), and after the foundation of OPEC in 1960 many other countries followed: Kuwait, Saudi Arabia, Algeria, Nigeria, Norway, Malaysia. It was the start of an inexorable trend in which ownership of oil reserves and production gradually passed from private companies to national companies. By the end of the century 90%

of the world's reserves, 75% of production and 60% of undiscovered reserves potential was in the hands of national companies.

This was a situation of which I fully approved. I had seen it at work in Algeria, and I believed this could (should) be the future in Britain too. Curiously this had been one of the areas of conflict between Wendy and myself. Her outlook was traditionally conservative, and she was totally opposed to the idea of public ownership. In June of 1976, whilst on vacation, I had an interview with the fledgling BNOC at Dorland House off Piccadilly in London, which was just beginning to come together, and another interview with British Gas, a long-established company which handled gas distribution, but had recently become involved in exploration. The BNOC interviewers told me that they were not yet hiring new personnel, but that I should apply again when they began recruiting staff the following year. British Gas made me an offer of a job on the spot (although I made my feelings about BNOC known to them), and I decided to accept. I resigned from my contact with CJB and Sonatrach and left Algiers for the last time in August. We had lived there for just over five years with pleasure and pain in equal measure. It was time to start a new chapter.

Wendy and the children moved in temporarily with her mother in Burton, a nice little village near Scunthorpe overlooking the River Trent, and she subsequently bought a house in the same village. I needed to find a place to live not too far from London. I checked out a few properties, including a pair of former gamekeeper's cottages at Wendover in an acre of lovely beech woodland on the Rothschild estate at the foot of the Chilterns, but the location was too far from London. At this juncture, good friends came to the rescue. Before going to Algeria, we had become friends with the Brown family who lived nearby and had children the same age as Susannah and Jonathan. They had since moved to a larger house in Crowthorne, two or three miles from Camberley, where we also knew a family called the Oxleys. Both were generous in offering me a bed until I found a place of my own, and I shall always be grateful to them. I eventually found a modest little house in the same town which I was able to buy, and I moved in there shortly afterwards. The house was very close to the Broadmoor Hospital for the Criminally Insane, a grim Victorian building, which held some of the most dangerous prisoners in Britain. Each Monday the alarm system would be tested with a siren, which was a forcible reminder of the proximity of the place. In the nineteenth century, it held an American doctor, William Minor, who had been a surgeon on the Union side during the Civil War who had murdered an innocent Irish workman in London who he believed was about to assassinate him. So he was incarcerated for life, but he was a scholar, and was allowed to purchase books, and was given an extra cell for his 'library'. He responded to a public appeal from the compilers of the Oxford English Dictionary to assist in providing references for words to be included in the dictionary, and he became one of their principal contributors. It was not until many years later that the editor discovered that one of his most important correspondents was confined to a mental hospital.

My divorce from Wendy was finalised in November of 1976. I was able to see the children regularly, usually in Halifax, but it must have been very confusing and heart-breaking for them. I justified it to myself by saying that they were better off with Wendy alone than with the two of us arguing and mistrusting one another, but I have to admit it was a hollow, self-justifying argument, and it is a spectre which has haunted me ever since. I had visited Irina once or twice and her divorce had also been granted. We were able to communicate by using innocuous postcards where the message was not in the text but in the picture. Big Ben meant one thing, Westminster Abbey or the Kremlin meant another. And so we began to plan our marriage which, of course, would have to be in Moscow.

Irina had an interesting background. She was born in Kazakhstan in Ust Kamenogorsk where her father was serving in the army, (now Oskemen, founded as a fort and trading post in the early 1700s). We used to joke that if she had been born any further east, she would have been Chinese. Her parents were both Russian, but her father had been born in Leninogorsk (now Ridder) in the Altai Mountains, where rich deposits of gold and silver, copper and zinc had been discovered in the 1780s. Her parents had an Alsatian guard dog. One evening in winter they put Irina to bed in her cot and went out, leaving the dog on guard duty. When they returned, there was no sign of Irina, the cot was empty and the dog was lying beside the cot, licking its lips. Panic! But then they found that Irina had fallen out of her cot, and the dog had curled up around her to keep her warm.

I started work with the British Gas Corporation at their offices in Bryanston Street in central London in September 1976. The publicly owned Corporation had been established in 1973 by the integration of many regional gas boards and the Gas Council to form a consolidated company responsible for all aspects of gas distribution and supply in Britain. The Corporation was headed by Denis Rooke, an engineer who had pioneered marine transportation of liquefied natural gas, and who did more than anyone to rationalise the company as an effective and integrated organisation. The Exploration Division though had a rather peculiar setup. It was a fairly recent creation and was regarded by many of the staff as an insignificant little group uncomfortably tacked onto the main business. The office was open plan and we shared a floor with Marketing, so that our exploration work was carried on against a background of telephone calls and chatter about contracts and sales and pricing which was seriously distracting. I was surprised one day to have a visit from an impressively uniformed Pakistan Airlines pilot with a box of mangoes, which caused a few raised eyebrows among the marketeers. They had been sent by my friend Qazi, following a visit by my mother to his family in Islamabad. The Exploration Department at that time was very small. The manager was Peter Hinde, a cultivated and knowledgeable man, who had battled against stiff opposition to drill the well which discovered the Morecambe Bay gas field. He lived in a manor house in Kent and had commuted to the office by train for many years. He claimed to have read all of Dickens' novels whilst travelling back and forth. The Chief Geologist was no-nonsense, outside-the-box kind of chap called Vic Colter who was enjoying much credit for the recent discovery of the Wytch Farm oilfield in Dorset in 1973, the largest field ever found onshore Britain. It is located in an area of outstanding natural beauty, so had to be very carefully developed and screened, so that once the drilling rigs had moved out the installation was almost invisible from the surrounding area. The field extends offshore into Poole Harbour and it was developed by drilling deviated wells from five onshore hub locations. It contains about 450 million barrels of recoverable oil, which compares favourably in size of some of the North Sea fields. The field came on production in 1979. Four years later British Gas was scandalously forced by Margaret Thatcher's government to transfer ownership of the field to BP following a bitter and acrimonious battle between Denis Rooke and Nigel Lawson, the Energy Minister.

At the time that I joined, British Gas was involved in appraising the Morecambe Bay gas discovery in the Irish Sea which was visible from the famous Blackpool Tower (think Coney Island and you will get a sense of Blackpool). Five wells had been drilled, two of which tested gas at shallow depth. Subsequent drilling proved the presence of a large gas field with 5 trillion cubic feet of reserves which at peak production provided

15% of Britain's total gas requirement. The field, as you might guess, ended up being sold to private investors in 1986.

The exploration staff was very small, with about a dozen geologists and geophysicists and a similar number of petroleum, reservoir and drilling engineers. Twenty years later I met one of the British Gas geologists by chance at a concert in London, and discovered that he had become an independent wellsite geologist working in Kazakhstan. We got together for a reunion after which he returned to Kazakhstan, where I subsequently heard he had been killed in an accident. I was given the assignment of conducting a regional study of the continental margin from Norway, west of Shetland, the Hebrides, Ireland, and the Western Approaches to the tip of Brittany, to examine these areas for their hydrocarbon potential. It was a fascinating project which involved collecting data from numerous sources and compiling geological maps and sections to illustrate the various basins and their potential. Some were partly explored and some were totally unexplored. The project involved tracing the history of sea floor spreading in the North Atlantic, which was a relatively new science at the time, and then trying to work out the subsequent development of each discrete area using evidence from the Deep Sea Drilling Programme, and from any other available wells, seismic data or reports. We had some good regional seismic lines over many of these areas, and two of the British Gas geophysicists picked the key horizons on these lines for me. With these data, I was able to build up a picture of the basins along the Atlantic seaboard and the sedimentary fill within them.

In the far north I did a quick-look review of the Norwegian offshore. Up to that time all the Norwegian exploration effort had been confined to the North Sea. Further north, the Norwegian continental shelf plunges into the Møre and Vøring basins which are very deep, but offshore Trondheim there is a terrace feature which steps down towards the trough, called the Halten Terrace, which appeared to me to have potential for hosting oil or gas in Jurassic fault blocks. No drilling was done in this area until the 1980s and 90s, but then a whole series of small gas fields was discovered, many of which have since been put on production. One of these is the Ormen Lange gas field which is located on the Storegga Slide, perhaps the world's largest landslide, when an area the size of Iceland collapsed into deep water at the end of the Ice Age resulting in a tsunami, with a run-up 10 metres high in Norway, but over 20 metres in Shetland and eastern Scotland, and which left evidence of its passage as far away as the Firth of Forth in Scotland. There was concern that gas extraction from the field could destabilize the slide, but it was judged to be a negligible risk since there was no evidence of any recent movement on the slide.

The continental margin west of the Shetland Islands steps down into the Faroe-Shetland Channel and has considerable hydrocarbon potential. It is terminated to the south by the Wyville-Thomson Ridge beyond which the continental margin becomes abrupt and plunges rapidly into deep water. In many places deep canyons have been cut into the slope by subsea turbidity currents. To the south-west of Ireland lie the Porcupine Trough and the Porcupine Bank, and further west still is a text-book example of a failed rift, where seafloor spreading began, and formed the Rockall Trough, but then aborted, to begin again much further to the west, leaving behind the Rockall Plateau, a 'microcontinent' of detached continental crust between two areas of oceanic crust. The 'fragment' extends for 400 miles from the latitude of Galway in Ireland to the latitude of the Orkney Islands, and on the Rockall Plateau there is a sedimentary basin called the Hatton Trough. These basins and platform margins all have theoretical potential for the presence of hydrocarbons, which it was my task to evaluate. Rockall, incidentally, is a tiny, sheer barren rock (formed of a particular type of granite, called – guess what - rockallite), located 280 miles west of the Hebrides which was 'annexed' by Britain in 1955 in order to extend the area over which Britain claimed jurisdiction (meaning rights of exploration and exploitation). A naval helicopter landed four sailors to symbolically plant the British flag. Since then the 'sovereign rights' have been extended even further west, to a point 550 miles west of the Hebrides, far out into the Atlantic Ocean. In 1971 it was claimed in a House of Commons speech that more men had landed on the Moon than on Rockall.

One of the main areas of interest was the West Shetland Basin, in which over a dozen wells had been drilled, but so far without commercial success. I rated this area highly. There was a ridge of tilted fault blocks with basins on each side, and the seismic and well data were very encouraging. I noted in my report that two wells were currently drilling, and one of these turned out to be the BP discovery well of the Clair field. Unfortunately, the oil is heavy and acidic and although the discovery was very large, only about 15% of the oil was judged to be recoverable. Development did not begin until 2001 with first oil produced in 2005, 28 years after the initial discovery.

I won't bore you with all the other basins I examined, but it was a rewarding study which revealed much about the origin and evolution of the Atlantic margin. Some of the areas I highlighted yielded discoveries in later years, others did not. The biggest disappointment has been the basins around Ireland. At the time of my study the only commercial discovery in Irish waters was the Kinsale gas field in the Celtic Sea Basin, 30 miles south of the Old Head of Kinsale (where the *Lusitania* was torpedoed in 1915), which was discovered in 1973 and came on stream in 1978. Since then a number of small satellite accumulations have been discovered, but they don't amount to much. There are many areas of platform margins, terraces and fault blocks on the western margin of Ireland which look promising, but in spite of much subsequent drilling only one modest sized gas field (Corrib) has been found in this area. Corrib was discovered by Enterprise Oil in 1996 and subsequently acquired by Shell, and has proved particularly troublesome to develop. Twenty-one years after its discovery the field is still not on production. No commercial drilling has yet been carried out on the remote Rockall microcontinent, although several Deep Sea Drilling boreholes and British Geological Survey test wells have been drilled. One problem here is the presence of thick Tertiary lavas which are extensively developed, but no doubt someone will drill there one day.

The Western Approaches (*Mer d'Iroise* in French territorial waters) and the entrance to the English Channel (*La Manche* to the French) have also proved disappointing, despite having good quality Triassic reservoirs and salt related structures. The problem may be a lack of source rocks.

There is an interesting corollary connected with the exploration for oil and gas in the English Channel. One specific area, called the Hurd Deep, 18 miles or so north-west of Guernsey, is off-limits for drilling and seismic activity. The Deep is a trench in the middle of the English Channel, cut by a catastrophic flood at the end of the Ice Age when the Dover Straits barrier was breached. This was used as a dumping ground after the First World War for unused shells, explosives, chemicals and other wartime *matériel*. It is also the last resting place of the German battleship SMS *Baden* which capsized after being used as a gunnery target by the Royal Navy in 1921. The Deep was also used after the Second World War for similar disposal of unwanted munitions, and the practice was not stopped until 1974. In 1951 the British submarine HMS *Affray* failed to return after a routine mission. After many weeks of searching the wreck was found in the Hurd Deep. Because of the lethal surroundings it was impossible to recover or enter the submarine, and the cause of her loss has never been satisfactorily explained. Given that seismic acquisition requires the use of explosive charges, it is not surprising that the Hurd Deep is off limits for oil exploration. The Hurd Deep was not the only place used for the disposal of noxious material. In 1946 an obsolete troopship the *Empire Woodlark* was filled with unwanted chemical weapons and scuttled north of the Shetland Islands, very close to the Benbecula gas field, discovered in 2000. Not the ideal place to shoot a high-velocity air-gun seismic survey. At the end of World War II, in a *reprise* of the First World War surrender, the Germans scuttled over 200 of their U-boats, but eventually about 150 were handed over to the Allies - and what happened to them? The Royal Navy towed them out to sea and sank them at various locations off the west coast of Scotland and the north coast of Ireland. There is an awful lot of wartime junk out there.

I submitted my final report on the geological history and hydrocarbon potential of the western seaboard to British Gas in August 1977, and I presented a summary to the Inter-Universities Conference in Swansea in September (and I gave a brief radio interview to a BBC reporter, who had heard my talk). I was pleased with the result. I felt that I had distilled the essential and most relevant points from a complex area, and I think by and large it has stood the test of time. I like to believe that it was of some use to British Gas in assessing which areas of the UK western seaboard held some prospectivity, and which did not. It also set me thinking about the contribution of people like me to a large company. No more than three or four percent of geologists ever win a prestigious award or reach the higher echelons of management. It is the ordinary, run-of-the-mill, competent geologists who do the work, interpret the data, identify the areas of interest, propose drilling locations, monitor the wells, and analyse the results, but it is usually the managers who take the credit. There is a name for this phenomenon, it is called the Matthew Effect, from the parable of the talents in the Gospel of Matthew. A fine example is the 'cabinet of curiosities' assembled by John Tradescant, gardener to King James I. It ended up in a purpose-built museum in Oxford, but it wasn't called the Tradescant Museum, it was called the Ashmolean, after the guy who bought it from Tradescant.

Letter 30
I went to Moscow for Irina's birthday in January of 1977, and took with me most of the documents I thought I would need to get married, but inevitably there were some documents missing. I returned in April still

unsure whether our marriage would be sanctioned, so I went alone. Fortunately, this time all went smoothly, and no obstacles were put in our path, and so on 21st April, (the Queen's birthday), we were married in Wedding Palace No 1 on the Ulitsa Griboyedov in Moscow. The wedding was attended by Irina's colleagues from Algeria, and of course by her family and friends. Irina looked stunning in a peach-coloured dress and carrying a bouquet of roses. Afterwards we drove to the Lenin Hills overlooking Moscow, and the reception, banquet, toasts and dancing continued long into the night. At Irina's parents' apartment, I was welcomed with the traditional offering of bread and salt, and in the following days met many of her relations and friends. Irina's mother was a hospital nurse and her father worked in local government. They had met during the war, when Irina's dad was convalescing in hospital after an injury. He had an impressive war record, being present at the defence of Moscow and in a tank regiment at Stalingrad, a titanic battle which proved to be the tipping point of the war. He ended up as a Captain in command of a squadron of T34 tanks, and went on to fight at Kursk, Warsaw, Berlin and Prague. His career had a curious symmetry with that of my uncle Dick. Both joined the army before the war, both married nurses whom they had met during the war, both fought their way through Germany in 1944, and both arrived in Berlin in their tanks in 1945. In the Tiergarten in Berlin to this day there is a Soviet War memorial flanked by two T34 tanks. Irina's brother Yuri was an engineer, recently married to Alla, who came from a very large family which included a Major-General who survived Stalingrad, and a KGB officer. I'm afraid my marriage to Irina did not help Yuri's career. Having a sister married to a foreigner was regarded with some suspicion.

Moscow at that time was still in thrall to the Communist Party, which controlled all aspects of life. Red banners with Soviet slogans were everywhere, all the hotels and stores were state controlled, and the arts were carefully monitored for any signs of deviation. The leaders Brezhnev, Andropov and Chernenko were sclerotic *apparatchiks*, like cardboard cut-outs from another age. Everyone could see that the system was ossified, but no one possessed the power to change it; the regime was designed to be self-perpetuating. That does not imply by any means that life was intolerable. People everywhere make the best of what they have. Irina's parents had a small flat about fifteen kilometres from the centre, plus a dacha at Malinovka, beyond the outer ring road. Yuri and Alla had a separate flat, and Alla's family had a dacha at Pravda, a small township, 45 kilometres to the north. They spent much of the summer there, and grew fruit and vegetables, arranged barbecues, collected mushrooms, sat around the samovar, and steamed themselves in the *banya*; not much different from Tolstoy, except that his estate was rather larger, and he had a few serfs to do the heavy work.

Moscow has some impressive sights, the Kremlin, Red Square, the Bolshoi, Tretyakov Gallery, Pushkin Museum, the famous marble halls of the Moscow metro, and the infamous yellow-brick façade of the Lubyanka Building. Just across the river from the Lenin Hills (as it was then) is the Novodevichy Cemetery, burial place of the elite of Russian and Soviet society: Bulgakov, Chekhov, Chaliapin, Gogol, Eisenstein, Khrushchev, Molotov, Prokofiev, Richter and Shostokovich. There are some attractive places out in the countryside too, elegant former palaces of the aristocracy like Arkhangelskoye and Kuskovo, and picture-postcard monasteries, like Zagorsk, with gilded or brightly coloured onion-dome cupolas. These places are particularly attractive in winter under a covering of snow, when you could take a ride in a troika, wrapped in furs, like some nineteenth century Russian nobleman. But in the late 70s Moscow was a dour place. There was not much traffic on the streets, people didn't smile a lot, and the officials treated the public with contempt, or at best with indifference. They tended to be abrupt, rude and hostile. Frankly I found it a bit intimidating, strangely silent, and rather oppressive.

So what about Irina? What were her thoughts and aspirations? She had a degree in librarianship, and had a wide knowledge of Russian and French literature. She spoke fairly fluent French but no English. However, on graduating she had taken a job in the state patent office, which involved contact with various trade delegations and trade fairs. She had met and subsequently married a man who worked in the exotic world of films, as a producer-designer for Mosfilm, Russia's answer to Hollywood, the largest and oldest film studio in the Soviet Union, famous for producing the classic films of Eisenstein, Tarkovsky, Bondarchuk and Ryazanov. In fact, we still possess a goblet from the set of *Till Eulenspiegel*, a film of 1977 in which he was involved. She was then offered the job of secretary-translator with the Russian technical mission in Algeria, which was a plum job, because it paid in 'gold' roubles which had much more buying power than normal roubles. And that is where I met her and fell in love. She reciprocated my feelings, and the rest, as they say is history. The next step was to apply for UK residence for Irina, and we knew this would take time, so I returned to UK alone, gathered all the paperwork, filled out the application forms and submitted them to the Home Office. At the same time Irina applied for an exit visa from Russia. I had a visit from a Home Office official in June, presumably to check that the application was genuine, and then we had to let bureaucracy take its course and hope that all would soon be resolved.

At our wedding I had met Irina's friend Svetlana and her affable and charming fiancée who was introduced to me as Joseph, a Greek who had been studying in Moscow. He came to London shortly before I left for Glasgow, and I showed him Crowthorne and Windsor and Henley and took him to meet my cousin Margaret and her Greek husband George. It quickly became apparent that he was not Greek, nor was his name Joseph, but I didn't find out the truth until later.

Oh, what a tangled web we weave
When first we practise to deceive!

In fact, he was Palestinian and called Imad. Why would a person lie about their origin? Did he have something to hide? He married Svetlana later that year, and we saw them often during the next few years. Imad, I was led to believe, was a safety engineer in the oil industry, and for a while they lived in London, but he led a peripatetic existence particularly in the Middle East, including with the Saudi Arabian Bin Ladin group of companies. They spent Christmas with us in 1978, but then, after one of his overseas trips he was refused entry back into the UK. Svetlana went to Sweden where she had friends, and Imad went to work in Abu Dhabi. I never felt comfortable with him, and it often seemed as though he had a hidden agenda. He would show up after a job with wads of money which he would blow very rapidly on gallivanting. Ostensibly this was money that he had made on his overseas assignments, but what he actually did was never quite clear. I didn't see him again until 1985 by which time his marriage to Svetlana had ended. He showed up out of the blue at an oil industry conference in New Orleans which I was attending and later he visited us in Glasgow, once again flush with money. We also went to visit Svetlana in Malmo with Susannah and Jonathan where she had remarried and settled down. I never saw him again. I heard that he ended up in Australia, but who knows? Maybe he did or maybe he didn't.

I still had my eye on the British National Oil Corporation. British Gas was fine, but the Exploration Department was small, and the working conditions far from ideal. BNOC started recruiting in the summer of 1977 and I had an interview with them as a result of which they offered me a position as a senior staff geologist. Due to some blatant political manoeuvring, it had been decided to set up the headquarters in Glasgow, which for me would involve a considerable upheaval, and the sale of the house in Crowthorne which I had only occupied for a few months. But apart from that it ticked all the boxes, and I accepted their offer. I visited the office in Glasgow and had a meeting with Mike Ridd, the Chief Geologist, who had recently moved to the company from BP, with similar motives to mine. I also found that the Exploration Manager was my old classmate from Durham, Tony Challinor, who had come up through the ranks of Burmah Oil Company. I tendered my resignation to British Gas and started work with BNOC in Glasgow in August.

The concept of a British National Oil Corporation had been under discussion ever since the first Wilson government, and at one stage BP was considered as a possible vehicle by which the aim could be achieved as it had been part owned by the government for many years, but it was judged politically unacceptable for an old-established international oil company to take on responsibilities such as compulsorily purchasing oil from other North Sea oil companies. In the end BNOC was assembled by cobbling together the North Sea interests of Burmah Oil, which had had to be bailed out by the Bank of England after incurring enormous debts on its tanker fleet operations in 1974 (an episode in which Denis Thatcher was involved), Signal Oil and Gas, a US company which had been acquired by Burmah in 1974, and NCB (Exploration), a division of the nationalised National Coal Board. The government deemed it inappropriate for the large parent NCB company to be involved in North Sea oil and gas exploration. Several employees from these companies transferred to BNOC, Don Shimmon from Signal became Production Manager, Dick Fowle from Burmah became Exploration General Manager, my former class-mate from Durham University, Tony Challinor, also from Burmah, became Exploration Manager and Iqbal Shoaib, from NCB, became one of the managers based in London.

BNOC was given wide and exceptional powers, which were strongly resented by purely commercial companies like Esso, Shell, Agip, Amoco, Gulf and BP which had to work under fairly rigid constraints. BNOC was allowed to borrow on the financial markets, had access to the National Oil Account, and was exempt from paying Petroleum Revenue Tax. It was given the right to buy oil produced from other operators (at market price) and resell it as it thought fit. It had the sole right to any licences issued outside the normal licensing rounds, and to 51% equity in all new exploration acreage, and had first refusal on all farm-in opportunities. Furthermore, it was given the right to have a representative on the operating committees of all producing fields. These were breath-taking powers which put BNOC into a totally dominant position. BNOC inherited the equity previously held by Burmah, Signal and NCB (Exploration) and before long had varying degrees of equity in over 150 blocks, mostly in the North Sea, but also west of Shetland, the Western Approaches, the Irish Sea and the English Channel. The coverage was comprehensive and immense, and included several producing fields and discoveries.

This amount of work involved in handling these responsibilities required a staff of hundreds if not thousands. The Head Office was established in a new building in St Vincent Street in Glasgow, with an operations office in Aberdeen, and two or three offices in London for senior executives, lawyers, negotiators and the like. The government appointed Lord Kearton as Chairman. Kearton had a degree in chemistry and before the war had worked for ICI. He was one of the British scientists who worked on the Manhattan Project (in fact he shared an office for a time with Klaus Fuchs the Soviet spy). He subsequently moved to Courtaulds and worked his way up to become chairman in 1962. (Interestingly Courtaulds had been forced to sell off its very profitable American business during the war as part of the price demanded by Congress for passing the Lend-Lease legislation). He had a reputation for being authoritarian and for deliberately keeping people under pressure, and he was intolerant of bureaucratic interference, but I was very impressed to discover that rather than go off to play golf at the weekend, he went instead to visit offshore drilling rigs and production platforms, and really made an effort to understand the oil business. The *Dictionary of National Biography* summed up his attitude neatly by saying 'he had no time for the mere redistribution of wealth. Rather, he believed in the creation of wealth by the application of efficient management and modern technology'. By the time he left in 1979, BNOC had a turnover of £3.2 billion, profits of £75 million and participated in almost half of the wells drilled on the UK continental shelf.

The Managing Director, Alastair Morton, was the son of an oil engineer and was famously abrasive and ruthless. He was committed to the concept of a national oil company and fiercely fought Nigel Lawson over Margaret Thatcher's plans to privatize BNOC. He subsequently became Chief Executive of Eurotunnel which he turned around from near disaster to successful completion. Another director, Ian Clark, noted for his shock of hair which had turned white at a young age, had been county treasurer and county clerk of Shetland Islands Council and was instrumental in obtaining an act which empowered the council to become port authority for Sullom Voe, a major oil terminal on Shetland, so that it could levy harbour charges on ships, and displacement charges on oil companies. These generated large revenues for the Shetland Islands, and allowed them to build up cash reserves of £400 million (for a population of 20,000). It has been suggested that the character of the canny accountant Gordon Urquart in the movie *Local Hero* was based on Clark. There is also a story that Shell executives regarded Ian Clark as a more difficult man to deal with than Colonel Gaddafi.

Letter 31
I started work with BNOC in August 1977 and was based for several weeks in the Royal Scottish Automobile Club in Blythswood Square, whilst I searched for a house in Glasgow. The Club was in an elegant Georgian building just across the square from where Madeleine Smith had lived in the 1850s. The barred area window through which she had passed the arsenic-laced cocoa was still there, and it didn't need much imagination to picture the scene. In the restaurant one evening, by mistake I ordered two portions of the same side dish. The response of the waitress underlined the fact that I was now in Scotland. 'Well', she said 'it's on your bill, you might as well have both'.

Scotland has a long and turbulent history. For centuries it was a proud and independent kingdom with its own laws, church, customs and commerce. Its relations with England have never been easy and there have been frequent invasions of one country by the other. Scotland was frequently allied with France against England, but few of its kings died peacefully in their beds. In 1603 Queen Elizabeth died heirless (and apparently hairless) in Richmond, and in order to preserve the Protestant succession Scotland's King James VI was invited to become king of England, thus becoming the first king of the three nations England, Scotland and Ireland. James's ambition was to unite not just the crowns of Scotland and England, but the two countries, nicely illustrated by his introduction of a coin called the unite bearing the inscription 'I will make them one nation'. The Scots however resisted, and the countries remained separate for a further hundred years, during which English trade prospered, but the Scottish economy collapsed following disastrous speculations in several dubious overseas enterprises. The Act of Union was promoted by Scottish merchants anxious to operate on equal terms with the English, but union was regarded by many Scots as a sell-out. Burns bitterly lamented 'We're bought and sold for English gold - such a parcel of rogues in a nation'. As a sweetener the Scots were allowed to keep their Presbyterian religion and their legal system, which they retain to this day. Nevertheless, Scotland embarked on a period of prosperity, it became a centre of the European 'enlightenment', producing a long line of eminent medical men, agronomists, political economists and scientists (including some pioneering geologists). Heavy industry grew rapidly and the Scots played a leading role in administering the British Empire. However, with the decline of heavy industry following the Second World War, Scottish nationalism re-emerged, and in 1979 a referendum was held to determine whether power should be devolved to a Scottish assembly. The referendum failed to gain the necessary 40% support, but Scotland finally got its parliament (with limited powers) in 1999. A further referendum on Scottish independence in 2014 failed to achieve a majority.

Glasgow in the late 70s was a city in transition, making the difficult adjustment from being 'the workshop of the British Empire' to a post-industrial modern city. At one time the Glasgow shipbuilding industry employed 70,000 people in 50 shipyards which extended for 11 miles along the river. Now all the major industries – shipbuilding, coal mining and steel making, were in terminal decline, which is the reason that BNOC was located there. It was a gesture to show that Glasgow was eager to embrace new technology in place of the old. Glasgow is a city of contrasts. There are areas of very affluent housing in the West End, Hillhead and Pollockshields, but there are also acres of dreary tenements in Maryhill, the Gorbals, Partick and Govan. Many tenements had been pulled down in the 1960s and replaced by anonymous tower blocks in places like Easterhouse, but Glasgow Council had belatedly realised that the tenements were part of the city's heritage and that they could be cleaned up and modernised and converted into rather desirable properties. Glasgow's George Square with its magnificent City Chambers compared favourably with anything in England, and the splendid buildings of the University and Kelvingrove Art Gallery were a legacy of Glasgow's proud cultural past. Glasgow was in the process of reinventing itself, and was doing so much more successfully than many post-industrial cities.

There is a curious and little-known geological fact about Glasgow. It is quite a hilly place, and the hills are actually drumlins left after the last Ice Age. London too has some significant relief, but there the slopes represent old river terraces of the River Thames. I once knew a chap who was studying the river terraces in London and he said that he was able to map them by recording each time he had to change gear in his car.

Another BNOC recruit, a young lad called Dave Pattrick, was also staying in the club. We were on expenses and we had tried many of the well-known Glasgow restaurants, but one late summer evening he said 'let's drive out to Gleneagles and have dinner there'. Gleneagles is a famous golf and country club in beautiful countryside between Stirling and Perth, but he had a fancy sports car, so we drove over there and parked among the Rolls and Bentleys, and had a delightful dinner overlooking the famous golf course. I discovered that Dave came from a well-off family who lived in the house near Sandringham in Norfolk in which Edward VIII had installed his mistress, the actress Lillie Langtry in the 1870s. It proved to be an expensive dinner (as you might guess), and when I submitted my expense account the Chief Geologist said 'Well I'll approve it on this occasion, but don't do it again'.

I found a modern house on an estate in the small town of Kilmacolm in the Renfrew Hills, 15 miles west of Glasgow, where a couple of my new colleagues were also living. The method of buying a house under the Scottish legal system is different from that in England. The vendor will set an 'upset' price (the lowest price at which he will sell), and then usually invites sealed bids from prospective purchasers. When a price is agreed 'missives' are exchanged, after which the vendor is committed to sell at the agreed price. The property is then 'disponed' (title is transferred) to the purchaser, and ownership is finally achieved when the sale is entered in the Register of Sasines in Edinburgh. It sounds complicated but the system works very well. Kilmacolm was served by a railway branch line which gave easy access to the centre of Glasgow. The town was located in pleasant countryside with a local recreational lake called Knapps Loch. It had a long history which, as in much of Scotland, was dominated by feuding between two local families. By the time I moved there it had become gentrified (the locals were known as Kilmacomics). It had a hydropathic hotel and golf club and a village for orphan children established by a Glasgow shoe-maker and philanthropist. A referendum in the 1920s led to Kilmacolm becoming a 'dry' town; the pubs and bars were shut down and shops banned from selling alcohol. If you wanted a drink you had to go to Bridge of Weir, four miles down the road. My house backed onto woodland where there were lots of fallen trees, ideal for sawing into logs for an open fire. I requested permission to collect them (like some feudal serf) and received a formal letter from Lady Maclay, the landowner, granting permission, but 'limited to fallen timber, and for one week only'. That's Scotland for you.

BNOC's head office, where I started work, was located in a glass palace at 150 St Vincent Street in Glasgow. I was assigned to a team headed by Brian Lee, who I had first met back in June of 1976. Brian was a nice chap who had previously been with NCB Exploration. It is a curious fact that companies have an ethos, a character, and a way of doing things that sets them apart from each other, and it is not easy in a new organization to mould them together. I had seen this in Algeria, and it was even more noticeable in BNOC where former loyalties were jealously preserved. Brian's group was responsible for monitoring a number of blocks in the South Viking Graben where several fields had already been discovered including the Bruce and Beryl oil fields, and the Frigg gas field which straddled the UK/Norwegian median line. BNOC operated block 9/14, and was a non-operating partner in two other blocks in the area, 9/18 and 9/19 which were operated by Conoco. I was given the task of evaluating the two non-operated blocks which were located in an area called the Beryl Embayment. This is where I first became seriously involved in calculating reserves of oil and gas. I had worked with oil reserves in Algeria, but at BNOC we had much more precise data to use, obtained from logs, core analysis and engineering studies, and carefully planimetered areas from detailed seismic maps. The calculations for oil are not difficult, but gas is much more complex. 'Reserves' need to be clearly defined. Do you mean the amount of oil originally in the ground, the amount remaining now, or the amount which is recoverable at the surface? There are acronyms like STOOIP (stock tank oil originally in place) and RRR (remaining recoverable reserves) and many others which have precise meanings. For undrilled 'prospects' there is the additional factor of risk, which involves a whole new set of parameters. But in the end, you produce figures which allow management to compare different fields and prospects and decide where they want to spend their money.

I left Crowthorne and moved to Kilmacolm at the end of September. Irina by now had obtained an exit visa, but we were still waiting for her British residence visa, which was taking longer than anticipated. Before leaving Crowthorne I had approached my local MP asking for his assistance in expediting matters with the Home Office. His name was William van Straubenzee, an old-school, one-nation conservative, not the type you would expect to intervene in granting a residence visa to a young woman from the Soviet Union. But within days of my request her visa was granted. I don't know whether it was due to his intervention or pure coincidence, but our last major hurdle was overcome.

I went to Moscow to collect Irina just before Christmas. Since my mother had not been there for the wedding I took her along to meet Irina's family. They were very welcoming and hospitable. We met her family, and they took us to the opera and showed us the sights of Moscow. We were invited to Svetlana and Imad's, and to Aristov's (to whose son I had taught English in Algeria), and they entertained us royally (perhaps not an appropriate word for Soviet-era Moscow). Irina held a farewell party, and we flew to London on Christmas Eve.

It must have been very difficult for Irina. She spoke only a word or two of English, she was going to a strange, unfamiliar country with totally different standards and values to those she was used to, and instead of living in London, she now found herself heading for Glasgow, and even worse to a small country town where she knew no-one. I assumed, naively, that she would love the freedom and affluence and ease of life in Britain, but failed to grasp that much of what she left behind was better than what she now faced. Irina's relations with my mother were not easy. Instead of seeing the best in each other they tended to be wary of each other's motives. My mother (like me), was not good at showing affection; she was not one for hugging and kissing and welcoming with open arms. Irina must have felt the lack of warmth keenly. My family's attitudes varied widely. My aunt Rhoda more or less disowned me after I left Wendy and the children. In the last eight years of her life she never once met Irina. My aunt Eva, on the other hand, was much more open and friendly and was very willing to welcome Irina into the family. My cousins and their spouses were rather bemused by the whole business, and waited to see how things would turn out.

The first few months were difficult. It took Irina a while to get used to the idea of bank accounts and savings, rather than collecting a salary and blowing it all before the next pay-day, which was the norm in Moscow. I assumed she would naturally enjoy the wide diversity of British food (!), but by and large she preferred Russian – *shchi da kasha pishcha nasha*, as they say over there. She picked up basic English surprisingly quickly, and fortunately made friends with Ziggy, a German woman who lived nearby, who in turn introduced us to two couples who were to become good friends, Bob and Jenny, and Kenneth and Yvonne. Bob had been a minister in the (Presbyterian) Church of Scotland, but had abandoned that for a career with IBM. Kenneth was an ambitious Financial Advisor who I suspect on occasion flew rather close to the wind. My mother, Susannah and Jonathan, Eva, and Svetlana and Imad (who were then living in London) all came to see us in our new surroundings. It was a particularly severe winter, and I recall driving to Aberfoyle and the Trossachs, and over the Rest and Be Thankful Pass to Inverary, all beautiful and picturesque under a covering of snow. In the Spring, we had a visit from Wayne and Billie, my Texan friends from Algeria, and we showed them the sights of Scotland: Edinburgh, Loch Lomond, Stirling, Robbie Burns' cottage. They had been good friends since my early days in Algeria, and they invited my mother to stay with them in Texas, where they took her all over the place. She loved to travel and, in the absence of a family life, it became her principal pleasure.

My uncle Leslie (Rhoda's husband) died that Spring of heart failure (a disease that runs in the family). He was what you would have to call a stick-in-the-mud, content to hang around home, lead a quiet life, work at a mundane factory job, undemanding and unenterprising. He liked to potter around the house, do a bit of DIY and cultivate his garden (and nothing wrong with that; contentment is not something that can be measured). They came with us once to Windsor, where my other aunt and uncle were living, but they usually confined themselves to trips close to home. This was in striking contrast to my mother, the woman who Rhoda had thought not quite good enough for her brother George. My mother had become something of a seasoned globe-trotter.

And so we settled down, Irina began taking driving lessons, started to make a few friends and became more fluent in English. That first summer I had an opportunity to go on a field trip to Orkney and Irina joined me. An island, in summer, it sounded great. But this was Scotland. On Orkney the wind is so unrelenting that there is scarcely a tree. It was cold, and she certainly didn't need her bikini. Instead we bought sweaters and a bottle of whisky to keep out the cold. But it is a fascinating place, full of archaeological wonders like Skara Brae (a stone-built Neolithic village), the Standing Stones of Stenness and the Ring of Brodgar, (Orkney's answer to Stonehenge), the Broch of Mousa (an Iron Age fortified round house), and the Neolithic burial mound of Maeshowe, which has graffiti from Viking times scratched in runic characters. We trekked across wild open moorland, where we were dive-bombed by Arctic skuas, to see the Old Man of Hoy, a dizzying, finger-like sea stack of Old Red Sandstone over 400 feet high, which was not successfully scaled until 1966. By the time you read this it may have collapsed, as erosion along this exposed coast is very rapid.

One evening we went to an arts-festival concert in St Magnus Cathedral, a building founded in the twelfth century by a Viking earl, which formed part of a Norwegian bishopric until 1472. It is built in a Romanesque style, reminiscent of Durham Cathedral, but in a striking polychrome mix of red and yellow sandstone. The concert featured new music by Peter Maxwell Davies who lived and worked on Orkney, and as we listened a thunderstorm raged outside, a fitting accompaniment to the wild and violent music.

But the most atmospheric place of all was Scapa Flow, an enormous natural harbour which was home to the Grand Fleet during the First World War. Now all is quiet, the wind blows over the empty anchorage, the buildings are gone, there is not a warship in sight. But at the end of May 1916 seventy ships of Jellicoe's fleet steamed out through Hoxa Sound to do battle with the German High Seas Fleet at the battle of Jutland. It was the greatest naval battle of World War I. Churchill had declared that Jellicoe was the only man who could lose the war in an afternoon, but the battle was inconclusive, mainly due to primitive communications based on flags and light signals, and to the unexpected vulnerability of the relatively lightly armoured battle cruisers. Britain lost more ships and more men than the Germans, but the High Seas Fleet never ventured out of their bases for the rest of the war. At the end of the war the entire German fleet of 24 capital ships and 50 destroyers was forced to surrender, and was interned in Scapa Flow. The ships were disarmed, anchored around the Flow and the crews reduced to a minimum, pending the outcome of the armistice talks. When, in June 1919, it appeared that none of the ships would be returned to Germany, the German admiral ordered the entire fleet to be scuttled. A few were beached, but all the rest capsized and sank to the seabed. Most were salvaged during the 20s and 30s, the larger ships requiring great ingenuity. Many lay upside down due to the weight of the deck armour and guns. Air locks were attached, divers sent down to create air tight compartments, into which compressed air was then pumped. The ships slowly emerged, still keel uppermost like great grey whales, and were then towed away to be broken up. Seven ships still remain on the sea bed including the battleships *Konig, Kronprinz Wilhelm and Markgraf*. The last to be raised was the battle cruiser *Derfflinger*, salvaged in August 1939, but with the Second World

War looming, she was beached and remained there throughout the war, a mute spectre of the previous conflict. In 1914 a German submarine had entered the Flow and Jellicoe had ordered blockships to be sunk in all but the major channels which were protected by nets and booms. Early in the Second World War another German U boat, the U-47, daringly penetrated the Flow between three blockships in Kirk Sound, and torpedoed the old battleship *Royal Oak*, killing over 800 sailors. Churchill ordered all of the smaller entrances to be permanently sealed, but this was closing the stable door after the horse had bolted. It was an eerie feeling to look out over the empty grey waters of the Flow and recall the echoes of the not-so-distant past.

But I digress. I continued my work on blocks 9/18 and 9/19, operated by our partner Conoco. They had already drilled five wildcat wells, either dry or with minor shows, which was frustrating because promising structures were present, the area was on trend with the Beryl field to the north, and the blocks had all the right ingredients. Another dry hole was drilled in 1978, but eventually, after I had moved on to other things, five small fields were discovered, three of which (Gryphon, Buckland and Maclure) were put on production.

I was sent on courses in core analysis and sedimentology, on field trips to see the onshore equivalents of the reservoirs and source rocks in the North Sea, and to a European Geological Societies meeting in Amsterdam, and I gave a lecture on North African geology at Glasgow University. I felt that I was settling in, and beginning to get my feet on the ground.

That first summer I thought it would be nice for Irina, me, Susannah and Jonathan to have a family holiday in Sicily. At the last moment, we discovered there was a spare place, and invited my mother too, since I thought she would enjoy being with the children, and hopefully begin to warm towards Irina. We stayed in a nice beach hotel at Noto on the south-eastern tip of the island, enjoyed the beaches and the historic small towns, and took a trip to the crater of Mount Etna, which was pretty awesome. But my efforts at peace-making were less successful, and there remained a lack of warmth and understanding between them.

In April 1978 production began from the Thistle Field, one of the most northerly fields in the North Sea, discovered in July 1973 by Signal Oil and Gas, then taken over in 1974 by Burmah and finally by BNOC in 1976. This was the first field to be brought on stream by BNOC and represented a major achievement. It contained 500 million barrels of recoverable oil in a water depth of 525 feet. A production platform was installed in 1976 on which two drilling rigs were mounted, which drilled very carefully controlled deviated wells into the reservoir 9,000 feet below. I was asked to do a comparative study with the Magnus field, the most northerly of all the North Sea fields, to see how it compared to Thistle. The field had been discovered by BP but was not yet on production. The two fields had much in common, they were of a similar size, Thistle had better reservoir quality, but Magnus had a thicker oil column. The main difference was that the oil at Magnus was in a slightly younger reservoir which was not present at Thistle.

At the end of October Irina's mother Tatiana came to visit us. It was her first visit to a non-communist country, and I think she was impressed by the affluence and ease of living compared with life in the Soviet Union. She did not speak English and my Russian was limited to just a few phrases, but somehow we managed to communicate quite well. She arrived in London, so we introduced her to Eva and Dick and showed her Westminster Abbey and the Tower of London, Madame Tussaud's and Harrods. We went to see my old house in Crowthorne and then to Windsor, before driving north to Glasgow. She stayed with us for a month so we were able to take her to Stirling and Loch Lomond, and we spent a weekend in Halifax with my mother and Susannah and Jonathan where she experienced the delights of Harry Ramsden's and the quaintness of York. By the time she left she had gained a real insight into what life was like in the decadent west. It's a pity that Irina's dad did not come too. He would have loved it. He never did visit the UK. By the time of Tatiana's next visit he was having heart problems and felt it unwise to travel.

And then, out of the blue, I was asked if I would like to go to Aberdeen as Chief Operations Geologist.

Letter 32
The offer of a job in Aberdeen posed a bit of a dilemma. It would be a significant promotion, the salary would be higher, the responsibilities greater, but against that I had to weigh a not inconsiderable reality. There was protracted and ill-tempered power struggle going on between the Glasgow and Aberdeen offices. The exploration management in Glasgow felt that it was their responsibility to select drilling locations and design the geological evaluation programmes, and that it was the responsibility of the operations office in Aberdeen to carry out those instructions. The

Aberdeen office felt that the Glasgow staff were a bunch of academics who knew nothing about the costs, the safety aspects or the technical difficulties of drilling in the hostile environment of the North Sea, and that they should have the last word in how the wells were drilled and evaluated. It became so bad that on one occasion they refused to visit each other's offices, and held a meeting half way between the two, in Perth. The Manager in Aberdeen was a burly, overbearing man called Mike Kelly, and the Drilling Manager a stubborn uncompromising Dutchman called Dirk de Neef. The Chief Operations Geologist was Jimmy Hay, a shrewd, canny, native Aberdonian. They formed a formidable triumvirate, heading up an office of about twenty or so engineers and geologists. Glasgow was determined to break up this coterie and more or less compelled Jimmy to move to Glasgow, which he was loath to do, and for me to go to Aberdeen in his place to represent the interests of the Glasgow office in Aberdeen. It was a poisoned chalice. I was in for a rough ride.

I accepted the offer, partly for the promotion, partly to gain experience at the sharp end of the business, and partly because I hoped that I would be accepted as a channel through which Glasgow's wishes could be explained, and Aberdeen's concerns could be transmitted back to Glasgow; a sort of peace-making role between the two camps. It was to prove a vain hope.

And so we packed our bags again. We searched the area around Aberdeen for a suitable house, and I remember being shown around one property by a young lad of seventeen or so, and I could barely understand a word he said. He spoke the local dialect called Doric, a variety of Lowland Scots which is very difficult for an outsider to understand. We eventually came across an impressive-looking house in Cove Bay, on the coast about two miles south of Aberdeen. It stood by itself at the top of a long drive flanked by trees, on a half-acre plot, with uninterrupted view over the sea, and was appropriately called Seaview House. We were very taken with the house, although it didn't have central heating and needed quite a bit of attention. We bought it in October and moved there in December, almost exactly a year after Irina had arrived in Britain.

It was a solid Victorian granite house built in 1855, and clad in white weather-proof harling. It needed to be. The trees on the drive were probably 50 years old, but only about 15 feet high, permanently stunted by the easterly winds blowing off the grey North Sea. The house had had a varied existence, being at one time a doctor's house and dispensary and later a café, and in the back garden there was a row of kennels for fox hounds, and the plot was surrounded by 'march-dykes', four-foot-high walls built of granite. The title deeds were hand written on parchment and contained the stipulation that the owner was 'not allowed to carry on any operations nauseous or noxious to the inhabitants of the neighbourhood such as iron works, breweries or distilleries'. Not that we had many neighbours. There was a farm behind the house owned by Hamish who, we discovered, had never been out of his native county, and on the corner, was a shop which doubled as a post office. To the south there was nothing but open fields. The village of Cove Bay was clustered around a tiny harbour, just large enough for two or three small fishing boats, where the fishermen would land crabs and lobsters which we would buy straight off the boats. Beside the port I found a large old discarded anchor which I retrieved and placed near the front door, a fitting symbol, I thought, of the house's maritime perspective.

In front of the house was a large area of lawn sloping down to the gate, and I inherited from the previous owner a large Atco mower originally from the Aberdeen Parks Department. We installed central heating and made a few alterations so that by Christmas the house was habitable. Svetlana and Imad came for Christmas, and my mother brought Susannah and Jonathan for Hogmanay, Scots New Year (although strictly Hogmanay means the last day of the year). We drove to Glen Shee, a ski centre in the Cairngorms near the Devil's Elbow above Braemar, but it was clear that there were no budding Eddie the Eagles amongst us. Later that month, while visiting Glasgow, we called in to see our friends Yvonne and Kenneth, who had a litter of new-born Labrador puppies, and what can be more appealing than sad-eyed, wet-nosed, silky black Labrador puppies? We placed an order, but of course the puppies couldn't leave their mother until they had been weaned. We collected her a few weeks later and named her Panther; she rapidly became a member of the family.

Aberdeen has several soubriquets – the granite city, the silver city, the city of roses, and the city by the grey North Sea – and all are appropriate. The buildings are constructed almost entirely from the famous Rubislaw granite, of which six million tons were extracted from the 450 feet deep 'big hole' to the west of the city. Whilst in production the quarry was kept dry by pumping, but since closure in 1971 the hole has filled with water. The granite is light grey in colour and sparkles as the sunlight reflects off the silvery flakes of mica. It gives the city a clean, cold appearance, which is entirely appropriate. Once the oil revenues began to flow in the city fathers chose to beautify the city by planting roses, and even the central reservations on the dual carriageways (freeways in American) are graced by colourful rose bushes.

Aberdeen has a long history, and owes its importance to its location between the estuaries of the Don and Dee rivers. Old Aberdeen, with the Cathedral and University, was built close to the Don estuary but the castle was constructed on higher ground towards the Dee. No trace of the castle now remains; it was destroyed during the Anglo-Scottish wars of the fourteenth century, but there is still a Castle Hill. As the city expanded the centre of gravity shifted southwards towards the river Dee. It became an important fishing and shipbuilding port, and in the nineteenth century an Aberdeen shipbuilder introduced the concept of the clipper ship, a sleek, fast sailing vessel for the India and China trade, which established the reputation of the city as an important shipbuilding centre. Docks and shipyards were constructed at the mouth of the river Dee, but although more than 3,000 ships were built there, most of them were small, and Aberdeen never rivalled the yards of Glasgow or Belfast, and the industry steadily declined after the Second World War. It was replaced by a much more lucrative industry, following the discovery of oil and gas in the North Sea, when it became the principal service base for offshore rigs and platforms. The harbour was full of oil service vessels, most of the oil companies had operations offices in the city, there were dozens of oil service companies based there, and it became the largest heliport in the world. The arrival of hundreds of oilfield specialists and workers employed by rich and powerful oil companies transformed the city, and when I arrived the boom was in full swing.

Aberdeen however is remote. It is 125 miles from Edinburgh and 150 miles from Glasgow, and is closer to Stavanger than London. It was (and still is) the third city of Scotland in terms of population, but until the oil boom, it was an inward-looking, provincial city, proud of its attainments no doubt, but wary of newcomers. It was as though the remoteness, and the hard, uncompromising granite of the city had permeated the souls of the people, and made them cold and proud, or so it seemed to me.

My hopes of a peacemaking role in the office were quickly dispelled. There was a meanness of spirit there which I found very unpleasant. My arrival was greeted with hostility and suspicion. Instead of being regarded as a mediator I was regarded as a stool-pigeon sent by Glasgow to keep tabs on the Aberdeen office, and as the man responsible for the unwanted transfer of Jimmy Hay to Glasgow. The North Sea and the continental shelf around the UK is a hostile and dangerous place to operate (as dramatically illustrated by the Piper Alpha disaster of 1988 in which 167 men lost their lives), and safety has to be the principal concern, but from the point of view of properly evaluating the wells the more information obtained the better. Cores and drill stem tests (where samples of fluid or gas are collected from the formation), and a good set of wireline logs are critical to interpreting what is present in a formation and in assessing the quality of the reservoir. It seemed to me that the main objective of the Drilling Department, invariably supported by the Manager, was to drill the wells as quickly and cheaply as possible. I came to feel that they used safety as an excuse not to cut a core, or run a DST or run an additional wireline log, and we had several angry confrontations on the matter. Perhaps I was naïve, but Mike Kelly thought it the business of the company to make money; I felt it was to find oil. We never did succeed in seeing eye to eye.

In spite of the unfriendly reception I had a job to do, and I determined to make the most of it, and become familiar with the operations side of the business. I was given a secretary, a nice local girl called Isobel who, when she flew off on holiday one time, told me she had never flown before. I had an assistant called Arthur Coleman, an experienced geologist not far from retirement, who maintained a very detailed and precise record of well tops (the subsea depth to key formations) for all the wells that we drilled. This was important for comparing the actual tops with the tops predicted prior to drilling to determine whether the prognosis was accurate, and exactly where the well had entered the oil-pool. In addition, I had a group of five or six young well-site geologists, who would rotate between the office and the rig. They were seconded from Glasgow for a six-month stint, as part of their training. It was a great opportunity for them to experience wellsite operations on modern, high-tech, state-of-the-art drill ships and production platforms. Some of them went on to very successful careers later in life and one of them, called Brian Maxted, became CEO and Chairman of the Board of an oil and gas company in Houston.

The Thistle Production Platform had been installed in August 1976, and was provided with two drilling rigs to drill the production wells. Unlike onshore oilfields where vertical wells are drilled all over the field, an offshore field like Thistle was developed from a single platform, by drilling deviated wells which were carefully steered to tap different parts of the field. Thistle had sixty slots from which individual wells could be drilled, and the wells snaked out from the sea floor to the reservoir like strings of spaghetti. The location of the drill bit had to be known to an accuracy of four or five feet and the process of steering a well demanded a high degree of skill. In order to plot the position and depth at which the well reached the oil pool drill depth in the deviated well had to be converted into true vertical depth, which also required some skill.

By the time I arrived, drilling of the Thistle production wells was well underway, and ultimately 24 producing wells were drilled plus a further 19 water injectors around the edge of the field (in order to maintain reservoir pressure). The injected water had to be very pure, as any bacteria or chemicals could seriously damage the reservoir. Production began in April of 1978, initially into a tanker moored alongside, but ultimately into a newly laid pipeline, and built up to a peak of 130,000 barrels per day (which is good by any standards). I was able to fly out to the platform and observe the operations at close quarters. It was a fantastic experience to see the platform, with its two drilling rigs, and smoke and flames belching from the flare stack, gradually appear out of the gloom, and become larger and larger as the chopper came in to land. I became quite familiar with the geology of the Thistle field and I was invited to give a talk on it at the Institute of Petroleum in London, which was subsequently published.

I was also fortunate to be able to fly out to the semisubmersible rig *Atlantic II* which at the time was drilling appraisal wells on the Clyde field, which had been discovered earlier that year. We also drilled several bold exploration wells in totally virgin areas. Well 72/10-1 was drilled on a block in the Western Approaches which had been granted exclusively to BNOC to evaluate the potential of the area. The well was dry but provided information which suggested that this region was not highly prospective. We also drilled two wells west of Shetland in an area that I had examined when I worked for British Gas. Well 205/10-1 was drilled west of BP's Clair field, but proved to be disappointing. Well 163/6-1 was a collaborative venture with 19 other companies, and was drilled as a stratigraphic test, to assess a deep-water area at the northern end of the Rockall Trough where it abuts the Wyville-Thomson Ridge. This well was drilled by *Discoverer Seven Seas*, a new drillship which I got to visit when she was provisioning at Peterhead. It was a much more sophisticated version of the Glomar drillships that I had worked on in Morocco, but this one was capable of operating in water depths of 7,000 feet and drilling to a depth of 25,000 feet. She was kept on station by a sophisticated dynamic positioning system using six thrusters, which constantly kept the ship on station. This well too was unsuccessful, but that is the nature of rank exploration drilling. If you are lucky, one well in five may be successful, the rest are 'dusters'. We also drilled well 9/14-1 for which I had written the prospect description and prognosis just before leaving Glasgow. In this case, appraisal drilling found the small Leadon field, which was brought on stream several years later. I met Lord Kearton on one of his visits to Aberdeen, and briefly reviewed the objectives of one of our wildcat wells. When I met him again six months later he remembered very clearly what I had told him. I was impressed.

There were two further success stories. The first was the discovery of a small field not far offshore, in an area called the Buchan Graben. The prospect had been worked up in Glasgow, and the discovery well was drilled in 1981. Seven appraisal wells were subsequently drilled which showed that the field extended onto an adjacent block. It was named Ettrick (after a river in the Scottish Borders) and was eventually brought on stream, producing directly into a floating storage vessel moored alongside.

The second success was on a much grander scale. In 1979 BNOC acquired an interest in the Beatrice field in the Moray Firth by buying out the equity of Mesa (UK) Ltd. The field had been discovered by Mesa, a US company founded by T. Boone Pickens, a colourful corporate raider and greenmailer, who had made his name by acquiring the Hugoton Production Company, which was fifteen times the size of Mesa. The field was named after Pickens' wife, the only North Sea field to be named after a woman. BNOC became operator of the field, which was much smaller than Thistle, but much closer to the shore, and in shallow water where the wells could be drilled from a jack-up rig. This was a mixed blessing as special precautions had to be taken in case of an oil spill so close to shore; instead of the cheaper tanker loading option, a pipeline had to be laid from the field to the shore terminal at Nigg. The oil had a high wax content and pour point which means that it begins to solidify below 18 degrees Celsius. If untreated it would, as someone graphically explained, 'produce the largest candle in the world'. The reservoir was also under-pressured and required water injection to flush out the oil. Fortunately, an additive was found which prevented the wax from solidifying, but this had to be introduced while the oil was still hot, using downhole electric pumps which at the time was untested technology. Once onshore it was not difficult to separate out the wax. This field presented major challenges, and was BNOC's first field to be developed from scratch. In the circumstances, they did a great job. The field came on stream in September 1981, and a lavish inauguration ceremony was held the following year (in the middle of the Falklands War) at the shore terminal of Nigg, near Inverness, to which Irina and I were invited. It was an impressive gathering, with banquet, champagne, string orchestra and speeches. Princess Alexandra, a cousin of the Queen, was the guest of honour, and Irina got to exchange a few words with her.

Letter 33
We made some good friends in Aberdeen, although none of them were Aberdonians. Through the office I met Tony Armistead, a senior rep for Gearhart-Owen, an oil service company which provided state-of-the-art MWD (measurement while drilling) equipment to monitor and guide controlled directional drilling. Tony's Iranian wife Laleh was a midwife at the Aberdeen Royal Infirmary. Tony and Laleh lived in Cults, a posh area of Aberdeen, five miles along the Deeside road, in an enormous house that they shared with their two sons and Laleh's parents. The house dated from the eighteenth century and was shown on maps from the time of General Wade's military roads in the Highlands in the 1740s and 50s. Tony was a very cultivated man, he played the piano and flute and sang in a local choir and was a fount of knowledge on all kinds of arcane subjects. We started playing squash together and soon became firm friends, a friendship that has lasted for forty years.

One of the icons of Aberdeen is Marischal College on Broad Street, a magnificent mid-Victorian building of Kemnay granite (the second largest granite building in the world after El Escorial in Spain – bet you didn't know that), resplendent with dozens of spires and pinnacles. The college was originally founded in 1593 and in 1860 was merged with King's College (1495) to form Aberdeen University. We discovered that the university had a Russian Department and, on making enquiries, Irina was offered a part-time job as a language assistant there. It was a small department headed by Jim Forsyth, an authority on ethnic minorities in the southern republics of the USSR and in Siberia, a subject on which he had written a large and scholarly book. His wife Jo was a lecturer in the same department. We were also introduced to an English couple, Judith and Jim Thrower. Jim was a lecturer in the Theology Department where, bizarrely, he lectured on atheism and the rejection of religion. He claimed to have fought on the barricades at Gdansk during the *Solidarity* demonstrations of the early 80s, reminding me of the line 'I have a rendezvous with death at some disputed barricade', although I don't believe, in truth, there was much danger of that. He was also warden of a student hall of residence which provided him and his family with a large house called Balgownie Lodge at Bridge of Don to the north of the city. We spent several happy Christmases there with them and their large family. There was another Russian language assistant called Masha, married to Peter, a Scottish architect, with whom we established a long-lasting friendship. One Christmas, I invited all my young wellsite geologists over for a party, along with Masha and Peter and a few others. Peter showed up with his Russian wife, wearing a black leather coat and looking like the archetypal KGB officer. My young geologists did not know quite what to make of him.

If you look at a geological map of the area around Aberdeen, you will see lots of red patches in a background of green (that's why they call geological maps 'sheets of many colours'). The green areas represent a group of heavily metamorphosed sediments called the Dalradian Group and the red represent granites which were intruded into the Dalradian sequence much later – bear with me, there is an interesting story coming up. In the Ordovician, say 480 million years ago, Scotland and England were separated by thousands of miles. Scotland formed part of Laurentia (North America) and England formed part of Gondwana (Africa and South America). Not only that but they were both located in the southern hemisphere. Gradually plate tectonics dragged them together and they finally slammed into each other at the end of the Silurian about 400 million years ago. The suture which marks the collision can be traced from the River Shannon in Ireland through the Isle of Man (where you can stand with one foot on what used to be Laurentia and the other on former Gondwana) to the Solway Firth and Lindisfarne on the North Sea coast. Subsequently a part of the Laurentian terrane of Scotland foundered between two faults to form the Midland Valley, the northern fault of which runs from Loch Lomond to Stonehaven, just south of Aberdeen. Now the interesting bit is that along this fault metamorphic zones have been found in the Dalradian which show a sequence from lightly metamorphosed (chlorite grade) to severely metamorphosed (sillimanite grade), each characterised by the presence of a particular mineral (similar to the ore minerals of the Northern Pennines). At Aberdeen (and Cove Bay) the most severe metamorphism is found, but close to the fault at Stonehaven the dominant mineral is garnet, representing less severe metamorphism, and there you can walk on the beach and pick up garnets which originated in 'America'. Fascinating.

I can't resist one more geological story. You will be aware of Meteor Crater in Arizona, 50,000 years old, and almost a mile in diameter and caused by a meteorite about 150 feet in diameter. There is another impact crater in Mexico which is supposed to have been responsible for wiping out the dinosaurs, 66 million years old, with a crater 110 miles wide, and a meteorite diameter of 6 miles. In Scotland there is evidence of another large impact. On the west coast there is a formation called the Stoer Sandstone, 1.2 billion years old which contains melted fragments of a green rock called suevite which have been forcibly embedded in the sandstone, smashing the sand grains in the process. Not only that but large broken blocks of Lewisian gneiss were found at the base of the sandstone. These were the ejecta from a large asteroid impact, and the bed thins towards the west suggesting an impact somewhere to the east. This area is covered by younger rocks but a gravity survey shows a big hole called the Lairg gravity low, 25 miles across, which probably represents the impact site. The meteorite has been estimated at 2 miles in diameter, putting it among the world's top 20 known impacts. Its effects would have been cataclysmic, but of course, apart from a few primitive organisms, there was no one there to witness it. Suevite was first described from Germany from an impact crater in Bavaria, and the cathedral at Nordlingen is largely built from rock produced by the meteorite impact.

Susannah and Jonathan came to Aberdeen for New Year in 1980. It was cold and snowy, we built snowmen and had Panther jumping for snowballs. We drove to the little town of Stonehaven, 15 miles south of Aberdeen, where there is a fire festival, to see in the New Year. As the bell on the Town House strikes midnight the locals light up fireballs which they swing around their heads as they process along the High Street to the accompaniment of a pipe band, ending at the harbour, where the fireballs are thrown into the sea. It is reputed to be an old custom designed to chase away evil spirits, and it has much in common with other New Year bonfires and boat burnings in other parts of Scotland, but in fact there is no record of the Stonehaven festival before about 1860.

We now come to a most remarkable episode. In 1980 the summer Olympics were held in Moscow in an atmosphere of animosity between the west and Russia following the Soviet invasion of Afghanistan (the Americans made the same mistake 20 years later). President Carter orchestrated a boycott of 65 countries in protest against the invasion, and the games were effectively spoilt. In the circumstances the Soviets moved known dissidents out of Moscow and rigorously discouraged fraternization between the visitors and the local inhabitants. To that end they published a few weeks before the games a series of articles in a weekly newspaper called *Nedelya* highlighting the dangers of fraternization. You will recall my account of Nilov, the Russian teacher in Algeria and his infatuation with the American girl. Now was the time that those chickens came home to roost. We learned from friends in Moscow that one of the *Nedelya* articles was about Nilov – and me! Jim Forsyth at Aberdeen University made a translation for me. The article claimed that on leaving Algeria Nilov had been recruited by the CIA as a spy and provided with clandestine equipment which he had used to pass classified information via a dead letter drop in Moscow to an American contact. It said that his activities had been uncovered by the KGB. He had been arrested, questioned and admitted his guilt. He was put on trial, pleaded guilty and was sentenced to 10 years in prison. During questioning he was reported to have told his interrogators that Donald Hallett, a rich British guy, posing as a geologist, had deliberately introduced him to Frank, and through Frank to the CIA in Algiers. The article included a photograph of me (taken at my marriage to Irina three years before), and went on to claim that I had visited Leningrad earlier that year, where I had been observed in a hotel restaurant having dinner with contacts from the US consulate.

Frankly I don't know whether Nilov had any contact with the CIA or not. What kind of information could he have provided? He was just a young teacher, not a nuclear scientist. Was he really guilty of spying, or just someone to be punished for having stepped out of line? The details of the equipment reportedly found in his possession, and the dead-letter drops seemed very James Bond-ish. He obviously used me as a convenient scapegoat at his trial, and the Soviet authorities seized on this opportunity to create a propaganda story to discourage other would-be dissidents. We asked one of Irina's friends in Moscow to check with the journalist who had written the story, who said that she had simply been told what to write. I contacted the Foreign Office and they sent someone to talk to me. They took notes and said they would investigate, and simply advised me not to travel to USSR in the near future. There was never any mention of any of this in the western press, Irina was never questioned, nor was her family, and nor was I. She had not been prevented from marrying me, nor from returning regularly to visit her family in Moscow. So we concluded

165

that the Soviet authorities themselves did not believe the story of my involvement. I did indeed visit Leningrad in 1980 (someone was obviously keeping tabs on me), but simply as part of an Intourist group to see the artistic treasures of the city, and I certainly didn't meet any Americans. Irina and her mother joined me there after the excursions and we all travelled back to London together. I was rather pissed-off by the whole affair. It must have raised suspicions among my former Russian friends, and there was nothing I could do about it. Several years later the Russian Consul in London informed the Foreign Office that the allegations against me were unsubstantiated and had been withdrawn.

Leningrad had a totally different feel to Moscow. It is a thoroughly European city, Peter the Great's 'Window on the West' or 'Venice of the North', built at great speed, on a muddy, disease-ridden estuary, and 100,000 people perished in the process. It was known as St Petersburg until the Russian Revolution, and it is redolent with history – the capital city of Russia for more than 200 years, the home of the tsars, and some of the best-known buildings in Europe, the city that saw the end of Rasputin, and the birth-place of the Russian Revolution. Peter the Great began it, and Catherine the Great (who was German, not Russian) completed it. She oversaw the construction of the Winter Palace and filled it with western art (including 200 pictures previously owned by Sir Robert Walpole). The collection expanded into six adjacent buildings to form the Hermitage Museum, which became the largest art collection in the world. The enormous Neoclassical St Isaac's Cathedral has a sumptuous, brightly-coloured interior with magnificent malachite and lazurite pillars framing the iconostasis, and the exterior is embellished with 112 red granite columns, each cut from a single block. During Soviet times the building became a Museum of Atheism and the white dove sculpture beneath the apex of the dome, symbol of the Holy Spirit, was removed. The Capitol in Washington is partly modelled on St Isaac's Cathedral. By the riverside is the Bronze Horseman, an equestrian statue of Peter the Great commissioned by Catherine the Great and bearing her name alongside his. It is mounted on a block of granite weighing 1,250 tons reputably the largest stone ever moved solely by muscle power. It is a city of superlatives. The Leningrad Metro was built after the war and encountered some major geological problems, particularly in the soft ground beneath the Neva estuary, necessitating tunnels 80 metres deep in some places. It was built in sumptuous style, some stations having marble halls, crystal chandeliers, mosaic artworks and columns covered in shimmering fragments of glass. The older stations, built during the Cold

War, also doubled as potential bomb shelters. Across the Neva is the famous cruiser *Aurora* which fired the shot which signalled the start of the revolution and the storming of the Winter Palace. Seven or eight kilometres south along the Moskovsky Avenue is Victory Square with a 50-metre obelisk commemorating the million soldiers and civilians who died during the 900-day siege of the city during the 'Great Patriotic War'.

One of the great stories of the siege, (apart from the fact that the inhabitants were reduced to eating the leather of their shoes), is that in order to encourage morale in the besieged city the Soviet authorities arranged for a copy of Shostakovich's newly completed Seventh Symphony, dedicated to the defenders, to be flown in by night, and Karl Eliasberg, conductor of the Leningrad Radio Orchestra, scraped together a scratch orchestra of professionals and amateurs, half-starved and housed in the shell-damaged hall, and broadcast the symphony on 9th August 1942, the day that Hitler had planned to celebrate the fall of Leningrad with a banquet at the Astoria Hotel. It was also broadcast by loudspeakers throughout the city and in a gesture of defiance to the besieging German troops. In the end, the defence held firm and Leningrad, like Moscow, was never taken by the Nazis. Leningrad, Petrograd or St Petersburg, whichever you chose to call it, is indisputably one of the world's great cities.

Catherine I, wife of Peter the Great, built a summer palace at Tsarskoye Selo (the Tsar's Village), 25 kilometres south of St Petersburg, later replaced by a much grander palace designed by Rastrelli. The Catherine Palace is a flamboyant Rococo building with a modest frontage of only 325 metres set in a large park with a lake. Many of the interiors were designed by the neo-Classical Scottish architect Charles Cameron, and the famous (or infamous) Amber Room was added by Peter the Great's daughter the Empress Elizabeth in the 1740s. The fate of this room will be recounted in a later letter. Suffice it to say that when I visited in 1980, reconstruction, following almost total destruction during the war, was still in progress, and the Amber Room was a bare shell. Now the palace has been restored to its former glory, complete with a new Amber Room, exactly replicating the original. The neighbouring Alexander Palace was the favourite residence of Nicholas and Alexandra, and the place where they were arrested and kept under armed guard in 1917. It survived occupation by German troops during the war, and neglect during the communist regime, but was saved from destruction in the 1990s and is now being slowly restored. In 1918, in true proletarian fashion, Tsarskoye

Selo was renamed Detskoye Selo (Children's Village), and in 1937 was again renamed, this time as Pushkin.

Peter the Great built himself a small summer palace called *Mon Plaisir* on the Gulf of Finland, from which he could admire his new city of St Petersburg to the right and the fortress island of Kronstadt to the left. Later he decided to build a 'Russian Versailles' on rising ground to the south which came to be known as the Grand Palace, and the entire complex as Peterhof. It was progressively enlarged by later monarchs who added gardens, cascades, statues and a canal connecting to the sea. The view of the palace from the Samson Fountain is curiously reminiscent of the view of Frederick the Great's palace of *Sans Souci* in Potsdam, but on a far more lavish scale. Frederick had a series of terraces planted with vines climbing up to the palace; Peter had a spectacular stepped cascade flanked by fountains and gilded statues. The palace was looted and destroyed by the Nazis during the war and left as a burning ruin in 1944 when they were driven out. The Soviets changed the name from the German Peterhof to the Russian Petrodvorets, but with the collapse of communism it has resumed its former name. Curiously the adjacent estate of Oranienbaum given to Peter the Great's army commander Prince Menchikov, and the nearby Kronstadt Fortress were never captured by the Nazis. Peterhof has been largely restored, and work is slowly progressing at Oranienbaum. In recent years a 25-kilometre flood barrier has been built across the Gulf of Finland between Oranienbaum, Kronstadt and the northern shore in an effort to protect St Petersburg from repeated inundations.

Letter 34
Whilst based in Aberdeen we had the opportunity to explore the area. Aberdeenshire possesses more castles, stately homes and ruins than anywhere else in the UK, perhaps a reflection of its turbulent history of clan warfare, feuds, and cattle raiding. Two Jacobite rebellions by Highlanders against the Crown in 1715 and 1745 brought harsh reprisals. After the battle of Culloden an Act of Proscription forced Highlanders to hand over their weapons, forbade them to wear Highland dress (tartan and kilts), and suppressed the power of the clan chiefs. Forts were built at strategic locations, linked by military roads, and the subsequence Highland Clearances dispossessed thousands of peasant tenant farmers (and replaced them with more profitable sheep, known as 'four-legged highlanders'), in a deliberate and successful attempt to destroy the Highland way of life. The resulting misery and bitterness led to mass

emigration from the Highlands which was on a par with that in Ireland after the 1845-1852 potato famine, and was graphically portrayed in Thomas Faed's famous painting *The Last of the Clan*. The British government's *laissez faire* attitude was a pathetic excuse for not interfering. Some measure of rehabilitation came about partly through the works of Sir Walter Scott which romanticized the Highlands, and partly through the carefully choreographed visit of a tartan-kilted King George IV to Scotland in 1822, the first ever visit of a Hanoverian monarch to Scotland. He was followed by Queen Victoria and Prince Albert in 1842 who bought the Balmoral Estate on Deeside. These events produced a sentimentalized view of the Scottish Highlands which has been carefully nurtured ever since. The royal presence at Balmoral popularised the area and a railway was built along the valley from Aberdeen to Ballater providing access to many of the castles along the valley.

Deeside is Scotland's answer to the Rhine valley between Bonn and Bingen, without the vineyards, and the Lorelei, but with a royal castle, salmon leaps, and a whisky distillery. There are impressive castles at Braemar, Balmoral, Crathes and Drum, and wonderful forest walks, which we often took with Tony, Laleh and Panther. Further north we visited more castles at Craigivar, Kildrummy, Castle Fraser, Fyvie, and the ruined Huntly Castle, former home of the Catholic Gordon clan. Catholic symbols on the gatehouse were carefully erased by the Covenanters, and after the 1745 rebellion the castle fell into ruin. Near Ellon, is the splendid neo-Palladian Haddo House, designed by William Chambers for the Earl of Aberdeen, whose great grandson was British Prime Minister at the time of the Crimean War. On the Moray Firth, an extensive area of coastal sand dunes called the Culbin Sands, were long held in place by marram grasses, until these were scavenged by local villagers for thatching. As a result, winter storms mobilised the dunes which overwhelmed the houses, and made the farmland unworkable. Not quite in the same league as Arcachon, but impressive nonetheless. Not far from Inverness is the battlefield of Culloden, site of the last battle to be fought on British soil, on a rather chillingly bleak moor, where Bonnie Prince Charlie and his Jacobite army were annihilated by the Duke of Cumberland in 1746, which brought the Jacobite cause to an abrupt and definitive end. And finally, I can't leave north-eastern Scotland without mentioning the famous Whisky Trail along the valley of the River Spey. Here is a different type of *dégustation*, where you can sample the 12-year-old single malts of Glen Grant, Cardhu, Glenfiddich, and Glenlivet. Each has its distinctive character. Glenlivet, for instance, obtains its water from Josie's

Well, a spring in glacial gravels behind the distillery, whereas Cardu takes its water from Dalradian quartzites. There is a pub in Glasgow which offers 50 different varieties of Scotch whisky and 50 different real ales. It is a very popular place.

And one more geological aside. The Grampian massif is terminated to the north by the Great Glen fault, which is now occupied by Loch Ness and Loch Linnhe. It continues northwards as far as the Shetland Islands, and southwards into Ireland through Lough Foyle to Clew Bay and can be seen again in North America in Newfoundland and the Gulf of St Lawrence. In Scotland, it is a wrench fault with a curious history. The fault cuts through the Dalradian metamorphic zones and through a distinctive granite intrusion, exposed at Strontian on the north side and at Foyers on the south side. These exposures are 65 miles apart implying horizontal movement in a sinistral sense (to the left with regard to the northern side). The Dalradian metamorphic zones are similarly displaced. Q.E.D. you might think, but scientists love to query long-accepted hypotheses, and more recently it has been suggested that the latest movement on the fault was dextral rather than sinistral (to the right with regard to the northern side), and another author claims a displacement of 1,250 miles on the fault, which in my view is nonsense, but that's geology for you. Strontian incidentally is where the element strontium was first discovered.

In the autumn of 1980 my aunt Rhoda died. I received a letter from her solicitor (lawyer) who said that she had not left a will and that I was legally her next of kin, but that she had left a handwritten note expressing the wish that her estate should be divided between a distant cousin and neighbours who had helped her towards the end. He asked what I wanted to do. I said that there were a few things, like a nicely bound set of Dickens which had belonged to my granddad, family photographs and a few bits and pieces that I would like, but it transpired that all these had already been disposed of by her relations or neighbours. I was seriously annoyed by that, so I told the solicitor that I could not accept the handwritten request as a valid will, and her estate came to me as her next of kin. It didn't amount to much, and I gave a token payment to those who had taken care of her, but the rest I split between Susannah, Jonathan, and myself – some kind of redress, I thought, for Rhoda's meanness to my mother when my granddad Alfred died.

In October of 1980 I received a call from Mike Ridd (the Chief Geologist) asking if I would be interested in an Area Manager's job in Glasgow. I did not need much persuading. I had gained invaluable experience in state-of-the-art operations geology during my two and a half years in Aberdeen, but I had made little headway in healing the rift between the two offices. The Area Manager's job represented a significant advancement, which would expose me to a new aspect of the oil business - joint venture activities with our licence partners. Furthermore, I preferred Glasgow to Aberdeen. It was a large bustling city, cosmopolitan and outgoing. Its days as the 'workshop of the British Empire', when a fifth of all the world's shipping was built on the Clyde, were long gone, and heavy industry – coal mining, steel making, ship building, locomotive works and iron foundries were in decline, but Glasgow still had dynamism. It was a city with a proud past, but also an eye to the future. The BNOC office had been established there largely for political reasons, but it also showed that Glasgow was prepared to embrace new industries to replace the old.

I was sent on an in-house Management course in preparation for the new posting, held at the rather fancy Peebles Hydro on the river Tweed in the Scottish Borders. It was quite an eye-opener. Up to then all of my experience and training had been technical, but here we did tests to determine whether we were skills-oriented or personnel-oriented, discussed motivational techniques and, since the main instructor had a degree in psychology, we also explored cognitive dissonance and socio-cultural influences and such like. I recall we watched the 1957 film *Twelve Angry Men* which examines the prejudices and preconceptions of a group of jurors, and discussed its theme and the powers of persuasion. It was all new to me, but a very useful insight in a totally new field. I invited Irina to come over at the end of the course so she could try out the pool and the spa and she brought Panther with her. The hotel of course did not allow dogs so we left her in the car, but then after dark brought her into our ground-floor room through the window. Unfortunately, when someone passed in the corridor she barked, so then we had to pass her back through the window and put her back in the car. For anyone watching it must have looked quite farcical.

We spent a long time looking for somewhere to live and this time we wanted something with more character than our previous house in Kilmacolm. We went to see Strathleven House, a splendid eighteenth-century country house in Dunbartonshire which had lain derelict for many years, but the costs of refurbishment would have been enormous, and it was surrounded by an ugly industrial estate. It was subsequently rescued and restored by Scottish Historic Buildings Trust and is now a conference centre. Next, we looked at Cleddans House, the former home of a coal owner near Monklands in Lanarkshire. It was a very impressive country house with crow-stepped gables over each of the four main windows. It was set in 24 acres of woodland and paddock. There was a cobbled courtyard with coach house and stables and a large walled garden with two greenhouses, and even an ancient tomb. The house was approached along a drive with gatehouses at each end, and was on the market for £75,000. Unfortunately, it was too big – what would we do with nine bedrooms and 24 acres? And it was located in the scarred and despoiled heart of the Lanarkshire coalfield. The last 'exotic' house we went to see was that of a gentleman farmer in the midst of rhubarb fields near Houston west of Glasgow, called Glebe House. It was a rather elegant place with a garden sloping down to a little stream. There were riding boots in the hall, a gun cupboard and fishing rods. The owner asked if I hunted – no. Did I shoot? – no. Did I fish? – no. We decided it was not for us.

Eventually we found what we were looking for, a very attractive and much more manageable house at Uddingston, ten miles east of Glasgow and just a short walk from the railway station. It was called Holmwood House and dated from 1836. It was built of red sandstone with ashlar facing, and it had a wonderful landscaped garden overlooking the rural Clyde Valley. The garden had a pleasant terrace for lounging or barbeques, but the weather in Scotland can be fickle and very wet, and I remember one year when we were only able to enjoy our terrace for about five days during the entire summer. What to the English is heavy rain, the Scots call Scotch mist. The previous owner had added a modern south-facing sun lounge, and a sauna in the loft, and there was a separate three-car garage. The entrance hall was panelled in yellow pine and the staircase had a large gilt mirror, salvaged from a ship (not quite the *Titanic*, but impressive nonetheless). It was perfect – it even had a monkey-puzzle tree in the garden - and we made an offer on the spot. We never regretted buying Holmwood House, but this was at a time when inflation was running at about 18%, so our mortgage repayments were shockingly high.

Just prior to moving to Glasgow I attended a meeting of the American Association of Petroleum Geologists in San Francisco. By some quirk of fate, I was given a business class ticket where my colleagues from Glasgow travelled economy. One up to Aberdeen, I thought. It was the first of several AAPG meetings that I attended. There was an extensive programme of lectures, social events and field trips, and I was surprised and pleased to encounter two of my former colleagues, my old friend Raymond Wright from UCL, now in a senior position with the Jamaican Geological Survey, and Dave Kingston my old boss at Esso in Walton on Thames. I went to see the geysers 70 miles or so north of San Francisco which have been exploited to provide geothermal power for northern California and produce more than 50% of the area's power needs. It is the world's largest geothermal project with over 300 geothermal wells and more than twenty power plants. The source of the heat is a magma chamber at a depth of about 17,000 feet. Output eventually began to decline and it is now necessary to inject water into the reservoir to maintain production. Interestingly this has led to an increase in the number and severity of earthquakes in the area, but unlike San Francisco the geysers are not located in an area of known faulting. We returned through the Napa valley, the best known wine producing area of California, and stopped off at a modern and prosperous winery, owned by a geologist. In the United States, unlike most countries, mineral rights, including oil and gas, usually belong to the owner of the land on which they are found. Furthermore, there is a legal requirement to deposit details of all wells or boreholes with the local authority which are then available for anyone to examine. There is a whole profession, unknown in other countries, of independent geologists making a living from constructing 'prospects' from public-domain data which may have been overlooked by commercial oil companies. If the geologist can persuade an oil company to drill his prospect, an override deal will be negotiated where he may receive perhaps 5% of the profits if the well makes a commercial discovery. The vineyard which we visited in the Napa Valley was developed on the proceeds of just such a successful override deal.

Letter 35
And so we returned to Glasgow. But the times were a-changing. In May of 1979 the Labour government of Jim Callaghan lost the General Election and Margaret Thatcher became Prime Minister. Despite her working-class origins, Thatcher chose to reject Keynesian policies which had been the backbone of government policy since the 1930s, and replace them with monetarist policies. She attacked the welfare state, nationalised industry and a regulated economy, and embarked on a plan to deregulate financial services, privatise national industries, break the power of the trade unions, and establish market flexibility. In this she found an ally in President Reagan with whom she developed a close friendship. Her

stubbornness and hubris were legendary and earned her the title 'the Iron Lady'. She started by selling 80 million shares in BP reducing the government's interest to less than 50%. Churchill would not have been amused since he had engineered the purchase of a controlling interest in BP, (then known as Anglo-Persian) for strategic reasons before the First World War. The remaining government shares in BP were sold off in 1987, many of them bought by the Kuwait Investment Office, which at one time owned 21% of BP. British Aerospace was the next to go, followed by Cable and Wireless and the National Freight Corporation in 1981. With each new disposal she became bolder, and the process gained a momentum which accelerated throughout her premiership (and beyond). Then, in cahoots with Reagan, she deregulated the financial markets, allowing banks and other financial institutions to adopt new and often dubious new practices in a mad scramble for increased profits. This policy would eventually come home to roost in the 2008 financial meltdown. This was the new political reality. Her crusading zeal alienated large sections of the population, inflation remained high and unemployment rose, and her approval rating sank to 23% in late 1980, the lowest ever recorded for a serving Prime Minister (until Theresa May).

She was saved by two factors. Production of North Sea oil, which had begun in 1975, was now beginning to generate significant tax revenue for the Treasury and production continued to rise until the bubble burst in 1986. By 1981, for the first time, UK oil production exceeded consumption (and continued to do so until 2005), and by 1983 North Sea production was twice that of Algeria or Libya or Nigeria. But whereas the Norwegians set aside excess revenues into an oil fund (worth $878 billion at the present time) which was (and still is) used to fund state pensions, the British government used their revenue for short term exigencies. The second factor was the Falklands War in the Spring of 1982, the last hurrah of Britain's imperial past. We won the war but it was a close-run thing. Britain lost seven ships, ten planes, twenty-four helicopters and over 250 servicemen. The principal Argentinian weapon was the Exocet anti-ship missile which caused great damage among the British ships. If they had hit the two aircraft carriers the result could have been very different. Margaret Thatcher derived great prestige from the victory, and in the end became Britain's longest serving prime minister of modern times. Her relation with the Queen was reputedly somewhere between cool and frigid, and there was a joke at the time which asked why the new one pound coin was nicknamed a Maggie, to which the answer was because it is brassy, two-faced and thinks it is a sovereign.

I took up my new role as Manager of Central Area (of the North Sea) in June 1981, a very active area in which over fifty oil fields have been discovered. I was primarily responsible for liaising with the many companies with whom we had joint operations in this area, whether BNOC was operator, in which case I would explain our thinking and review our progress in evaluating the blocks, or as non-operator when the procedure was reversed. I had a small team of geologists and geophysicists who did most of the data interpretation, and kept track of the wells being drilled on our blocks. One of them admitted that his main reason for joining BNOC was to be able to climb the Munro's (the 282 peaks in Scotland over 3,000 feet in height first listed by Sir Hugh Munro in 1891). By the time he left he had climbed most of them. At any one time we would have five or six wildcat or appraisal wells to monitor. As a typical example, picked at random, in November 1981 we had five wells to keep track of: 20/3-2 (Conoco), 13/28-1 (Occidental), 20/9-2 (Phillips), 16/21b-4 (BNOC), and 13/13-1 (Zapata). Occidental was always known as Oxy, and an in-house joke of the time wanted to know whether an oxymoron was an inept Occidental geologist. Activity on producing fields was handled by a Production Geology and Geophysics Department, headed by Colin Maclean, a dynamic and ambitious man who later became something of a legend at BP. He came up with an ingenious method to develop marginal fields by spreading the risk and development costs between BP and their associated contractors, and he later became head of BP's Grangemouth Refinery, which he turned around from loss-making to profitability. Later still, as one of Lord Browne's trouble shooters, he was brought in to head BP's Texas City oil refinery following a devastating explosion and fire in 2005. He was reported at one time as owning five homes in four different countries.

Exploration concessions on the UK Continental Shelf were offered periodically by the Department of Trade and Industry. Companies, or usually consortia of five or six companies, would bid for the blocks they believed held some potential, and select one company to be operator (i.e. acquire the seismic data, drill the wells, and handle any eventual field developments), so the operator was usually the company with the largest resources. Each concession was granted for a specific period and had a minimum work commitment. Once the work programme had been completed and the time limit reached the licence holders had to either relinquish the block, or in the event of a discovery apply to the DTI for a production licence. Individual companies had differing priorities and would often dilute or augment their equity in a block, or 'farm-out' all or part of their interest to another company, giving other companies an opportunity to 'farm-in' to a block, so the make-up of the partnerships was constantly changing. Equity changes, fulfilment of work programmes, well abandonments, and concession relinquishments were initiated by the licence holders, but had to be approved by the DTI – hence the need for negotiators, accountants and lawyers at the operating committee meetings.

My new job brought me into contact with these people at BNOC, and with the DTI in the person of John Brooks, the 'gate-keeper' at the Department of Trade and Industry, when we needed to discuss drilling or relinquishment obligations. He was a rather abrupt and humourless man who always gave the impression that BNOC would receive no favours from him. Most of our partners were American (who rather resented BNOC's privileged position) and I soon got to know my counterparts in companies such as Amoco, Conoco, Chevron, Phillips, Texaco, Mobil, Tenneco, Arco, Sun, Occidental, Ultramar, Union, Monsanto, Zapata (a company founded by George Bush senior) and Norcen (a Canadian company). But we had British partners too like Tricentrol, Lasmo, GOAL (of which Dick Stabbins, a contemporary of mine at UCL, was Exploration Manager), Charterhouse (where I met my future colleague Danny Clark-Lowes), and Cluff Oil, founded by 'Algy' Cluff, another legend of the UK oil scene.

Cluff was an entrepreneur. He earned his nickname Algy from the wealthy, irreverent bachelor Algy Moncrieff in Oscar Wilde's *The Importance of Being Earnest.* He attended Stowe School, where he claims to have learnt nothing, then served in the Grenadier Guards with whom he spent some time in Malaya, where he made some serious money by investing in undervalued rubber estates. In 1972, when North Sea licences were being handed out cheaply, he teamed up with an American company, Transworld Petroleum to bid for North Sea acreage. They were granted several blocks, on one of which the Buchan field was discovered in 1974. Cluff subsequently sold his interest in the field making a spectacular profit, and then diversified into minerals, particularly gold, in Tanzania, Zimbabwe and Ghana, again with remarkable success. He was owner of *The Spectator*, an old-established conservative political magazine from 1981 to 1985, and he remains a director of eight of his many mineral companies. As a country retreat he bought Furzey Island in Poole Harbour, ironically unaware that Britain's largest onshore oilfield (not yet discovered at that time) was located beneath it. Later British Gas built a production facility just a stone's throw from Cluff's former mansion. He was always innovative and late in his career he became interested in underground coal gasification, and acquired several licences on which to test the viability of this process. This is still in its infancy but if successful it could revolutionise Britain's energy industry. He now lives in a converted windmill overlooking the white cliffs of Dover.

All our partners had offices in London, so I would often take the 'red-eye special' early morning flight from Glasgow to London and return the same evening. British Airways used to serve a rudimentary meal on the return flight where the wine was labelled 'product of more than one country'. The mind boggles. In November most of the technical staff moved from the glass palace in St Vincent Street to a much more workaday building in Gordon Street which had formerly been a dance hall. It was a bit of a rabbit warren, and had rather decrepit lifts (elevators). On one shocking occasion one of our geologists pressed the lift button, the door opened and without paying attention he stepped forward and fell down the lift shaft. The car was not there. He had very serious injuries and the accident changed his personality from a jocular ebullient character to someone much more reserved and serious.

The new job also carried additional responsibilities which enabled me to utilise some of the lessons learnt on the Peebles management course. I became a member of the contracts committee which oversaw the evaluation and award of contracts, and I was responsible for evaluating the performance of the people in my group. This marked a distinct shift away from purely technical work (which I enjoyed) towards management (at which I was a novice), underlining the often false assumption that successful technical employees make successful managers. But in practice managers invariably outstripped technical employees in pay and status, so no one turned down the offer of promotion. BNOC attempted to establish parallel staff and line ladders which were supposed to offer equal opportunities to both, but this was 'more honoured in the breach than the observance'.

We moved into Holmwood House in September, but Irina continued teaching in Aberdeen two days a week, since she was able to fly on BNOC's chartered Bandeirante aircraft, which saved the hassle of having to drive or travel by train. We renewed our friendship with Bob and Jenny in Kilmacolm but they soon moved to France, and eventually decided to remain there. They bought an old water mill on the Loire where we visited them several times, and they very quickly integrated into the local community. We also got to know a Glasgow lawyer, Stewart Dallas and his wife Eugenia. Stewart was a director of Rangers football club and he took me to see several matches from the director's box, and to see their trophy room stuffed full of the silverware which they had won. They had a country house at Strachur on Loch Fyne which we visited once or twice for summer picnics. Just along the road was the house of Fitzroy Maclean (a famous wartime commando, who assisted Yugoslav partisans during the war, and was rewarded with another house on the island of Korcula in Dalmatia by a grateful Marshal Tito). We also formed a close friendship with Mike Ridd, Chief Geologist at BNOC, and his wife Song. We found that we had very similar views on politics, geology, culture and travel, and we have remained firm friends ever since. Mike bought a derelict cottage near Drymen to the north of Glasgow and converted it into a delightful country retreat. The house came with several acres of land on which at one time he kept a herd of Highland cattle. Song was a delightful, gentle, cultured woman from Thailand of whom we became very fond.

My uncle Dick's final job was caretaker of the Westminster City Library in Great Smith Street in London, and when he retired in 1981 Margaret and George took over their flat. George became library assistant and caretaker, while Margaret worked as a PA in a firm of property developers in Belgravia. Eva and Dick then moved into a nice flat in Eccleston Square in Pimlico but in March of the following year my uncle Dick died. He had been good to me, particularly when I was young. He enjoyed the camaraderie of army life, both during the war and afterwards, and he maintained his army connections well into retirement. He was a larger than life character, physically big, but with a sense of humour and a hearty laugh, and he was the life and soul of every sergeant's mess where his battalion had been based. He was very proud of his scarlet guard's uniform and busby, and had done sentry duty at Buckingham Palace and Windsor Castle, and been a standard bearer at Trooping the Colour, when the regiment paraded their colours (flag) before the Queen on her birthday. He had a good war record, and had seen action all the way from Normandy to Berlin. Later in life he became a Freemason, which gained him further influential friends and I suspect that they helped him to find suitable jobs after he retired from the army. I went to his funeral in London, sad to see the end of a man for whom I had much respect.

Letter 36
Margaret Thatcher's first Energy Secretary was David Howell who succeeded Tony Benn in 1979, He took the view that British Gas should

not be involved in oil exploration, and against the fierce opposition of Sir Denis Rooke, hived off that part of the business as a privatized oil company called Enterprise Oil, and their Wytch Farm oilfield in Dorset was sold to BP in 1984. The main target of the government however was BNOC which, despite its impressive record and highly profitable status, was regarded as anathema to monetarist policy. In preparation for this event the government appointed Philip Shelbourne as Chairman and Chief Executive of BNOC in 1980 in succession to Lord Kearton, provoking the immediate resignation of Alastair Morton, who hated him. Morton famously said that the Treasury had done more damage to this country than the Luftwaffe ever did. He was replaced by Malcolm Ford, another rough and overbearing man who joined BNOC from Shell. Shelbourne was a merchant banker, former Chairman of Samuel Montagu who knew nothing about the oil industry, and unlike his predecessor showed little inclination to learn. His job was simply to oversee the privatization of the company.

The plan to privatize the exploration and production activities of BNOC was announced in November 1981 and the sale took place in November 1982, with the newly privatized company being named Britoil. In an effort to win over employees the government offered a preferential share deal which most employees took advantage of, including myself. The government retained a golden share which gave them the power to reject any future take-over bid which they judged to be undesirable. On the initial flotation only 27% of the offered shares were taken up. Shelbourne nevertheless collected a knighthood for his services. He did not live in Glasgow of course. At one time he had a house in the prestigious close of Salisbury Cathedral near to that of Edward Heath, and later he lived in an elegant Victorian house in Highbury filled with valuable antiques to complement his gay, but very discreet life-style. The adjective used most commonly to describe him was feline. His office was in a mansion overlooking Green Park, where of an evening he would walk his black Labrador, Brit. It was rumoured that when he finally left Britoil not only did he receive a magnanimous golden handshake, but he was given the additional bonus of free business lunches for the rest of his life, giving the lie to the expression there is no such thing as a free lunch.

Britoil lost most of the special privileges which had been granted to BNOC and thereafter operated as a normal commercial entity. Privatization changed the entire ethos of the company and was much resented by most of the staff, but at my level we soldiered on. We were still busily engaged with our partners, selecting drilling locations, obtaining agreement on work programmes and budgets, and evaluating the results of wells as they were completed. Exploration was boosted by the 1983 budget which for the first time allowed exploration and appraisal costs to be offset against income in the calculation of Petroleum Revenue Tax. This had the effect of making several of the smaller fields economically viable, such as BNOC's Deveron and Don fields. By this time tax revenues from North Sea oil and gas contributed £10 billion per year to the Treasury, equivalent to one third of the revenues obtained from Income Tax. Calls were made for an oil fund to be set up, as had been done in Norway, but they were ignored.

Glasgow had a darker side, and in the not-too-distant past the city had a reputation for violence. The militancy dated back at least to 1919 when a major demonstration escalated into violence in 'the battle of George Square'. Churchill, who was Secretary of State for War, sent 10,000 English soldiers armed with machine guns, tanks and a howitzer to restore order, and power stations, docks and public buildings were guarded by armed soldiers. Machine guns were set up on the roofs of the General Post Office and North British Hotel. The government was terrified that the violence might develop into revolution, as had happened in Germany and of course in Russia. Thereafter the district became known as Red Clydeside. During the 1920s and 30s when unemployment rates were high, a gang culture developed, particularly in the Gorbals and Bridgeton areas of the city, when there were reckoned to be six times as many gangs as in London. Their preferred weapon was the cut-throat razor. The gangs were largely sectarian since a large number of Irish Catholics had migrated to Glasgow to work in the shipyards and foundries, who were strongly resented by the existing Protestant population. It took the combined efforts of a new Chief Constable, Percy Sillitoe, the recruitment of big, strong policemen, the introduction of police radios, and paid informers to break the gang culture. By the time we came to the city these events were a fading memory, except for the continuing rivalry between the Rangers (Protestant) and Celtic (Catholic) football clubs, but communism and militancy lingered on into recent times, and the name Gorbals still has a sinister ring to it. We met a couple who represented a legacy of that era, called Greg and Ira, who lived in a typical Glasgow tenement in Partick, in the heart of Glasgow's docklands, an area noted for its militant communist shipyard workers. Greg was a serious communist, and his book shelves were full of the collected works of Lenin, Stalin's speeches, Gorky, Marx, Herzen, all in Russian, which he had taught himself to read. He had visited Russia several times, where he met his wife Ira. She was not a serious communist. In fact she was a cheerful, good-time girl, who wanted nothing more than to party, enjoy the bright lights and buy what she liked. She was more of a socialite than a socialist.

By this time Mike Ridd had been appointed General Manager of Exploration, and Alan Parsley, formerly with Shell, was promoted Chief Geologist in his place. Mike was instrumental in setting up a Laboratory for the company in Glasgow, and a Data Services Group was established at about the same time, headed by Graham Baxter (son of Raymond Baxter, a well-known BBC sports commentator), who subsequently went on to become Chief Operating Officer at Hutchinson 3G (UK). Under the terms of its charter BNOC was not limited to operations in the United Kingdom, so the search began for opportunities elsewhere. Ian Clark became responsible for overseas ventures whilst Malcolm Ford principally handled UK operations. Arco had approached BNOC suggesting a joint venture designed to increase Arco's North Sea holdings, and in return give BNOC the opportunity to join some of Arco's overseas operations. Mike identified two areas of interest, a concession in the UAE called Margham on which a significant gas and condensate discovery was made, which now functions as a strategic gas storage facility for Dubai's electricity generation and desalination. The second area was in Indonesia on the Kangean block north of Bali on which the Terang gas field was discovered in 1980 and the Pagerungan gas field in 1985, and a Britoil geologist was relocated to Jakarta to monitor the work of the Arco team. Both fields are now on production. The other significant overseas venture, negotiated by Ian Clark, was the acquisition of a 25% interest in the gas reserves of a US company Freeport-McMoRan for $73.5 million, and in exploration acreage in Montana, Colorado and Texas. Another Britoil geologist was sent to monitor activities at Freeport's office in Denver. Britoil ultimately acquired 18.6 million barrels of proven reserves in the US for $160 million, equivalent to about $8.60 per barrel, at a time when the finding cost was around $10 per barrel.

Alan Parsley was a clever and innovative geologist trained in the hi-tech world of Shell in The Hague. His principal contribution was to introduce the Monte Carlo method of prospect assessment to Britoil (possibly inherited from Shell, but perhaps I do him a disservice). The method was designed to calculate reserves so that one prospect could be compared with another. In order to eliminate human bias, each parameter (size, net pay, porosity, water saturation etc.) was given a range of values which were then subjected to repeated random sampling to produce a bell curve for each prospect showing the largest potential reserves (but with only a low chance of finding this amount of oil or more), the most likely (with a an evens chance of finding this amount or more), and lowest (but with a high chance of finding this amount or more). The method was adopted as standard for all Britoil's prospects.

During the early eighties the oil price was high as a consequence of the Iranian Revolution, the Iranian hostage crisis and the Iran-Iraq War. When I joined BNOC in August 1977 the oil price was $14.85 per barrel, but by June 1980 it had risen to $39.50 which had the effect of making many previously marginal discoveries viable for development, and with the price remaining above $30 a barrel until November of 1985, activity in the North Sea boomed. This in turn emboldened the government to accelerate its privatization programme. Britoil's privatization was followed by Associated British Ports, Enterprise Oil (subsequently bought by Shell), Jaguar, British Telecom, British Gas (against the bitter opposition of the Chairman Denis Rooke), British Airways, Rolls-Royce, British Airports Authority, British Steel, and the regional water, electricity and power companies. Not for nothing was this dubbed 'the sale of the century' and 'selling the family silver'. Many of these companies are now in foreign ownership. Furthermore, the Thatcher government authorised the sale of council houses where tenants were given the right to buy their houses at a discounted price from the local authority. This was great for the tenants, but led to a rapid rise in property prices which, in places like London, has made it impossible for young people to get onto the 'property ladder'.

Susannah and Jonathan usually came to stay with us during the school holidays and for New Year, and my mother would often come to visit, when she would take the train from Leeds on the famous Settle to Carlisle line which crosses some of the wildest and most dramatic scenery in England. We made excursions to places like Arthur's Seat in Edinburgh (the relic of an ancient volcano), or the Trossachs, a lovely area of lochs and glens, where you could take an old steamboat on Loch Katrine. Nearby is the Lake of Menteith, the only lake in Scotland (all the rest are lochs), with a ruined priory on an island in the middle, and Jonathan and I (and Panther) finally conquered Ben Lomond. On another occasion, Jonathan was playing Tarzan in Bothwell Castle policies (woods) when he fell off a tree and broke his arm. We also went to New Lanark, a model village on the Clyde, where David Dale established a cotton mill in the

1780s with power provided by the Falls of Clyde. Along with his son-in-law Robert Owen, a philanthropist and reformer, they built workers houses, a school, a chapel and an Institute for the Formation of Character, (but no pubs), an early example of utopian socialism. But there was a darker side. The village also had Nursery Buildings to house orphan children supplied from the poorhouses of Glasgow to work in the mill.

In January of 1983 I was invited to a Burns Supper in Hamilton, by Charles Westwood, a Britoil lawyer who also lived in Uddingston. Curiously, during communist times, Burns' suppers were popular in Russia where his memory was much revered, probably because of the sentiments expressed in poems like *A Man's a Man for a' That*, though how the Russian's coped with Burns' dialect is a mystery. The suppers were (and still are) formal black-tie or Highland dress, men-only events, held to celebrate the birthday of Robert Burns. Burns Clubs exist in many towns and cities and their only function is to organise this one annual bash, and the Hamilton Club was one of the most famous. The tables are charged with bottles of whisky, with not a wine bottle in sight. There are a series for formal speeches 'to the Queen', 'to the lassies', 'to the immortal memory', and an invited speaker gives a witty and erudite speech with numerous references to Burns' poetry. The highlight is the piping in of the haggis (by a Scottish bagpiper if full Highland dress) when the lights are dimmed and a flaming haggis is ceremoniously carried to the top table, and the President declaims Burns' famous *Address to a haggis*, which begins 'Fair fa' your honest, sonsie face, great chieftan o' the puddin' race'. There is a story that this poem was translated into German and then back into English where the last phrase was transmogrified into *'mighty Führer of the sausage people'*. Haggis, ye'll ken, is minced, spiced, sheep offal mixed with oatmeal, slow-cooked inside a sheep's stomach, served with tatties and neeps and a wee dram. The season for wild haggis hunting is officially from 1st October to the end of February.

In February 1983, whilst attending an operating committee meeting in London, I was invited by my opposite number in Charterhouse Petroleum to see *Tosca* at Covent Garden. Apart from the opera I had seen at the Bolshoi in Moscow, this was my first experience of opera in England, and I was seduced – not simply by the opera itself, but by the whole ambience of the experience, the champagne and smoked salmon, the Crush Bar, the splendid ornate auditorium, and dinner afterwards at one of the traditional old restaurants in Bow Street. It was an evening to remember, and opened my eyes to a whole new genre of music.

Late in 1982 the post of Chief Geologist became vacant when Alan Parsley was promoted to head up International Ventures, and following company policy the job was advertised in the press. I decided to apply for the position, since by now I had a good background in all the major aspects of petroleum geology, but I anticipated stiff competition from both inside and outside the company. In fact only a handful of applications were received, and none of those from outside were found to be suitable, so in March 1983 I found myself appointed to the job. It was the peak of my career, a prestigious job, Chief Geologist of a major integrated oil company employing over two thousand people, and still holding a dominant position in the North Sea, despite the removal of BNOC's former privileges. The position was ungraded and was on the Chief Executive's List. It came with a fancy new car, an impressive salary, share options, and special pension rights. It also gave me access to the Executive Dining Room that allowed me to meet and get to know many of the senior managers in the company.

There were however two negative factors. Under Britoil management staff and line functions had become more strictly segregated. No longer was the Chief Geologist involved in seeking overseas ventures as in Mike Ridd's day, nor was he directly involved with partners or with operations. His function was basically recruitment, training, ensuring standards were met, performance appraisals, promotions, remuneration and representing Britoil at conferences and public events. The second factor was that Mike Kelly, my former boss in Aberdeen, had now been transferred to Glasgow as Exploration General Manager, and was once again my boss. Not good news.

Letter 37
In the summer of 1983 Irina and I took a brief trip by hovercraft to France. Hovercraft were great, they glided over the sea on a cushion of air, like a Jesus bug (which walks on water). They were four times faster than the ferry, and could be run up onto a beach, but they were noisy and expensive to operate. We overnighted at a little hotel in Calais, on the first floor, and during the night heard someone trying to open our window. As soon as we got up to investigate, we saw the head of a youth, who immediately jumped down and ran away. Irina's bag was right under the window. Our trip had two objectives, one for Irina to visit the champagne houses around Reims, and for me to visit the battlefield of Verdun, so we spent a day sampling champagne at Ludes, Ay, Epernay and Reims, and then headed off to Verdun.

Verdun holds a unique position in the battles of the First World War. The city guarded the valley of the Meuse and was protected by a star-shaped citadel built by Vauban in the 1670s, and after the Franco-Prussian War by a ring of 19 outlying forts. By 1915 these forts were considered semi-obsolete and were stripped of many of their heavy armaments. It was a mistake. In early 1916, the German army launched a major offensive against Verdun, in an attempt to 'bleed France white'. The French were determined to hold the city at all costs (just as the Germans had anticipated) under Nivelle's rallying cry *'Ils ne passeront pas'*. The defence was entrusted to General Pétain who organised a supply route along the so-called *Voie Sacree*. Initially the Germans made progress and captured two of the outlying forts, Vaux and Douaumont, after ferocious hand to hand fighting. But progress slowed and it developed into a slogging match of attrition along a very narrow front resulting in enormous casualties on both sides. The advance was halted in the summer when the Germans were forced to transfer large numbers of troops to the Somme, but the battle ground on until the winter, by which time the meat-grinder had consumed 143,000 German lives and 158,000 French. The French called it a victory.

There are some interesting sights at Verdun. The *tranchée des baionnettes*, a French trench in a very exposed position in front of Fort Douaumont was obliterated by shell fire and after the war was found to contain a line of rifles with fixed bayonets projecting from the trench, with the remains of a French soldier beneath each rifle. Forts Vaux and Douaument are large bunkers built of enormously thick concrete, and excavated deep underground, faced with a sloping *glacis* swept by machine guns from the fort. The forts housed about 150 men, and inside there were firing steps, gun emplacements, corridors, barracks, a communications room and ammunition stores. The corridors are silent now but imagine when the Germans broke in with their grenades, rifles, bayonets and smoke bombs.

There is a swath of country north of Verdun designated after the war as part of the *Zone Rouge* a region defined as 'Completely devastated. Damage to property 100%. Damage to agriculture 100%. Impossible to clean. Human life impossible'. Driving through the forest north-east of Verdun there are shadowy traces of six former villages which were simply wiped off the map. At Douaumont there is a large French war cemetery, and an ossuary, a sinister looking building, something between a cathedral and a bunker with a tall tower containing a death bell. In the nave of the building are niches containing the bones from 130,000 corpses recovered from the battlefield – both French and German, in untidy, disorderly piles, as though thrown in at random. It presumably has some symbolism to the French mind, but to me underlined the gulf between Gallic and Anglo-Saxon attitudes. Such a macabre memorial would never have been built in Britain.

In 1983 the Conservatives defeated Labour (again) and Margaret Thatcher began her second term of office. Labour was in disarray. The leader, Michael Foot, was a sincere socialist intellectual but he looked like the Scarecrow in *The Wizard of Oz*, not the kind of image likely to wow the voters. The Labour Manifesto of that year has been described as the longest suicide note in history, but ironically many of its proposals have since been achieved. It advocated unilateral nuclear disarmament which was considered to be totally irresponsible by the majority of voters. Today there are calls for the abandonment of the Trident missile deterrent by people not regarded as particularly extreme. It called for withdrawal from the European Economic Community; this was achieved following the Conservative-sponsored referendum in 2016. It called for greater control over the banks; following the 2008 melt-down the government was forced to take control of several banks, and to limit some of their more irresponsible practices. It advocated a national minimum wage, a policy adopted in 1998 and now supported even by the Conservatives, and finally it proposed a ban on foxhunting and hare coursing, a measure which became law in 2004. It just goes to show that what goes around comes around.

So the Conservatives were free to continue their crusade against national companies, and now they began to wage war against the trade unions. The principal – perhaps the only – purpose of the Trade Union movement is to protect workers from exploitation and ensure that they receive 'a fair day's pay for a fair day's work'. Unfortunately, by the 1970s and 80s trade unions in Britain were abusing their power. They operated 'closed shops' where workers in a particular industry were obliged to become union members and contribute to their funds. They used block votes instead of allowing members to vote individually, and they abused the right to strike by calling out their members for the most trivial reasons, and by using flying pickets bussed in from areas far removed from the scene of the dispute. They caused grave damage to the British economy and they were perceived to be dominating the Labour Party, and as representing a serious challenge to the authority of the government.

Industrial relations became ever more bitter in the early eighties. Both the steel and coal industries in Britain were operating at a loss, so Thatcher determined to sell them off. She poached Ian MacGregor, from Lazard Brothers (paying them £1.8 million compensation) to become Chairman of British Steel. Like Beeching in the 1960s, he closed down dozens of plants and reduced the workforce by more than half in preparation for privatisation, which was completed in 1988. Thatcher was delighted, and in 1983 appointed him Chairman of the National Coal Board, where he was given a similar task. MacGregor ordered the closure of scores of deep mines and ultimately reduced the work force from 240,000 to 50,000. But it led to the most bitter and violent confrontation in recent history, eclipsing even the General Strike of 1926 in violence. The miner's strike lasted for almost a year from March 1984 to March 1985, and culminated in the Battle of Orgreave in June 1984 where mass-picketing of a coke plant in Yorkshire was broken up by 6,000 police in full riot gear, supported by over 40 mounted police, who made repeated baton-wielding charges into the crowd. The cases against all the miners who were arrested were dismissed due to unreliable police evidence, and the police eventually paid out almost half a million pounds in compensation. Afterwards a police chief admitted that it 'had come dangerously close to the police being used as an instrument of state'. Thatcher won the war and destroyed the power of the unions, but at enormous cost. The NCB was privatised in 1994, but by then there were only 15 deep mines left (compared with 180 in 1980). Both the steel and coal industries were sacrificed on the altar of monetarism, and in future Britain would be dependent on foreign imports for its needs. The rifts in British society were huge and persist to the present day.

It was against this background that I began my new job. I now had a fancy office with oil paintings from Britoil's art collection, someone's idea of a sound investment, and a new secretary. Margaret was an efficient and conscientious PA who had previously worked for Alan Parsley. She made my appointments and arranged my meetings, and typed my letters and memos. Best of all she was discreet and dependable.

During the boom years it was easy to attract high quality geologists and specialists. Britoil salaries and benefits were tempting and we were able to entice a number of very well qualified geologists to join us from universities. We also had a programme of graduate recruitment and each year I would organise promotional tours of the major universities, emphasising what a satisfying, glamorous, and well-paid job it was (which was true, as my own career had proved), and the applications flooded in. We would receive maybe three hundred applications for five or six places, so we could really pick and choose. I winnowed them down to a short list of about twenty who we invited to Glasgow for interview. We put them in a nice hotel and invited them for dinner on the evening before the interviews. We wined them and dined them, which allowed us to observe them in a relaxed setting, but at the end of the evening, I would say, 'by the way during the interview tomorrow I would like each of you to give a five-minute presentation on a geological subject of your choice'. This caused some consternation. It was now about 10.30 in the evening, some had imbibed rather too freely, and they were suddenly faced with having to plan a presentation for the following morning. To me it was a legitimate test of how they could cope with stress. The interviews were conducted by myself with the assistance of a second experienced geologist, and a senior representative from Personnel Department. Most of them coped surprisingly well, but many were more attracted by the pay and conditions than the job itself, and very few had read any books or articles on petroleum geology. Nevertheless, we were able to recruit some outstanding candidates, some of whom went on to very successful careers. We put them onto a one year training programme where they attended technical courses, worked in one of the exploration districts, and gained experience of wellsite work. How different to my introduction to petroleum geology; in those days you were thrown in at the deep-end – without a life-jacket.

I also found it interesting to conduct staff appraisals and tease-out the strengths and weaknesses of individual geologists. I was then able to suggest suitable training courses for them, and send others to conferences or field trips relevant to their jobs. Salary increases had two components, a cost of living increase applicable to all, and a performance increase determined by the result of the annual appraisal. That way those who contributed most could be suitably rewarded. Similarly promotion to the next grade was determined partly by length of service and partly by performance (and at more senior levels by the availability of a suitable vacant slot on the organisation chart). There was a certain satisfaction in witnessing the flowering of a persons capability and usefulness to the company.

Mike and Song introduced us to several of their friends, Iain Harrison and his wife Fabienne, Frank and Joy Quinn, and Anne and Sandy Mackintosh. Iain was director of a family-owned shipping company, which split from a larger consortium in 1956. They had (and still have) a charming and comfortable house near Drymen to which we were invited for several New Year parties (which happened to coincide with their son Patrick's birthday), at which the champagne was served from Jeroboams. We took Susannah and Jonathan to one of these parties, where Sue developed a taste for champagne - rather to excess on that particular occasion. They also organised musical evenings at which the celebrated violin and piano duo of Isabelle Flory and Robin Colvill would often perform. Frank was a technician and piano tuner at the Royal Scottish Academy of Music and Art who was married to Joy, a kind and warm-hearted Burmese lady. Each Christmas the choir from the Academy would visit their house in Langside to sing carols. It made for a magical evening. Anne Mackintosh lived at Killearn, just down the road from Drymen. She was a plant physiologist who took up painting in her spare time. She painted portraits and we asked if she would paint one of Irina. She produced a wonderful, luminous oil portrait, which has hung in our lounge ever since. She became well known, and in 1990 she was invited to paint Margaret Thatcher, and subsequently painted a host of well-known people like John Major, David Cameron, the Duchess of York, Nelson Mandela, King Hussein of Jordan, Donald Dewar, Danielle Steel and Arnold Palmer.

Mike resigned from Britoil in September 1984, frustrated by the attitude of the two directors Ian Clark and Malcolm Ford. Neither had had any involvement in the exploration side of the business and they lacked an understanding of its risks, subtleties and implications. They had a domineering, intolerant style of management, which contrasted badly with the days of Lord Kearton and Alastair Morton (who, although abrasive, were at least open to reason). Mike decided to become an independent consultant. He had a wide knowledge of the petroleum geology of the UK continental shelf and of the potential of the onshore UK, plus many areas overseas. He then went one step further and formed a partnership with Iain Harrison, his ship-owning friend, and together they set up a small independent oil company which they named Croft Exploration. Mike established a consulting company called Liberty Exploration, which offered geological expertise to oil companies, and of course particularly to Croft. Oil and gas exploration is an expensive business, so they developed a strategy of acquiring the lease of one or two concessions (which is reasonably cheap), either alone or with partners, work up its potential, and then offer equity in the licence to an established oil company, on condition that they drilled the first well. It was a variation on the over-ride system which operates in the USA. So, for the time being, we went our separate ways, but we kept in regular contact, and later in life became very close friends.

And what of the family during these years? Irina was by now working part-time as a language assistant at Strathclyde University in Glasgow. Susannah and Jonathan were growing up, and at sixteen Sue went to a sixth-form college in Scunthorpe while Jonathan continued at secondary school in Winterton. We took them to Agadir one year which had grown enormously since we had lived there, and it took us a long time to find our former house. We also went with them to visit our old friend Svetlana, now living in Malmo with her new husband, and a pair of love-birds. I recall some delightful picnics and a visit to the southernmost point of Sweden. My mother continued to travel, particularly on a cruise ship called *Aquarius*, based in Piraeus, where I think she rather fancied the radio officer. She visited Corinth and Delphi, Knossos, Istanbul, Petra and Aqaba, Jerusalem, Beirut and travelled up the Nile to Karnak and Abu Simbel. The cruise ships always had experts on board to talk about the history, archaeology or natural history of the places they visited, and she often suggested that I should offer my services as a geologist, a rather nice if impractical idea, given my situation. She often visited us in Uddingston, and she was very happy to look after Panther when we were away.

Most of the family visited at one time or another – Eva, Dick, Margaret and George with their son Paul. My other cousin John and his wife Julia had spent a couple of years in Kimberley in South Africa with the Diamond Corporation, where their daughter Sarah was born. John loved it, but Julia missed home and her family and so they returned to England. Their son Andrew was born shortly afterwards. In 1985 John did a parachute jump for charity sponsored by the Diamond Corporation, which was a very impressive achievement for someone with polio. Maybe he had inherited something of the bravado of his mother's cousin who won the VC in the First World War. It was decided that he would have to land in water so they selected the largest lake in south-east England, called Bewl Water in Kent. They arranged for a pick-up boat to be standing by, gave him some basic training, kitted him up, and off he went. He jumped from 2,000 feet (in order to be sure of landing in the lake), and raised over £2,500 for the premature baby unit at University College Hospital, and then gave an interview on local radio. He later took up gliding and became quite an expert glider pilot.

Irina's sister in law, Alla, came to visit in early 1986, but January is not a good time to visit Scotland. She enjoyed Edinburgh and Glasgow and Panther, and we took her to the old homestead in Halifax, which I remember was deep in snow. We told her next time to come in summer but, as far as Scotland was concerned, there was to be no next time.

Glasgow by this time was just beginning to get to grips with confronting its industrial past. The soot-blackened buildings were being cleaned, tenements were refurbished instead of being demolished, and in 1974 the building of tower blocks was abruptly terminated following completion of the totally inappropriate brutalist blocks at Sighthill and Balornock. Architectural masterpieces like Alexander Thomson's astonishing Greek Revival churches in Caledonia Road and St Vincent Street were saved from demolition but only at the eleventh hour. The policy of redevelopment zones was abandoned and replaced by one of infilling vacant sites with new buildings. Britoil occupied one of these in St Vincent Street, but in 1985 commissioned the construction of a new headquarters building at the top end of St Vincent Street. Beware the building of prestigious new headquarters buildings; it is often a harbinger of impending disaster.

Glasgow is rather proud of its art and artists. It was the home of the architects 'Greek' Thomson and Charles Rennie Mackintosh, and the Kelvingrove Art Gallery has two entire rooms devoted to the Glasgow Boys and the Scottish Colourists (not to mention Dali's famous *Christ of St John of the Cross*, and numerous Impressionist paintings). We often used to go to viewings of contemporary Scottish painters, and even Britoil arranged periodic exhibitions in the foyer of their St Vincent Street office.

Pollock House is an impressive Georgian country house built by William Adam located in parkland three miles south-west of Glasgow. It is famous for its collection of Spanish paintings particularly those of El Greco, Goya and Murillo. It also had an arts society (still in existence I believe) which hosted black-tie concerts in the splendid library, followed by dinner in the former kitchens. We were members, as were Mike and Song, Iain and Fabienne and Frank and Joy, and we regularly attended concerts with artists such as John Ogden and the Cuban pianist Jorge Bolet who, on one occasion, almost demolished the piano whilst playing Liszt.

But the jewel of Pollock Park is the Burrell Collection. Burrell was a Glasgow ship-owner who along with his brother, developed the practice of buying ships during economic recession and selling them at a huge profit when the economy recovered, amassing a large fortune in the process. Burrell developed an eclectic interest in antiques and began collecting a wide range of items – Chinese porcelain, Persian carpets, Indian furniture, and pictures, employing agents to scour the market for him. During the First World War he sold his entire fleet of ships for three times the building cost and thereafter devoted his time entirely to collecting. He widened his interests and began to acquire architectural items – medieval gateways, carvings, Greek, Roman and Egyptian antiquities. He bought Hutton Castle in Berwickshire where he housed his collection and when he died in 1958 he left the entire collection to Glasgow Corporation, but with several stipulations. These included housing the collection far enough from Glasgow to avoid industrial pollution but close enough for working people to be able to visit. He also requested that the dining room, drawing room and hall from the castle should be re-created as part of the display. The collection was crated and stored for twenty-five years, but eventually the Corporation did Burrell proud. They built a modern gallery in Pollock Park with a wall entirely of glass facing onto parkland in which to house the collection. It was beautifully designed and expertly displayed, and they even recreated the three rooms as he had requested. It opened in 1983 and became one of the principal tourist attractions in Scotland. There is a delightful story about a visitor to Hutton Castle who admired the Persian carpet in his bedroom, to which Burrell replied 'aye, but you should see the one underneath'.

Letter 38
One of the perks of the Chief Geologist job was to attend conferences and meetings and fly the flag for Britoil. I attended several conventions and field trips of the American Association of Petroleum Geologists, and wrote reports on the meetings for Britoil's in-house magazine. One year it was held in San Antonio, Texas, and I took the opportunity to visit the Alamo where James Bowie, William Travis and Davie Crockett made their last stand against Santa Anna's Mexican army. The site is now a shrine jealously protected by the Daughters of the Republic of Texas. In later life, we became good friends with a Texan who invariably inscribed the back of his party invitations the rallying cry 'Remember the Alamo!' He was also very proud of the fact that Texas had existed under six flags: Spain, France, Mexico, Republic of Texas, United States and Confederate States. I leave you to guess which one he most revered. The field trip on that occasion was in northern Mexico around Saltillo and Monterrey, where I recall seeing some spectacular folds within Jurassic rocks which

were not simply overturned, but overturned twice, which is decidedly unusual. Following the meeting I flew to Denver for a review of our US holdings with the resident BNOC geologist who was monitoring the activities of our US partners.

Another year the conference was held in New Orleans where I got to sample the local dishes of gumbo, po-boys, jambalaya and mountainous platters of sea-food. It has always puzzled me why American restaurants serve such enormous portions. Is it to demonstrate that they have an abundance of food? Americans presented with *nouvelle cuisine* in Europe must be quite shocked. The French Quarter of New Orleans originated before the Louisiana Purchase, but in fact the architecture is more Spanish than French. The Quarter has a certain reputation, but I had a conference to attend, so of course business before pleasure. I went on a field trip to the very southernmost tip of the Mississippi Delta which is extending seawards at a rate of about 250 feet a year. You can almost see it growing. From 1885 to 1935 the delta expanded enormously, but since then the sediment supply has been drastically reduced, largely due to interference by man, and now the area above sea level is decreasing. Birdfoot deltas like the Mississippi form important oil and gas reservoirs, and their characteristics have been exhaustively studied. The Mississippi delta has it all, delta plain, braided streams, lagoons, swamps, natural levées, over-bank deposits, crevasse splays, distributary channels, distributary mouth bars, barrier islands, tidal creeks, and those are just the surface features. As one delta lobe becomes silted up the distributaries migrate to form another lobe, and over time an enormous pile of deltaic sediments accumulates with lobes overlying each other like a series of superimposed hands.

From New Orleans I flew to Columbia, South Carolina to visit the offices of ESRI, a research group attached to the University of South Carolina. The group was established by Bill Kanes, a former Esso geologist, who came up with the clever idea of combining academic and commercial interests to the benefit of both. He got to use the university facilities and in return he attracted research students, mostly from overseas, to study at USC. He also furthered his own interests by amassing a large amount of data from the foreign students which he was able to use for his own purposes. We discussed whether there were any areas of mutual interest between ESRI and Britoil, but I was rather distrustful of his *modus operandi* and I was unable to find any potential benefits to Britoil. I found Columbia a curious city. I only had one day and no time to explore, but much of it was typical US ribbon development with endless advertising signs. What a surprise then to turn into the university campus and find a lovely tree-shaded quadrangle surrounded by classical buildings. It has suffered a fire, an earthquake and the Civil War, but has survived as an architectural gem.

It was accepted practice to send senior staff to attend a week-long Management Course at the Manchester Business School and my turn came in 1985. It was attended by people from all walks of life and focussed on MMM, men, money and materials, and SWOT, strengths, weaknesses, opportunities and threats, and there were lots of exercises usually in teams where challenges were set and conclusions analysed. It was an interesting exposure to a commercial world I barely knew, but I felt that the participants came from such a diversity of backgrounds that it was difficult to find much common ground between them.

I had become interested in remote sensing as a means of providing information useful to oil and gas exploration, and I visited Hunting Surveys at Borehamwood in Hertfordshire to see the current state of play. Hunting had begun as a photo-reconnaissance company which had expanded into methods applicable to mineral exploration. They showed me astonishing results obtained from ground penetrating radar, air-borne gravity and magnetic surveys, instruments for measuring radioactivity, and temperature differences (useful for assessing the health of growing crops), ultraviolet and infra-red cameras for identifying rock types and minerals, instruments for detecting oil seeps and escaping gas in the sea. They also showed me evidence of a bulge on the surface of the ocean which was indicative of a low-density sedimentary basin below. I wrote a summary of the meeting for Britoil management highlighting methods which might be applicable to our activities, but received no feedback. It is probably still buried in the files somewhere. Huntings also had a military division which operated worldwide (and manufactured Trident missiles), and at one time Osama bin Ladin was the company's agent in Sudan. Sadly, for such an innovative company, it went into liquidation in 2003. You might think the government would have offered some support, but Blair was no more sympathetic to ailing companies than Thatcher had been.

As far as flying the flag for Britoil was concerned, I gave talks and lectures at seven or eight universities across the country including University College Dublin (established principally as a Catholic rival to

Trinity College), and Durham (where it was a strange feeling to lecture in the same lecture theatre I had known as an undergraduate). I attended the bicentennial symposium of the Royal Society in Edinburgh, and the sesquicentennial anniversary of the British Geological Survey. I wrote two papers, one on the history of oil exploration in Scotland, in collaboration with two colleagues, which was published in a Scottish geological journal. It dwelt on the career of James 'Paraffin' Young, a chemist who invented a way of extracting oil from cannel coal (an oil shale formed from decaying land plant debris), which he patented in 1850. He expanded his prototype operation into an industrial scale plant at Bathgate in Scotland which began producing oil in 1851, (ten years before 'Colonel' Drake drilled his famous oil well in Pennsylvania). The second paper was on subtle methods of oil exploration, which drew on the information I had gathered from my visit to Hunting Surveys and on information gleaned from investigating a number of remote sensing techniques. This was presented at a meeting of the British Association for the Advancement of Science, a venerable organisation formed in the days when science had a much wider appeal than it has today, which could boast as former presidents such names as William Buckland, Adam Sedgwick, Sir Roderick Murchison, Prince Albert, the Duke of Northumberland and so on. It must have generated some interest as I was invited to give a radio interview on the BBC *Today* programme.

In March of 1986 Halley's Comet appeared and although viewing conditions were not great I was able to watch its progress across the sky for about a week. The comet appears every 75 years or so and has gained a reputation as a harbinger of disaster (but as someone neatly put it, I'm Taurus, and Taurans are not superstitious). But in this case it was true. The oil price had remained over $30 a barrel from October 1979 through the period of the Iran-Iraq War and in November 1985 it still stood at $30.81. But demand for oil was falling and some exporting countries began discounting their sales below the official OPEC price which had a major impact on Saudi Arabia since for many years it had acted as a swing producer, adjusting its sales to maintain a steady world oil price. Saudi revenue fell from $119 billion in 1981 to $26 billion in 1985 and they decided enough was enough. They introduced an ingenious system called netback dealing which did away with the fixed OPEC price and deliberately allowed the world oil price to fall in the belief that high cost areas like the North Sea would become unprofitable. They were ultimately wrong, but their action threw the oil industry into panic as the price tumbled. This effectively marked the end of OPEC as custodian of the world oil price, as oil exporting countries scrambled to look after their own interests. By July 1986 the world oil price had fallen to $11.58 per barrel. Companies found that projects which were profitable at $30 a barrel were no longer profitable at $11 a barrel and many projects were shut down, and the people associated with them laid off. Britoil was particularly badly hit since 95% of its activity was in the high-cost North Sea. Companies tend to panic in these situations, concentrating on the immediate crisis and ignoring lessons from similar events in the past (which suggest that dramatic price falls are usually followed by recovery). So it was with Britoil. In May of 1986 they announced that 400 jobs were to be axed, about a fifth of the entire workforce. I had just been on a visit to Cambridge promoting Britoil to students about to graduate, following which I was scheduled to give a one-week course to our young geologists on methods of reserves calculation. Redundancies were to be announced at the start of that week. I was just about to begin the course when Margaret came and said 'Mr Kelly wants to see you'. I had believed my job to be secure but, for whatever reason, it was deemed non-essential. After ten years of working my way up through the ranks I was no longer needed. It was a bitter blow. Many others of course were in the same boat, including several senior managers, but that didn't make the blow any easier. The redundancy package was generous, a year and a half's salary, £10,000 relocation allowance, the right to keep my share options, and to take my pension from the age of fifty, but of course it would almost certainly mean having to sell our house and leave Scotland and all the friends we had made there.

Britoil set up a redundancy office with phones, faxes, secretarial services, information on jobs, counselling and so on, and once I had recovered from the shock, I began to think that my wide experience in virtually every aspect of petroleum geology would be attractive to many companies. But they were all in the same boat. Companies involved in the North Sea were not hiring, they were firing, just like Britoil, so I found I had to cast my net wider. I looked for opportunities with foreign oil companies, oil service companies, scouting organizations, (scouting is a service which monitors the activities of oil companies and provides a digest for its subscribers), and universities, and eventually I received indications of interest from two oil companies, Kufpec (Kuwait) and Sirte Oil (Libya), a service company called Robertson Research, a management consultancy outfit called Arthur D. Little, and Imperial College and Royal Holloway in London and Curtin University in Australia, plus an offer from a Government Office in Fiji which was responsible for issuing oil concessions. It was a very mixed bag so, with the exception of Fiji, I went to visit most of these organizations. It was a major decision, because it would determine the future course of my career.

Sirte Oil Company had an office in Weybridge which I visited, and found a group of Brits plus one or two Libyans, who were surprisingly positive about the company. Like Sonatrach and BNOC it was a large integrated national company which operated about twenty fields and was actively exploring a dozen or so concessions. The company drilled three or four wildcat exploration wells a year plus five or six appraisal and development wells, and produced about 170,000 barrels of oil per day. There was a vacancy for a Senior Staff Geologist based in Libya, with free housing and an attractive tax-free pay package. It beat all of the other job offers by a considerable margin, but the problem was that Libya was something of a pariah state. The leader, Muammar Gaddafi, actively supported revolutionaries, dissidents, terrorists and freedom fighters of many different persuasions. He was rabidly anti-Israeli and he had sent troops into Chad to seize areas rich in minerals, reputably including uranium, and he was believed to have nuclear ambitions. Following the bombing of a night club in Berlin frequented by American servicemen, thought to have been planned by Libyans President Reagan ordered a bombing raid on Tripoli and Benghazi which was carried out in April 1986, so by the time of my discussions relations between Libya and the west were at a low ebb.

I suggested that they fly me out to Libya to take a look at the company and the location. They agreed and in January 1987 I flew to Tripoli with an American geologist, Frank Palen, who had been with the company for several years. We then flew to Marsa al Brega on the Gulf of Sirt, which was the main operations base of the company. Sirte Oil Company had been formed to take over the interests of Esso Libya following their departure in 1982. The town had been built to house Esso's American personnel back in the 60s, so all the houses, offices and recreational facilities were American in design and construction. I met the Exploration Manager, an American-educated Libyan, who was very welcoming, and I went to see a typical family house, and was quite impressed by the accommodation. Each house was on its own plot with small gardens back and front surrounded by a high wall for privacy, and there was a very nice beach nearby. It reminded me of Algeria, but with the advantage that the office was only five minutes' drive from the house. I returned to UK to report on my trip to Irina.

Irina's mother came on her third visit to the UK in September and stayed for a month, but this time the mood was subdued. It was obvious that we would not be in Holmwood House for much longer. Irina and I faced a tough choice, but in the end opted for the Libya job. It was the kind of technical work I liked and which I knew I could handle, the pay was good, we were familiar with ex-pat life, the climate was excellent, and there was a thriving ex-pat community. Little did I realise what an important role Libya would come to play in my future career. We decided we would give it a try for a couple of years and see how things stood at that time.

I had a farewell bash with Bob Duke, one of my Britoil colleagues who had also been laid off. We went to a famous Glasgow restaurant called Rogano's for lunch and didn't leave until evening. Just how many bottles of champagne we consumed I do not recall, but I don't remember much about the journey home. We put our lovely house on the market and sold it in December, and moved into Bob and Jenny's house in Kilmacolm, as they were now living Paris. My mother and Sue and your dad came for a rather melancholy Christmas and New Year. We spent Christmas Day with Tony and Laleh at their enormous house at Cults, and I remember Tony accompanying a group of visiting carol singers on the piano, with the snow falling outside, and we went to Harrison's for our final New Year celebration. It seemed like my high-flying career had come to an abrupt end. This was my bonfire of the vanities.

Britoil did not long survive me. In 1988 it was acquired by BP with many additional job losses for Britoil personnel. BP's takeover gave them a large addition of proven oil reserves for a cost of about $5 a barrel at a time when the finding cost of new reserves was about $7 a barrel. BP didn't want the fancy new headquarters building that Britoil had built on St Vincent Street, and sold it to an insurance company which in turn got into financial difficulties and had to shed a large proportion of its workforce. BP had a further shakeout of personnel in 1992, by which time there were very few former Britoil employees left. It was a sad end to what had been (and could still have been) a great company.

There was a tragi-comic coda to the Britoil story. Just before the BP takeover a couple in Britoil's accounting department in Aberdeen, Alison Anders and her lover Royston Allen, attempted to steal £23 million in a heist worthy of Frederick Forsyth. Anders found details of a girl of about her age who had been killed in a road accident, she applied for and obtained a birth certificate in the girl's name, which she then used to acquire a false passport. She used an agent to set up a Swiss Bank account using her false name, and in June 1988, as a routine part of her job, she was asked to prepare a bank instruction to the Bank of Scotland in Aberdeen authorising the transfer of £23 million (for the lease of a drilling rig) to Lombard North Central in London. She changed the destination of the transfer to the account in Switzerland and handed it to the bank. All would have been well, but she added a hand-written note on the authorisation which was highly unusual. A sharp-eyed cashier queried the transfer, and the fraud was uncovered. Anders and Allen holed-up Abu Dhabi and later Anders went to the USA, but Allen's estranged wife found evidence of the plot and 'shopped' her husband by handing over the evidence to the police. Anders and Allen got five years in jail apiece.

Letter 39
We left Scotland in the middle of April in a snow storm, and since we had no intention of returning we decided to buy a *pied a terre* in London, for use during our vacations. We found a very nice flat in Warwick Square in Pimlico, close to Eva who lived in the adjacent square. It was small but nicely situated in the middle of the square with a south-facing view over the gardens. The Pimlico garden squares had been very fashionable in Victorian times, and our house, number 12, had been owned by Viscount Cross, who was Home Secretary in Disraeli's government in the 1870s. Churchill too had once lived in the square. The entire area had belonged to the Grosvenor Estate, but after the war the houses were converted into apartments of all manner of shapes and sizes. Our flat was on the second floor and had high ceilings and original sash windows and shutters. It must originally have been the principal bedrooms. We filled it with some of our furniture from Scotland, and what would not fit in we put into store.

I was not scheduled to start my new job in Libya until May, so we decided to take a vacation in Thailand. We spent a few days in Bangkok and then flew to the resort of Phuket, an island off the west coast in the Malacca Strait. We went to Phang Nga Bay to see the astonishing needle-karst limestone pinnacles made famous by the James Bond movie *The Man with the Golden Gun*, and by boat from Krabi to the exotic Phi Phi islands. We saw the locals collecting birds' nests with long poles from caves, to be used in making the very expensive bird's nest soup. The nests are made from the hardened saliva of swifts which inhabit the caves - not everyone's idea of a tempting appetizer. We hired a car and visited a rubber plantation, and we ended our trip with a few days in Singapore which was impressive, but so efficient and clean that it seemed to have lost all traces of its oriental origins. Of course, we dropped into Raffles Hotel for a Singapore sling, and I remember Irina and I had an argument about Britain's fighting capability during the war. Singapore was not a good example of British military prowess.

We left for Libya, with Panther of course, in May 1987. Panther travelled in a large animal crate and was rather traumatized by the experience. We overnighted in Tripoli and walked her near the hotel, but she was used to taking a pee on grass, and there is very little grass in Tripoli. Next day we flew to Brega in the company's F28 plane. The dog box would not fit in the hold, so eventually they put it in the cabin. The plane was only half full so it was not a big problem, but the steward was subsequently reprimanded for disobeying company rules. Poor old Panther, she did not enjoy the experience. We arrived to find that our allocated villa was not yet ready so we were put into temporary accommodation on 'Secretary's Row' – a street used to house most of the ex-pat secretaries. Panther cheered up when she found the beach and the sea, but it took her a while to adjust to the temperature, and she shed a lot of hair in the process. Eventually she acclimatized and became part of the community.

I think Irina must have been rather shocked with Brega. It was an oil camp with a population of about 10,000 people, half ex-pat and half Libyan. There was a port for loading oil tankers, an airport, oil refinery, gas plant, office buildings, a large housing area, supermarket, mosque, school, cinema, bowling alley, football field and an ex-pat club. The whole complex was surrounded by a boundary fence to provide some measure of security. Some of the housing was for bachelors, with six men to each house, each with his own room and bathroom, and with a communal recreation room. The family houses were three-bedroom villas with two bathrooms, a large living room and kitchen equipped with American appliances (since the town had been built by Esso). There was a car-port, shed and garden, all surrounded by a wall. The houses were refurbished and re-equipped for each new arrival and were actually quite comfortable – and free. The company provided the houses, the furnishings, electricity and water, and even spare light bulbs, at no cost. Palm trees and oleanders and camel grass had been planted in an effort to make the place attractive, but there was no disguising the fact that the surrounding area was desert.

The original inhabitants of Libya were Berbers (which is the origin of the word barbarian), who still form a significant part of the population, especially in the area bordering Tunisia. Further south the nomadic Tuareg (a word apparently signifying 'abandoned by God') form the major ethnic group in the south-west, and the Toubou (the rock-people) in the south-east. Libyan society is still dominantly tribal in character with loyalty to the tribe outranking loyalty to the state. Libya has had a turbulent history, being under the control at various times of Phoenicians, Carthaginians, Persians, Greeks, Romans, Vandals, Byzantines, Arabs, Normans, Spanish, Maltese Knights and Ottomans. Texas with its six flags pales into insignificance. The Ottomans were ejected by the Italians in 1912, and the Italians by the Allies during World War II. Libya became independent as a federal kingdom in 1951 and was seized by 'Colonel' Gaddafi in a military coup in 1969. The country is dominated by the Sahara desert which reaches the coast along a 800 kilometre stretch between Misratah and Zuwaytinah. Modern Libya is something of an historical accident. With the desert separating the two principal areas of population there was little to hold the country together. That is until oil was discovered in 1957, in the desert, between the two areas of settlement. Within a few years it became clear that Libya was a major petroleum province with many giant fields of both oil and gas, and Libya was transformed from an insignificant pastoral backwater into a rich, oil producing nation. That, more than anything else, held the country together, at least for a while.

Morocco had been a feudal dictatorship, Algeria a dour socialist state, but Libya was different. Gaddafi, of Berber origin, had his own brand of government, which was expounded in his 'third universal theory' and published in his *Green Book* (what the first two theories were, is something of a mystery). In practice, it was a mix of socialism, nationalism and pan-Africanism. Ostensibly the country was governed by People's Committees which fed ideas and suggestions upwards to the central administration. But this was not the reality. Representatives of the People's Committees were coerced into endorsing policies handed down from above, and the regime was in fact a military dictatorship. The country was formally called the *Socialist People's Libyan Arab Jamahiriya* (reminiscent of Algeria's optimistic epithet *'Démocratique et Populaire'*), which was usually abbreviated as SPLAJ. The word *Jamahiriya* was coined by Gaddafi and means something like 'state belonging to the masses'. Gaddafi himself was not President or Prime Minister, but simply Leader (shades of Hitler), there to advise and guide. But Gaddafi's active espousal of freedom fighters (or terrorists and rebels, depending on your point of view), and his invasion of Chad, and border disputes with Algeria, Tunisia, Malta, Sudan and Egypt made him highly unpopular on many fronts. He once turned up for a heads of government meeting in Paris with his tent, white horse and a posse of female bodyguards (known as Amazons).

There were some other bizarre facets. Moslem countries base their calendar years starting from the *hijira*, the flight of the prophet Muhammad and his followers from Mecca to Medina. Gaddafi's government decided that it was more logical to date the years from Muhammad's death, so they became ten years out of step with every other Moslem country. Not only that, but for business purposes the Libyans used the Gregorian calendar, but Gaddafi did not wish to use the western names of the months, so they changed them, rather along the lines of the French revolutionary calendar. October became At *Tumur* (the dates), November *Al Harth* (the tillage). July however was *Nasir* (after President Nasser of Egypt) and August *Hanibal* (the Carthaginian general). It became a considerable chore when writing an official document to put the Islamic date, and year-according-to-Gaddafi, followed by the Western calendar date, but with the new name of the month. The timing of the start of Ramadan and the Eid el Fitr (which are determined by sighting the new moon), also produced problems. Does sighting in Mecca become binding on all other countries or should each country rely on its own observation? The Libyans of course adopted the latter view but on occasion it would be cloudy at the critical time, which gave the authorities a real headache. Libya also experimented for several years with summer time, when the clocks were advanced by one hour during the summer months. It would be announced at the last moment on Libyan radio and TV, but such was the ponderous nature of SOC organisation that it might not be introduced in Brega for a couple of days, during which SOC time was out of sync with the rest of Libya. You may scoff at such confusion, but England too has had its calendar problems. By the reckoning of the time King Charles I was executed on 30th January 1648. In Europe, where the Gregorian calendar was in use, that date was 9th February 1649.

The oil industry had developed rapidly in the pre-Gaddafi era, with companies like Esso, BP, Gulf and Agip making major discoveries, but after King Idris was overthrown, Gaddafi and his associates imposed majority state participation on all oil and gas concessions and producing fields. Four or five of the oil majors pulled out (or were forced out), and Libyan state companies were formed to take over their interests, so that BP's former concessions were taken over by a Libyan state company given the ironic name of the Arabian Gulf Oil Company (Agoco), in retaliated for Britain handing over a number of islands in the Arabian Gulf to Iran when British forces pulled out of the area. Esso accepted state participation for a number of years, but in 1982 they pulled out too, handing over their interests to Sirte Oil Company, the company I had now chosen to join. This was a different model to that of Algeria where one large oil company handled all the state's interests. In Libya new state companies were set up to replace each of the majors, but all under the umbrella of NOC, the National Oil Corporation.

Relations between Libya and the US began to take a downward path shortly after Gaddafi came to power. In 1973 Libya unilaterally established the so-called 'line of death', claiming the entire Gulf of Sirt as Libyan territory. The US insisted on the freedom of navigation within the Gulf and periodically conducted naval exercises, which led to armed conflicts in 1981 and 1986. Following the 1981 incident the US embargoed the importation of Libyan oil, which led to the departure of Esso from Libya, and they banned the sale of aircraft and electronic equipment to Libya. In 1984 a policewoman (who was ironically protecting the Libyan Embassy in London from a hostile demonstration) was killed by a bullet fired from the embassy, leading to an immediate severing of diplomatic relations between Britain and Libya. Two weeks after the 1986 conflict in the Gulf of Sirt a discotheque in Berlin frequented by US servicemen was destroyed by a bomb, for which the Libyans were held accountable, and President Reagan ordered punitive bombing raids on Tripoli and Benghazi. Gaddafi's compound in Tripoli and military airfields were attacked. Sixty tons of bombs were dropped and about seventy people were killed and twenty Libyan planes destroyed on the ground, but the Libyans shot down an American F-111 fighter bomber with a Russian-made missile, following which Gaddafi renamed the country the *Great* Socialist People's Libyan Arab Jamahiriya. It was into this unsettled environment that we arrived one year later.

There is a bizarre corollary to the bombing story. *En route* to Tripoli the US warplanes encountered an unexplained shape just below the waves between Sicily and the island of Pantelleria which they mistook for a Libyan submarine, and attacked with rockets. In fact it was the *Empedocles* sea-mount, one of several active volcanic features in the area. (Empedocles was a Greek philosopher, born in Sicily, who threw himself into the crater of Etna in the belief that he would be reincarnated as a god – not too wide of the mark, since he was reincarnated as a volcanic sea mount, which I thought rather fitting). Empedocles – the volcanic feature, not the man - periodically emerges from the sea as a temporary island but is eventually eroded back below sea level. It rose from the sea in August 1831 when it reached a height of 60 metres. The British navy immediately planted a flag and claimed it for Britain under the name of Graham Island, but two weeks later representatives of King Ferdinand of the Two Sicilies cut down the British flag and replaced it with the Neapolitan flag and renamed the island Ferdinandea. In September a French geologist arrived and planted the *tricolore* and named the island Julia, to be followed by the Spanish, who called it Corrao. The island was visited by numerous tourists including Sir Walter Scott, and Fenimore Cooper used the story in his novel *The Crater*, and someone even proposed building a hotel. Alas, the ash and pumice were soft and friable and within two or three months the island had disappeared. In 2001 the Sicilians lowered a marble plaque weighing 150 kilos reasserting Sicilian claims, but within six months it had been smashed to smithereens. How that happened is still a mystery. As the *New York Times* commented it would make a great plot for a comic opera.

We moved into a family house and began to get a feel for the place. Sirte Oil Company had a small administrative office in Tripoli, but the main base was in Marsa el Brega on the coast of the Gulf of Sirt, 770 kilometres from Tripoli and 240 kilometres from Benghazi. It was an integrated company employing about 6,500 people with facilities and personnel handling exploration, drilling, engineering, production, refining, gas transmission and processing, petrochemicals and shipping. There was a Libyan school and an expat school, a hospital, mosque, fire station, private airport, supermarket, restaurants and recreational activities. In addition, there was a Management Building and departments for personnel, training, safety, and security. It was a veritable town, all operated and run by Sirte Oil Company.

The expat community covered a wide range of skills and nationalities. There were geologists and engineers, teachers and doctors, firemen and pilots, divers and welders, cooks and physiotherapists. All the management positions were held by Libyans, but the vast majority of the technical work was done by expats. Many expats, especially those on bachelor status, were from the UK, but there were also Canadians, Yugoslavs, Poles, Germans, Czech, Maltese, Indian and a handful of Americans. We once attempted to list the number of different nationalities and reached 80 without any great effort. All departments included some

Libyans who were being trained so that eventually they could replace the expats. Libyan oil companies were obliged to offer attractive packages to expats in order to recruit suitably qualified personnel, but the Libyans themselves received very low salaries (my Libyan manager claimed that his British secretary was earning more than he was), so many of them had a second source of income, a shop or perhaps a farm, and they were not highly motivated to work eight hours a day for 1,500 Libyan dinars a month (about $750). The dinar was pegged to the US dollar, but at a totally unrealistic rate, so there was an active black market where expats could buy dinars at far below the official rate. After a year or two most of the young Libyan trainees would be sent overseas to study for a master's degree in Britain or Canada and they would go to great lengths to extend their assignments for as long as possible. It was not unusual for them to be away for two or three years, and one or two of them never returned.

In the early days of Esso's operations Libya had been a highly desirable posting. The overseas pay was good, the climate very attractive, alcohol was available through the company store, and there was the presence of the US airforce at Wheelus Field near Tripoli (formerly the Tripoli Grand Prix circuit, 'the Ascot of motor racing', which functioned from 1925 to 1940). After the 1969 revolution the sale of alcohol was banned, so employees took to making their own. Indeed, Esso produced a booklet called *The Blue Flame* which explained in some detail how to construct and safely operate a still to produce alcohol. The neat alcohol was cut with 50% water and flavoured with whisky, gin or vodka essence with added oak chips or juniper berries, to produce 'flash', a rough imitation of the real thing (in both senses of the word), and of course it is not difficult to make beer or wine. The tradition continued after the departure of Esso, but was technically illegal. The Libyan authorities knew perfectly well that alcohol was produced in the camp, but turned a blind eye so long as it was kept under wraps. Occasionally Industrial Security would raid the bachelor quarters and seize their booze, but they would never raid a family house in case a woman was present. Almost all the expat family houses had a bar which became the focus of the many social occasions within the community. We used a baby-food supplement called Biomalt, produced in Switzerland, to make beer, and the consumption of Biomalt in Libya was so high that the company sent investigators searching for all those healthy, bouncing Libyan babies.

The Libyan economy was eccentric. At one time private enterprise was forbidden, at another small shops and businesses were permitted, and at other times there was virtually free enterprise. The company had its own store which sold basic items, but also such luxuries as chimney sweeps brushes (none of the houses had chimneys) and snow shovels (it doesn't snow much in Libya). There were more shops outside the compound at a place called Crossroads where there were traders selling fruit and vegetables, dry goods, chickens, goats and sheep, usually on the hoof, but they would kill them for you for a small fee, and at one time there was a Government store which was something of a joke. It stocked 20 kilo bags of sugar and rice, 10 litre cans of cooking oil, rack upon rack of bottled water, plastic shoes in only two sizes, and not much else. Fortunately, the idea of government stores was quickly abandoned. Many basic staples were subsidised, bread was ridiculously cheap, and petrol was cheaper than bottled water. I could fill up my 35-litre gas tank for less than $5. Towards the end of our time in Brega the Libyan authorities decided that the free enterprise of Crossroads could not be allowed to continue: it was unregulated, unauthorised and was beginning to make certain people quite rich, so they used the excuse that the buildings were too close to the road and sent in heavy military bulldozers to demolish the entire site. Such was the arbitrary way of life in Gaddafi's Libya.

Letter 40
We quickly got to know a large number of expats, and there were some very interesting characters. One was an Australian marine pilot called Danny Gunn, who guided tankers in and out of Brega port and anchorage. He was originally from Tasmania and conceived the idea of building a boat himself and sailing it around the world with his family, stopping off at various places *en route*, where he would get a job and earn enough money for the next leg of his journey. Three years in, and they had reached Europe, where he deposited his family with his mother-in-law in South Wales, while he came to Brega to work as a marine pilot for six months, before continuing his circumnavigation. He said he expected to be away from Australia for the best part of ten years. When I last heard of him he was approaching the Panama Canal. I hope he made it.

One of the Petroleum Engineers, with whom we established a long-lasting friendship, was Iranian by origin. In his student days in Fasa and Shiraz he had become a communist and participated in student protests against the Shah's regime. His activities came to the attention of Savac, the Shah's secret police, and he had to be constantly on the move to avoid arrest. Eventually he decided to escape across the border into the Soviet Union. He made his way there and was able to cross into Turkmenistan without problem. He went to the nearest watchtower to request asylum, but it was unmanned. They went to the nearest village, but there was no police post, so he continued to the capital Ashgabat where he was held in jail for three months whilst his story was checked out and verified. After spending a year in Tajikistan, he left without permission and travelled to Moscow where he applied to study Petroleum Engineering. He was sent to the Petroleum Institute in Grozny, the capital of what was then the Soviet Autonomous Republic of Chechnya in the north Caucasus near the Caspian Sea, and the centre of a major petroleum province. At that time the region produced 7% of the Soviet Union's oil, but it had a troubled history. During the Second World War the Chechens attempted to revolt against Soviet domination, and in retaliation Stalin forcibly deported the entire Chechen population to Kazakhstan and other areas. They were eventually allowed to return, but found their homeland occupied by large numbers of Russians. The Chechen wars which raged from 1994 to 2000 were the direct legacy of these events. After a year he returned to Moscow and completed his studies at the Gubkin Institute, and he married a charming, homely woman called Natasha. Some years later he obtained a tourist visa to visit Paris with his wife, and decided not to return to Moscow. He applied for political asylum, which after a long battle was granted, and they found a place to live in St Denis (a noted communist quarter of Paris at the time, with its two principal roads named Avenue Lénine and Avenue de Stalingrad). Mohamed joined Total and was sent to East Pakistan where he found himself in the middle of a civil war when the Pakistani army invaded to prevent the establishment of Bangladesh as an independent state. He managed to find a boat in Chittagong and escaped to Kolkata (Calcutta). Eventually he left Total and came to work in Libya. He arrived in Tripoli two days before the American bombing, and was immediately sent back to France where he waited for six months until it was safe to return. He eventually repudiated his name Mohamed in favour of Michel. I told him he should write his memoirs.

In Brega I was given a company car but this could only be used in and around the camp, so I bought an old Fiat 124 from a Canadian engineer who was leaving, which then gave us the means to drive to places like Benghazi, where there were hotels and restaurants, a gold souk, shops and, best of all the port, where we bought freshly caught fish and lobsters, which we then took home to barbeque.

The Exploration Department had a staff of about thirty and was headed by a Libyan, Taleb M. Taleb, who had been charming when I met him on my reconnaissance trip in January. Advancement in Gaddafi's Libya (for Libyans) depended on having the right connections, the right family, the right tribe and the right provenance. In spite of Brega being much closer to Benghazi than Tripoli, virtually all of the senior posts were held by people from the Tripoli region. The chairman of the company, Mansur Benniran, had connections to Gaddafi, and his brother was reputed to be the head of the Security Services. Gaddafi would occasionally visit Brega with his entourage, and his tent would be pitched on the beach amidst very tight security. He had a communications truck and a mobile hospital in case of emergency. In later days, a deep bunker was built for him beneath the Brega Guest House. In Sirte Oil Company (abbreviated to SOC in future) the Libyan department heads had to walk a very fine line. Any slip up and they could be out within days. They had to demonstrate (and keep on demonstrating) loyalty and commitment to the regime, and a certain degree of success and achievement in their posts. In the case of Exploration Department this meant not too many dry holes, producing new and innovative exploration ideas, and making business-like presentations at the periodic review meetings with the National Oil Corporation (NOC).

I began work as Senior Staff Geologist but within a month was promoted to Subsurface Geology Coordinator, reporting to the Geology Superintendent, Bob McCrossan, an experienced Canadian geologist who had been recruited largely for one purpose. In Canada he had produced a very well regarded collaborative atlas on the *Geological History of Western Canada* published by the Alberta Society of Petroleum Geologists, which had been widely used by oil companies and universities in Canada and beyond. Taleb had obviously seen and been impressed by this report and had tempted Bob to come to Libya for a few years with a lucrative pay deal and benefits package to produce a similar atlas of the Libyan offshore. This area was very lightly explored. Esso had made a couple of small offshore discoveries just prior to leaving in 1982, and SOC had subsequently found two more small fields nearby, but the principal success in the area was the discovery of the Bouri field by Agip in 1977 which proved to be one of the largest oil fields in the Mediterranean. Large areas were unexplored, and Bob was hired to fill this gap. No expense was spared. A regional seismic programme was designed, acquired and processed, well data was purchased, laboratory studies were commissioned and a special team set up to coordinate and assemble the atlas. The project was close to completion when I joined, and had cost a mint of money. Unlike normal company reports the atlas was professionally typeset and printed with colour illustrations, just like the Canadian atlas, but only five or six copies were produced all of which were jealously guarded and kept under lock and key. It was an excellent piece of work, and on the basis of the report SOC drilled three further offshore wells, all based on sound reasoning, but unfortunately all dry. Taleb received a great deal of kudos for the report, but curiously little use was made of it after the three dry holes. Like a great expensive ship for which there was no further use, it was relegated to being used by NOC as promotional material to encourage foreign companies to apply for offshore concessions.

Bob McCrossan was a larger than life figure. He was a good windsurfer and he could often be seen in very brief swimming trunks heading out into the bay with a cigar clamped between his teeth. On one occasion he went to Malta with his wife Janette, and decided to windsurf to the small adjoining island of Gozo, about 5 kilometres distant. When he arrived there the wind dropped and he was marooned on the island with nothing but his surf board and swimming trunks. He was a very good geologist, but his commitment to the offshore project (and the frequent calls to his stockbroker) meant that he had little time for anything else, and much of the onshore work devolved on me.

This gave me the opportunity to become familiar with the petroleum geology of the eight onshore concessions held by SOC and with the discoveries which had been made on them. Almost all of the discoveries had been made by Esso in the 50s and 60s, but following the Gaddafi takeover in 1969 Esso stopped exploring and devoted all their attention to extracting as much oil as quickly as possible. We operated some very large fields. Zelten (subsequently renamed Nasser) covered an area of 50,000 acres (three times the size of Manhattan), and had 250 wells, which reached a peak production rate of more than half a million barrels per day, but by my time production was down to 50,000 barrels per day. Altogether the field has produced about 2.5 billion barrels of oil, of which 2 billion were produced by Esso before they left. Esso had made many discoveries which they had not had sufficient time to fully evaluate, and SOC adopted the rather cautious strategy of devoting much of their effort to following up leads which Esso had regarded as either too small or too difficult to produce.

One of my first jobs was to look again at the Nasser field. Esso had established that the structure was a buried reef, shaped like an upturned shallow dish, but we had a few deep wells which showed a more complex picture. One of our geophysicists, a 'plastic' Canadian of Russian origin, came up with the concept that beneath the reef were probable lagoonal sediments surrounded by a fringing reef (as you can see today, in the Maldives for example). I examined the geological data and found evidence to support his idea. We eventually secured funding from NOC to test the concept and sure enough we found the fringing reef, which was oil bearing, and was soon contributing to the production of the Nasser field.

SOC had acquired another concession which had been relinquished by the French company Aquitaine which also had oil shows in some of the wells, but in a very tight reservoir, which they regarded as non-commercial. SOC deepened three of the Aquitaine wells and found a very thick oil zone in fractured quartzite. We commissioned a specialist on fractured reservoirs to come and advise us and then proceeded to drill a further fifteen or so wells which all flowed oil, and the field which was named Wadi, was brought on production at a very respectable rate.

There are some fascinating stories about the early stages of oil exploration in Libya. Occidental was a small California independent until Armand Hammer took control. He transformed the company into a major international player largely through his decision to go international. He submitted bids for two concessions in Libya in 1966, partly on acreage relinquished by Mobil as non-prospective, but unlike the established majors he theatrically presented his bid written on sheepskin, rolled and bound with silk ribbons of red, black and green, the national colours at that time. Not only that but he undertook, if his bid was successful, to explore for water in addition to oil, particularly in the parched areas of southern Libya where King Idris had been born. Much to the chagrin of the majors Occidental obtained its concessions, and they proved to be a goldmine. Concession 102 was found to contain the southern half of one of the largest fields in Libya, and Concession 103 contained half a dozen pinnacle reefs, full of oil, one of which flowed at 74,800 barrels per day – a rate unheard of in Libya, and matched only by the oilfields of the Arabian/Persian Gulf. It was jokingly said that the Arabs must be a favoured race for Allah to endow them with so much oil. One of the Occidental reef discoveries was made on the site of a former Mobil camp. The fields were named *Idris* in honour of the king but renamed *Intisar* when the king was overthrown. Occidental drilled production wells, laid a 130-kilometre pipeline to the coast, built an export terminal and shipped out the first cargo in less than a year from the initial discovery. It was the

Grandparents

Alfred Hallett, 1890-1949

Mary Hannah Hallett, 1892-1945

Walter Riddlesden, 1882-1959

Ada Riddlesden, 1882-1959

Parents

George Hallett, 1916-1944

Mary Hallett, 1917-1999

*The last photograph of my dad
1944*

*George's grave in the
CWG cemetery in Florence*

4 Westbury Terrace, Halifax, where it all began

Panorama of Halifax

St Cuthbert's Society, my college in Durham, 1960

The Rookhope Borehole, 1961

University College London,
'the Godless institution in Gower Street'

Fieldwork area,
Semer Water, Yorkshire, 1963

Fieldwork area, Malham Cove, Yorkshire

Esso Bletchingley No 2, gas discovery well, Surrey, 1966

Morocco

Our house in Agadir, 1969

Glomar Grand Isle, drilling Esso's MO-2 well, offshore Morocco, 1969

The busy drill floor on Grand Isle, 1969

Zagora, starting point, for the desert crossing to Timbuktu, 1969

Algeria

*Susannah and her friend Nipa,
Boumerdes, 1973*

*Jonathan gets mobile,
Boumerdes, 1974*

Irina, 1975

*Hammam Meskhoutine,
the remarkable travertine hot springs, 1975*

Back in the UK

*Jonathan and Susannah,
Crowthorne 1976*

*Seaview House,
Cove Bay, Aberdeen, 1980*

*Atlantic II,
semi-submersible drilling rig, 1979*

*BNOC's Thistle Field,
production platform, 1981*

Scotland

Holmwood House, Uddington Glasgow

*BNOC headquarters,
St Vincent St, Glasgow, 1981*

Irina and Panther, 1981

Jonathan, master of Rubik's Cube, 1981

Libya

The garden of our villa in Brega, 1995

Sunfish sailing, Brega, 1992

Duck Rock on the road to Zelten, 1995

Drilling rig on SOC's Meghil Field, 1997

Libya

Three of the half-million 4-metre diameter pipes for the Great Man-Made River project

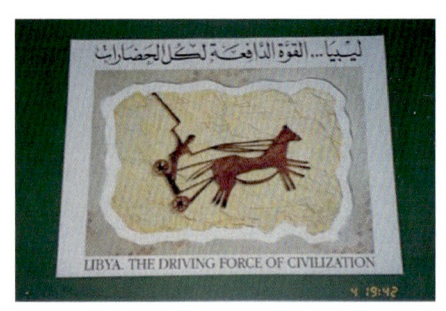

Gaddafi-era poster in Ghat

The natural arch of Fozzigiaren, Acacus mountains

The theatre at Sabratha

Irina and I,
Dubai, 2017

The terrible trio:
James, William and Eddie, Dubai, 2014

Sue and Alex, 2014

Jonathan, Jenny and guess who, Mystic Seaport, 2013

Geology

The Aletsch Glacier, Switzerland

Uluru, Australia

Hoodoos, Lake Iseo, Italy

Lake Bogoria, in the Great African Rift Valley

making of Occidental. After the revolution however, Occidental was vulnerable, and when the new Gaddafi government opened discussions on state participation they were able to threaten to shut down Occidental's production unless they agreed terms. Hammer sought support from the majors but they refused to help, and Oxy was forced to accept 51% state participation. The government then took on the majors one by one and forced them to do the same. Hammer's discussions were held with Major Jalloud, Prime Minister of the new regime, who was very courteous, but placed his belt on the table, complete with a holster and revolver. As Hammer says in his memoirs 'He smiled. I smiled', but rather than overnight in Tripoli he flew back to Paris in his private jet every night whilst the discussions were in progress. As a postscript, the search for water in Kufrah in south-east Libya was successful and discovered an enormous aquifer, which led to the construction of the *Great Man-Made River*, one of the most ambitious civil engineering projects ever undertaken in Africa. Occidental have continued their involvement in Libya to the present day.

Armand Hammer was a remarkable man. He was born in New York of Russian immigrant parents from Odessa, and his father was a doctor, drug manufacturer and a committee member of the Socialist Labour Party. Armand Hammer was named after the arm-and-hammer emblem of the party. The family ran a business called Allied Drug and Chemical which in 1919 began manufacturing a tonic containing ginger with a high alcohol content, which became very popular during Prohibition and made the company a million-dollar income in the first year. Hammer went to the Soviet Union in 1921 taking with him pharmaceutical supplies to combat a typhus epidemic, and arranged shipments of millions of bushels of American wheat, which at the time was almost valueless in the US, in exchange for furs and caviar. He became a confidant of Lenin who allowed him to set up Russia's first pencil factory (and monopoly) in Moscow in 1926, which made him a second fortune. With this he bought scores of works of art which he shipped back to the US by the train-load and sold at his gallery in New York. Hammer left Russia when Stalin became all-powerful in 1930, but during the Gorbachev era he paid another visit, and when asked what he would like to see, he said 'my old pencil factory'. He was taken there and found it virtually unchanged. His desk was exactly where it had been in 1930.

At the age of 58, rich and ready to retire, he bought the small, almost bankrupt California oil company Occidental, 'just for fun'. He turned the company around and, being an inveterate wheeler-dealer, he acquired several other companies, and by 1966 Occidental recorded annual sales of $700 million. That was when his Libyan adventure began. When he died in 1990 he had a personal fortune of $200 million, and he left his large collection of Impressionist and Post-Impressionist paintings to UCLA in California.

Letter 41
Libya as a country is a modern creation, not formally named until 1934. The name however is ancient and was used by the Greeks to refer to the entire coastal area of north Africa, in fact the name is mentioned in Homer's *Odyssey*. 'Libyans' was the Greek name for the Berbers, the indigenous people of this area. In Hellenic times, Greek settlers arrived in Cyrenaica (north-east Libya) and established the Pentapolis, the five cities of Cyrene, Apollonia, Ptolemais, Tauchira, and Berenice. A thousand kilometres to the west the Carthaginian Empire flourished, and three cities, Sabratha, Oea and Leptis Magna, were founded in a well-watered fertile area, known in more recent times as Tripolitania. Between the two lay 700 kilometres of inhospitable desert.

When the coastal highway from Tripoli to Benghazi was completed in 1937 a grandiose 30-metre commemorative Fascist arch was built over the highway, 70 kilometres west of Brega marking the boundary between Tripolitania and Cyrenaica, and the lavish opening ceremony was attended by Mussolini and the Governor General, Italo Balbo. (During the war Balbo suffered the unusual fate of being killed when his plane was shot down by Italian 'friendly' fire). Forty years later the arch was blown up on Gaddafi's orders as an unwelcome reminder of Italian colonialism. The site however was close to a much earlier boundary called the *Arae Philaenorum* after the legend of the Philaeni brothers. During the power struggle between Carthage and Cyrene, it was agreed that the boundary between the two states should be determined by two runners leaving their respective cities at the same time, with the boundary being drawn where they met. The Carthaginians, represented by the Philaeni brothers, covered much more ground, but were accused of cheating by the Cyreneans who insisted that either they allow the boundary to be established much further west or that the Carthaginian pair should be buried alive at the meeting point, in which case the Cyreneans would accept the boundary. The Philaeni brothers of course, being selfless, patriotic Carthaginians, naturally accepted the live burial option, and so the boundary was established. An elaborate altar was erected by the

Carthaginians to commemorate the event (recorded by the Roman historian Sallust, although all traces of it had disappeared by his time), and this formed the inspiration for the 1937 arch. The ancient boundary has remained unchanged for 2,500 years.

Evidence of the extent of Carthaginian and Cyrenean influence can be seen in some of the archaeological sites around the Gulf of Sirt, although the picture has been confused by later Roman and Byzantine settlements. The modern town of Sirt is located in a small fertile area surrounded by desert, and it is likely that this represents the most easterly outpost of Carthaginian settlement. A little further east, 60 kilometres from Brega, are the ruins of a small village called Automalax which marked the westernmost limit of Cyrenean territory. In Hellenic times, there was a fort on the headland at Brega and a port called Kozynthion, which was (and still is) one of the few natural harbours on the Gulf of Sirte. Following the destruction of Carthage, the entire north African littoral was incorporated into the Roman Empire, but Tripolitania and Cyrenaica were separately administered. To the east of Brega are the ruins of Boreum, a tiny settlement with a landing place, which grew up around a castle on a promontory. It was simply a way-station during Roman times, but in the sixth century it became the westernmost outpost of Byzantine Cyrenaica (suggesting that by this time Automalax and Kozynthion had been abandoned), and extensive fortifications were constructed to protect it – the only walled city between Berenice (Benghazi) and Leptis Magna. It became a Byzantine bishopric in the time of Justinian, and the surrounding area is dotted with the remains of small fortified farms from this period, reinforcing the view that this was a dangerous frontier zone guarding the western extremity of Cyrenaica. The situation is not much different today.

In 642 AD Arab invaders attacked the poorly defended coastal cities of the Byzantine Pentapolis which were quickly overcome. They rapidly spread westwards, demanding tribute from local tribes, and Tripoli was taken in 647. Arabic control tightened, and the Berbers slowly adopted Islam, but North Africa proved difficult to administer and in effect became more or less independent under a succession of shifting dynasties (some of which were Berber), whilst acknowledging the spiritual authority of Baghdad.

By the sixteenth century the power of the Arabic caliphate was in decline and most of the southern Mediterranean coastline fell the Ottoman Turks in the 1550s. They installed local governors called pashas who ruled on behalf of the Sultan in Constantinople, but by the early eighteenth century the Ottomans too were losing their grip, and a Turkish janissary called Ahmed Karamanli staged a coup and seized power in Tripolitania. He then persuaded the Sultan to recognise him and his successors as hereditary pashas. Karamanli power was based largely on the employment of corsairs to attack shipping in the Mediterranean, but countries wanting immunity from attack could pay a tribute instead. In 1801 the pasha demanded an immunity payment from the United States. President Jefferson refused, and the pasha cut down the flagpole outside the American consulate and made it clear that US ships would be open to attack by the corsairs. Jefferson send a task force into the Mediterranean and blockaded Tripoli. One of the ships was the frigate USS *Philadelphia* which while attempting to chase a corsair ship was lured onto a reef near Tripoli. The ship was captured as a prize and taken in triumph into Tripoli harbour. Accounts vary but the ship ended up being burnt to the waterline. One account says this was done by the Turks, but another claims that the Americans organised a daring night raid in which they boarded the *Philadelphia* and set fire to it. During my time in Libya I remember seeing a large painting of the ship on the gable of a building overlooking Tripoli harbour.

The Karamanlis did not long survive. Their corsairing days were brought to an end in the aftermath of the Napoleonic wars, and in 1835 the Ottomans reasserted control over Tripolitania and exiled the last Karamanli pasha. Ottoman control lasted until 1911.

The Ottoman Empire became known as 'the sick man of Europe' at the end of the nineteenth century. It suffered defeat at the hands of Russia in 1878 and lost many of its European provinces as a result. Spurred on by the acquisition of Tunisia by the French and Cyprus by the British, the Italians decided to make a bid for Libya (having already obtained an undertaking from France not to interfere). They declared war on Turkey in September 1911 and sent an expeditionary force to Libya to take and occupy the main cities. The Ottoman army in Libya had been greatly depleted by the transfer of large numbers of troops to put down a rebellion in Yemen, but despite superior equipment and much larger forces Italian progress was slow. What they assumed would be a walk-over turned out to be a costly and protracted war. The conflict was notable as the first use of aircraft for reconnaissance and bombing. Curiously Mussolini, later a great champion of Italian colonialism, opposed the war in Libya. On the

Turkish side Mustafa Kemal, then a major in the Ottoman army, held the city of Derna against a far stronger Italian force until the end of the war. But when the Balkan wars began, Turkey sued for peace, and Libya was ceded to the Italians.

Italian control was extended over much of Tripolitania and Cyrenaica, but during the First World War Italian troops were needed in Europe, so their presence in Libya shrank to the coastal areas. After the war, Italian dominance was slowly reasserted against strong local opposition, and complete control was not established until the 1930s. Resistance effectively ended when Omar Mukhtar was captured and executed in 1931, a campaign dramatically portrayed in the Libyan-sponsored film *Lion of the Desert*. Under Mussolini's dictatorship Italian settlers were brought in, reaching over 100,000 by 1939, the towns and cities were transformed with European architecture and a number of short railway lines were built, but not until 1934 was the country finally unified and named Libya. Even then the two provinces of Tripolitania and Cyrenaica were maintained, and the boundary between them, unchanged from the time of the Philaeni brothers, was still the *Arae Philaenorum*. The new colony was dubbed by the Fascists the Fourth Shore (the other three being the Adriatic Sea, the Tyrrhenian Sea and the Dalmatian-Albanian coastline). During the war Tunisia was added, completing the short-lived Fascist dream of a *Mare Nostrum*, a latter-day recreation of the Roman Empire.

There is a darker side to the story of the Italian occupation of Libya during the 1930s which still casts a shadow over the area. During this period the Italians built concentration camps to house dissident Libyans and Bedouin tribesmen. One of them was located at Brega, and another near the Italian fort of Al Aqaylah (now Bishr). 21,000 prisoners were held at Brega and 11,000 at Al Aqaylah. Many hundreds died from starvation, some were shipped to prison camps in Italy and some simply disappeared. Even as late as the 1990s the Libyan authorities were still demanding to know the fate of *'les disparues'*. The Esso complex at Brega was built on the site of the former camp, and some of the more impressionable residents claimed to have seen strange cloaked figures at night. To misquote a famous poem:

And still of a winter's night, they say, when the wind is in the trees,
When the moon is a ghostly galleon, tossed upon cloudy seas,
The spectral men come tramping, tramping, tramping,
The spectral men come tramping,
Through the darkened town.

Libya saw a great deal of fighting during the Second World War and once again the wide expanses of sand and sabkha of the Sirtic Gulf played a crucial role. In 1940 the Italian Tenth Army invaded Egypt (then under British control), but halted their advance at Sidi Barrani. The British launched a counter attack and forced the Italians back into Libya. The Tenth Army retreated to Benghazi and continued southwards, but was intercepted by the Allies at Beda Fomm north of Brega, and was annihilated. 130,000 prisoners were taken (including 22 generals) and the Allies advanced to Al Aqaylah. There was a wry joke at the time that the shortest book ever written was 'The Book of Italian War Heroes'. However, Hitler immediately sent reinforcements to prevent complete collapse of the Axis forces in North Africa, under the command of Lieutenant-General Rommel, (given the nickname the Desert Fox by the British press). Rommel was an able strategist, and his tanks and field guns were far superior to the Allies at that time. Churchill made the grave mistake of transferring three divisions from Libya to Greece, and the remaining forces dug in at Brega where they established a defensive line. This was attacked by Rommel, and the Germans broke through at the end of March 1941. Rommel's forces advanced to Tobruk in eastern Libya, but were unable to capture the city, and his rapid advance outran his supplies. He had no alternative but to retreat, and by December 1941 was back in Al Aqaylah.

But the respite was brief. History repeated itself. Rommel quickly re-grouped and re-equipped and in January 1942 he attacked again, catching the Allies by surprise. He advanced against weak opposition all the way to Tobruk. After outmanoeuvring the Allied forces at Gazala, Tobruk was besieged, and on 21 June 1942 the garrison surrendered, with 32,000 Allied soldiers taken prisoner. It was the worst British Commonwealth disaster since the fall of Singapore. Rommel was promoted to Field Marshal, and with the Allies in disarray he invaded Egypt, intent on reaching Alexandria and the Suez Canal. He requested more troops and equipment, but this was refused by Hitler who was unwilling to divert troops from the Eastern Front. Like Churchill's before him, this decision was a crucial strategic mistake. Rommel chased the Allied forces all the way to El Alamein, just 110 kilometres from Alexandria, where the panicking Brits were desperately burning documents (on a day dubbed Ash Wednesday). But the Eighth Army held, and was able to repulse Rommel's attacks. General Montgomery was appointed commander and large supplies of equipment began to arrive. In contrast, Rommel's supplies were dwindling. After a titanic struggle at El Alamein Montgomery broke through. Rommel had no means of resisting and was chased all the way back through Tobruq, Benghazi and Brega to Al Aqaylah. Beyond Al Aqaylah there was no place in which to establish a defensive line, and he was chased back to Tripoli and eventually into Tunisia. Axis control of Libya was at an end. Churchill said 'Before El Alamein we never had a victory, but after El Alamein we never had a defeat'. Fine words, but not entirely accurate in either case.

There was one area in which Britain did have some notable successes. The Long Range Desert Group was formed in Egypt in 1940 utilizing a small group of volunteers with experience of driving and navigating in desert conditions (as portrayed in the movie *The English Patient*), to operate behind enemy lines in Libya, gather intelligence on troop movements, organise covert patrols and conduct opportunistic raids. They assisted the French (coming from Chad) to wrest Kufrah in south-east Libya from the Italians early in 1941, and launched a successful raid on the town and airfield of Al Marj (Barce in those days) in 1942.

In November 1958, a BP seismic crew was operating in the Great Sand Sea, south-east of the Sarir oilfield when they came upon the wreckage of a US Liberator bomber. The plane was broken into two pieces but was quite well preserved. The radio and machine guns were still operational, but there was no sign of any parachutes or of the crew. On the forward fuselage was hand-painted the name *Lady Be Good*. The logbook showed that the plane had flown from its base at Soluq, south of Benghazi on a bombing mission to Naples on 4th April 1943. The US records showed the plane as missing, and it was assumed to have been lost over the Mediterranean. An investigation was carried out and a search was mounted for the bodies of the crew. Eventually the bodies of eight of the nine crew members were found, and their story was pieced together. This was the first operational mission for both the crew and the plane. Contact was soon lost with the rest of the flight due to strong winds and poor visibility, but *Lady Be Good* flew on to Naples. Visibility was so poor they jettisoned their bombs without finding their target, and headed back to Soluq. They apparently overflew the airfield without realising it and continued on the same heading for a further two hours. Either they could not see the ground or the flat featureless desert resembled the sea. They eventually parachuted from the plane which flew on for a further 25 kilometres before crashing in the desert. The parachute of one of the crew did not open properly and he was killed. The rest made it safely to the ground and began to walk northwards towards the coast, not realising that it was 640 kilometres distant. They had only one canteen of water between them. Notes which they left behind show that after five days the group had travelled 125 kilometres, but five of them could go no further. The remaining three continued for a further three days, and the last body was found an astonishing 175 kilometres from the parachute location. This plane remained where it lay until 1994 when the wreckage was taken to a Libyan Air Force base.

Millions of mines were laid in northern Libya during the Second World War, and when SOC began to acquire seismic data in some of its northern concessions it was necessary to send in mine clearance crews to remove them. It seemed oddly appropriate that SOC employed a German company to do this work. They would stack them up in great heaps and periodically blow them up.

Sadly this area became a battleground yet again in 2011 when Gaddafi sent his forces to destroy the 'rebels' in Benghazi *zenga zenga* (street by street). They shelled the city and sent in a squadron of tanks, but they were halted in the suburbs by the defenders, and following a UN resolution French fighter planes attacked and destroyed several tanks and armoured vehicles, forcing the Gaddafi forces to withdraw. They never came back. Enormous damage was done by the warring factions and the rival militias, and the installations at Brega were extensively damaged.

Letter 42
There was a very nice beach in Brega on a bay protected by a reef that had been used as a harbour since classical times, where we were able to swim, wind-surf, snorkel and scuba dive, and there was a sailing club which I joined shortly after arriving. I had never sailed before and it seemed like an ideal opportunity to learn. The club was equipped with Sunfish boats, small dinghies with a single lateen sail, easy to handle and not too difficult to set upright in the event of a capsize. They could be sailed single-handedly or with another person to handle the sheet (the rope which controls the sail). We had the use of an old Land Rover to tow the boats from the boatyard to the sea, but it could be very tricky launching

the boats through the surf before jumping aboard. Every Friday races were organised on varying courses marked by three or four marker buoys, arranged so that each race tested the skills of tacking, gybing (jibing in American) and running with the wind. There was a safety boat on standby and a control tower from which the starting signal was given. We would often sail outside the reef and around the remains of an old offshore loading gantry. Usually there were about a dozen boats in each race, and they were frequently very heated affairs, with numerous protests about overlap and shouts of 'starboard' and earnest consultations of the rule book after the race. Some of the sailors were very good, in particular Mario, a Maltese doctor who regularly won, or ended among the top two or three.

I also acquired a wind surf board which was a bit of a beast. It was actually a racing board (not best suited to a beginner) and it could be very difficult to handle against the combination of wind and waves, especially when gybing. Irina was much better at it than me, but even she sometimes had difficulty in returning to the beach, particularly in a light wind. There was also a Scuba Diving Club with a small membership since there were not many qualified divers. But there were some interesting things to explore. The bay was protected by a stony reef which was home to some large Moray eels, and there were lots of broken *amphoras*, some of which we were able to recover. Whether they were Roman or Byzantine, or more recent we were unable to discover, but it was very satisfying to bring them to the surface after so long under water. But the main attraction was the remains of a Hurricane fighter plane from World War II which had ditched into the sea in the gap between the reef and the shore. Its broken remains lay in about thirty feet of water but it had long since been stripped of its guns and most of its instruments. I managed to acquire some bits of the armoured windscreen and one or two other fragments. On one memorable occasion, we took the Scuba Club Zodiac to the beach with a company Land Rover. The launching ramp was occupied by another vehicle so we drove to the water's edge and launched it from the beach. We were out for about an hour, searching for amphora and exotic fish (and Irina got sea-sick in the rescue boat), but when we got back we found the Land Rover sunk into the sand. We struggled to extricate it but only made it worse, and it sank to the axle – and the tide was on the turn. We were afraid that Industrial Security would arrest us for misuse of company property, but we rounded up about twenty strong guys and after a lot of effort lifted it out. We managed to get it safely onto the ramp, but as I put it in gear to drive it away the gear lever broke off in my hand.

There were all kinds of recreational clubs in Brega – golf, darts, bridge, even a cricket club, but it was the Tennis Club in which we became most involved. The committee organised tournaments, a ladder, and even a coach who was brought over occasionally from Malta. But usually we just played social tennis, and after the game gravitated to someone's garden or bar to analyse the game, or just to put the world to rights.

On one of our visits to Crossroads that summer we were rather shocked to see a column of about fifty tanks and other military hardware being transported along the coastal highway. These were clearly destined for the conflict with Chad along the southern border of Libya. It involved the disputed ownership of the Aouzou Strip on the border between the two countries which was believed to contain uranium minerals, and Libya's claim was based on an agreement between France and Italy in 1935, which was never ratified. It became known as the Toyota War since Toyota Land Cruisers were extensively used by both sides. Initially the Libyans advanced deep into Chadian territory, but with the aid of the French, the Libyans were ultimately forced back and heavily defeated at Fada and at the Libyan Air Force base of Maarten as Sarra resulting in 7,500 Libyans killed and the loss of hundreds of tanks and other *matériel*. A cease fire was agreed in September 1987 and ultimately the International Court of Justice awarded the Aouzou strip to Chad.

In August of 1987 Susannah and Jonathan came to visit us in Brega. Jonathan soon made friends with the other kids visiting for the summer vacation, and before long was into late night poker sessions in the garden which he would usually end up winning. Sue was a little more circumspect, especially after she and I were hauled in by Industrial Security for taking photographs on the beach. There was an old, ineffective-looking anti-aircraft gun on the headland, but it apparently constituted a military installation and photography was forbidden. It illustrates just how paranoid they were, but having been bombed by the Americans and defeated by the Chadians, their nervousness was not surprising.

Irina returned with Sue and Jon via Amsterdam where she tried to interest them in the Rembrandts and Vermeers in the Rijksmuseum, but I think they would have preferred the Heineken Experience. Sue started university shortly afterwards at Thames Polytechnic (later the University of Greenwich), initially studying Sociology, which was fashionable at the time, but rather quickly switched to Estate Management which offered the prospect of a much better job. She was based first at the Woolwich campus and later at Dartford. It gave her the freedom to stretch her wings and meet new people, and it was there that she met Alex who she married a few years later. During two of her summer vacations she worked as an assistant at a students' summer camp in Fryeburg, Maine, just down the road from your place in Shelton.

Irina visited her family in Moscow and returned to London in October. She arrived to find a state of devastation. The evening before the Great Storm had raged across southern England. In Warwick Square and Eccleston Square, where Eva lived, hundred-year-old trees had been toppled and were lying against the houses round the square, there were crushed cars in the street, and debris lying everywhere. Winds of 120 mph were recorded, 18 people were killed and 15 million trees brought down. In Brighton the music room of George IV's exotic Pavilion had just completed a multi-million pound restoration following a fire in 1975. The storm caused a massive stone ball to crash through the roof, and the restoration had to start all over again. Even today in Richmond Park there is a plantation with fallen trees lying where they fell, all aligned in one direction, the direction from which the storm hit.

Now and again we drove to Benghazi for the weekend. There was a hotel called the Tibesti overlooking the inner harbour where you could have a suite for the weekend for a ridiculously low price, as there were very few visitors. The building looked like the Hanging Gardens of Babylon with cascades of vegetation covering terraces on the lower slopes. It had originally been the offices of Agoco, the first of the Libyan nationalised oil companies, but Gaddafi decided that it should be converted to a hotel so Agoco was kicked out. It was used by some of the Agoco expats as they shuttled between Europe and Libya, and I remember asking the porter if they ever had any tourists. He shook his head, but then said 'Yes, we did have one – a European cyclist heading towards Egypt'. They didn't serve alcohol of course, but they served Bin Ghashir mineral water and Fanta in very elegant carafes. Like all Libyan cities Latin script was regarded as unpatriotic, so all the road signs, street names, shop signs and such like were in Arabic only. It was a very good incentive to learn at least enough Arabic to be able to read the signs. There is an urban myth that the Algida double-heart logo was originally the other way up, but it had a superficial resemblance to the Arabic word for Allah, so was hastily turned upside down. The Libyan flag under the Gaddafi regime was unadorned green (a colour of special significance to Islam), so every door, shutter, fence and decoration was green. Such slavish conformity was depressing, like the Chinese all wearing Mao jackets, but Bengazi was an interesting city. It had a large port, with an excellent fish market which sold fish and lobsters straight from the boat, an interesting souk where we could buy gold and jewellery with our black-market dinars, and we found an excellent clean and modern butcher, who sold *filleto* for a quarter of the price in the UK, and a nice, shaded, open-air fruit market with grapes and pomegranates, oranges and *chammari* (the fruit of the strawberry tree) from the Jabal Akhdar (Green Mountains). Before the Gaddafi regime there were vineyards and a small wine industry in the area, and we found that the grapes sold in the market were very suitable for home-made wine making. The element that was missing in Benghazi was the complete absence of any kind of entertainment, or of any half-way decent restaurants. Benghazi had a well-known football club called Al-Ahly which won many trophies in the years before 1995 and thereby earned the envy (and enmity) of Gaddafi's son Saadi, who considered himself something of a football star. In 2000 the stadium was demolished by the government, allegedly for insulting Gaddafi by dressing a donkey in a football shirt bearing Saadi's number. To call someone a donkey is one of the worst insults an Arab can utter.

To me Cyrenaica is to Libya what Catalonia is to Spain. It is different, it stands apart, its origin was Greek rather than Carthaginian. It resents the pre-eminence of Tripoli and it is proud of its heritage. Cyrenaica is dominated by the Jabal Akhdar between Benghazi and Darnah, an area of limestone hills, gorges, caves, forests, and a fertile coastal plain. This is where Omar Mukhtar's guerrillas held out for many years during the struggle against the Italian occupation. One of the deep gorges is spanned by the spectacular Wadi el Kuf Bridge, 170 metres above the valley floor, just a metre or two less than the Pont Sidi M'Cid in Constantine. We visited many places in Cyrenaica including Al Marj (the Meadows) which suffered a catastrophic earthquake in 1963. The town was abandoned, and looks for all the world as if it has been bombed. A new town was built three kilometres away. The main city of the ancient Pentapolis was Cyrene and its prosperity was largely based on a plant called *silphium*, long extinct (but perhaps related to giant fennel), but so important that it appeared on many Cyrenean coins. Its resin was claimed to have remarkable medicinal and contraceptive properties and to be worth its weight in silver denarii. It was apparently not amenable to cultivation, and was harvested to extinction. The last stalk is said to have been presented to the emperor Nero as a curiosity.

The gospels of Matthew, Mark and Luke all mention a Simon of Cyrene who carried the cross at Christ's crucifixion, and the Coptic church claims that St Mark was born in Cyrene and became its first bishop. There was also a sizeable Jewish population in Cyrenaica which rebelled against Roman rule in 115 AD and massacred a large number of the Roman inhabitants until order was restored by one of Trajan's commanders. By 1912 5% of the population of Libya was Jewish and a commission was set up to examine the feasibility of establishing a Jewish homeland in Cyrenaica. The proposal was rejected, but by 1940 a quarter of the population of Tripoli was Jewish with 44 synagogues in the city. However the rise of fascism, World War II, the Holocaust and Gaddafi's anti-Semitic policies led to the removal of the entire Jewish population one way or another, and today there is not a single Jewish citizen remaining in Libya.

Cyrene is a spectacular place, perched on the edge of an escarpment overlooking the sea. It was founded by Greek colonists in about 630 BC. Legend has it that the Greeks were guided to this spot by local tribes who told them that 'in this place there is a hole in the heavens'. Not only was there abundant rainfall but there was also a natural spring which provided a copious and constant water supply. The city was named after a legendary Greek huntress called Kyrene. It grew into a large and important independent city-state with its golden age from about 450 BC to 320 BC, after which it came under the control of Ptolomaic Egypt and from 74 BC it became a province of Rome. The site of Cyrene is complicated since the city is bisected by a deep ravine, the Roman city was extensively built over by the Byzantines, and the modern village of Shahhat occupies the middle of the site. It was sacked during the Jewish revolt and devastated by an earthquake in 365 AD. Cyrene was built on several different levels with the agora and forum at the top, the acropolis hill to the west, the Sanctuary on a sheltered terrace to the north, and an almost unexcavated eastern hill with the Temple of Zeus and a hippodrome. From the northern gate a road winds down to the port of Apollonia, 10 kilometres away flanked along part of its length by elaborate mausolea and tombs. In fact the necropolis is larger than the city itself, containing over 1,200 crypts and rock-cut tombs and thousands of individual graves, all of which have been broken into and robbed. Cyrene has only been partially excavated, and in the 1930s the Italians laid a narrow-gauge railway track to facilitate the removal of fallen stones. One of our party said 'I didn't know the Romans had railways'!

Cyrene is very impressive. In the agora quarter the Roman forum was partially restored by Italian archaeologists in the 1930s who carefully re-erected more than 60 columns and connecting architraves to form a very striking enclosure. There are two theatres, law courts, a public records office from which thousands of seals have been recovered, the walls blackened by fire, probably from the Jewish revolt of 115 AD, and the palatial house of a prominent citizen called Jason Magnus, full of mosaics (some now removed) including one of a labyrinth with the inscription *'Epagatho'* (good luck). A colonnaded street 130 metres long connects the forum to the *agora* (market) and contains the memorial tomb of Battus the founder of the city, and at the end of the street there are the remains of a gymnasium. The acropolis was a military citadel surrounded by walls and watchtowers, but now much obscured by Italian defensive works built during the colonial period.

Down the hill, on a narrow terrace looking out towards the sea, is the Sanctuary, the most hallowed part of the city, entered through a restored *Propyleum*, with the original spring - the Fountain of Apollo, surrounded by drinking troughs for watering animals. Nearby is the great Temple of Apollo with some re-erected columns, but the colossal statue of Apollo which once dominated the building is now in the British Museum in London. Nearby the *Strategheion* is a sort of victory temple where captured war trophies were displayed, which has been rebuilt and re-roofed to show what it would have been like 2,500 years ago. In Roman times a large bathing complex, the Baths of Trajan, were built on the terrace, and one of the pools has been restored and filled with water. At the western-most end of the terrace the Greeks built a theatre awkwardly shoehorned into a narrow space between a cliff face behind and a cliff edge in front. The Romans converted it to an amphitheatre where they held gladiatorial combats and spectacles with wild animals. Just to increase the excitement the animals could be admitted from any of ten entrances into an arena adorned with bushes and tree branches to resemble a jungle. On the eastern hill, yet to be seriously excavated, are the partly restored remains of the Temple of Zeus, larger than the Parthenon in Athens. It contained a gigantic statue of Zeus, twelve times life size. The temple was destroyed during the Jewish revolt, but partially rebuilt afterwards. The head of Zeus was smashed into over a hundred pieces, but has been painstakingly reassembled in recent times and placed in the nearby museum. Next to the temple are the remains of a hippodrome where chariot racing took place in Roman times. The Greeks and Romans took their sports seriously. In Cyrene they had a gymnasium, a blood-sports arena and a chariot racing circuit.

Returning home one evening with friends from one of our weekend jaunts in Benghazi the water pump on my car failed, and we ground to a halt on the desert highway miles from anywhere. We managed to flag down a truck and the driver agreed to tow us to Ajdabiya, the next town along the road. We eventually talked him into towing us all the way to Brega, probably a hundred kilometres from where we had broken down, and I offered him a sizeable tip for his trouble. He refused to accept it, and it was only when we insisted that he take it for his children that he reluctantly agreed. The Libyans are proud people.

Another of our expat friends used to visit his cousin in Benghazi in his M series BMW, which he liked to drive fast. The road undulates over ridges and troughs through scrubby desert, with the occasional camel visible in the distance. One evening he was returning to Brega driving at perhaps 130 kph when he crested a ridge to find a flock of sheep crossing the road in the dip. He had to make an instant decision whether to plough through them or drive off the road. He chose the former and smashed through the sheep, with heads and legs flying in all directions. In all he killed 22 sheep. There was a shepherd with the flock, the police were called, my friend was arrested and taken to the local lock-up, where they kept him until someone produced $1000, which was given to the shepherd in compensation, upon which he was released. The car stood up to the carnage remarkably well, but cost him another $1000 to repair. We said at least he should have brought the carcasses back with him for a barbecue. He became something of a local celebrity and was often greeted by the single word *'baa'*.

Letter 43
We took advantage of the ability to buy airline tickets with 'cheap' dinars (a loophole which was closed shortly afterwards) and had some exotic overseas vacations. Our first trip was to Antigua where we stayed at the St James's Club, and it was there that we learned to scuba dive. It was a marvellous experience to swim in the warm waters of the Caribbean and look up to see keel of the diving boat thirty or forty feet above. We dived on the wreck of a ship that had caught fire whilst carrying a cargo of asphalt from Trinidad. The asphalt had melted and run out through the seams of the wooden hull. There were lots of brightly coloured tropical fish and I was bitten by one, which I thought was distinctly unfriendly.

Back in the office we continued the policy of appraising the discoveries which Esso had considered uneconomic and one of our principal targets was the Attahaddi (meaning unique or miracle) gas field discovered by Esso in 1964. The gas was contained in a quartzite reservoir which was only productive where the quartzite was fractured. Esso had drilled six wells and following their departure SOC continued the appraisal. Some wells flowed at very high rates, where others were tight, but we finally figured out that the fractures were best developed close to faults which traversed the field so we concentrated on those areas. The reservoir was deep and well costs were high, but it was necessary to evaluate each fault compartment before committing to a development plan. During my time we drilled fourteen appraisal wells, and the field was not finally brought on stream until 2004, forty years after its discovery. However it lived up to its name and proved to be one of the largest gas fields in Libya, capable of producing 300 million cubic feet of gas per day. I took a trip to Attahaddi with one of the wellsite geologists which reminded me of the drilling operations in Algeria – the same desert surroundings, the same type of rig, the same logging trucks, the same accommodation trailers. Another guy had made quite a collection of finely worked flint arrowheads which he had found in the neighbourhood. This geologist, a Dutch-Canadian called Gus van Heusden was a remarkable man. He always received far more mail than anyone else, often from places like Kenya and Tanzania. He was a bachelor and I discovered that he spent most of his vacations in these areas arranging for water wells to be drilled and school rooms to be built, and for surgical operations to be carried out on deformed children, all at his own expense. That was real philanthropy and it impressed me no end. The letters were from his grateful beneficiaries.

Irina got a part-time job in Aviation Department which allowed us to broaden our circle of friends. Two of the pilots were Argentinian, and both had served in the Argentinian Air Force during the Falklands conflict. One of them was a good guitar player, and we often met them socially. We discussed the Falklands War and the effectiveness of the Exocet missile, which they said could have won the war for Argentina, but they only had eight of them.

Virtually every expat house had a bar which became the centre of social life. We played darts and we had barbecues. There were black-tie casino evenings, discos and drag parties, but the highlight of the year was the Hallowe'en party where prizes were given for the most inventive costumes. One year I went as the sinister messenger from *Amadeus* in a cloak, Janus mask and tricorn hat, while someone else went as the Phantom of the Opera, and another as Miss Piggy. People would go to enormous trouble to create impressive costumes. One day we had a wine tasting evening where people brought a bottle of their best home produced wine, to be blind tasted by the rest. One of the samples was rated by a woman as 'disgusting rot-gut'; it turned out to be her own brew.

We became friendly with several of the Turkish cooks from the Mess Hall, Erol, Muharram, Hasan and Turharn, who would drop by for a drink or a game of darts, but principally to escape from the tedium of their own accommodation, which was little more than a modified shipping container. They would often arrive with a *çanta*, a bag containing a chicken or two or *filleto* or eggs, as their contribution to the evening, and Hasan proved to be an excellent mechanic when my car needed a new clutch (he had previously worked in a BMW repair shop). Naturally we also got to know many Libyans, particularly through the Tennis Club, and while some were reluctant to mix socially with expats, many were eager to do so. They would come to our barbecues and parties and they would drink alcohol with no qualms at all. Alcohol, after all, is an Arabic word.

There were many Gas Plant and Maintenance guys on bachelor status, mostly British, who worked a 28:28 schedule, that is 28 days without a break, followed by 28 days' leave. They lived in communal houses, but with individual bedrooms and bathrooms, and a separate recreation room – which was invariably a bar. These were tough, hard drinking, earthy types, with great wit and humour but with language that would make a roughneck blush. They were great raconteurs, and got involved in all kinds of extraordinary episodes. I used to visit 1010, which over the doorway had the inscription *'Action not Words'*. One time I was travelling to the UK when I met some of them at a hotel in Djerba. They asked me if I could speak French, and when I said yes, they asked me to ask the manager for the loan of a spade. I said 'Why do you need a spade?', and they explained that a month before they had arrived in Djerba *en route* to Libya with a bottle of whisky. They knew they couldn't take it into Libya so they had buried it in a hole in the mud-brick wall of the hotel garden. Unfortunately, in the meantime the wall had collapsed, so they got the spade, and with lots of curious onlookers, started digging for the bottle, but someone had beaten them to it, and the bottle was gone. They would start imbibing as soon as the plane left Amsterdam or Frankfurt, and drink as though there was no tomorrow, whereas in reality their houses in Brega were stuffed with booze. They would occasionally be raided by Industrial Security who confiscated the alcohol (and drank it themselves as likely as not). I went over there not long after a raid and the guys were sitting around drinking as usual. I said 'I thought you had been raided'. They said 'Yes, but they didn't find our stock in the roof'.

But life was not all beer and skittles. There were some major frustrations. They were paranoid about industrial, political or security information being leaked to the outside world, so although the company was equipped with the latest German telecommunications equipment, if you wanted to make an overseas call you had to take a written request to the telephone exchange (this was long before the days of mobile phones and Skype and WhatsApp). Sometimes they would get it for you, sometimes not. Incoming calls were all routed through the exchange, and my mother and Sue and Jonathan would often spend hours trying to get a connection, and I suspect that even then the calls were routinely monitored. The only way to circumvent the system was to go out to Crossroads and call from there.

The other major headache was exit visas. When you arranged an overseas trip you had to submit your passport through the SOC passport office to Tripoli for the issue of an exit/re-enty visa. Frequently it would not appear until the day before departure, and on one occasion a guy went to the Passport Office to enquire about his missing passport only to find it wedged under a table leg where it had been placed to prevent the table from wobbling.

Tripoli airport could be difficult too. I recall arriving at the very door of the plane only to be told that the security guard could not read the exit stamp, and I had to go all the way back downstairs to get it re-stamped then dash back before they closed the plane door. All the departure boards and announcements were in Arabic only, so you had to listen very carefully. One of our American engineers arrived there one day heading for Zurich with a group of Brits who were flying to Frankfurt. He was there in good time for his flight and waited with the others in the departure lounge. The Brits left and after a while he went to check his flight, just in time to see it taxiing down the runway for take-off.

Brega had a very pleasant climate. It is located on the coast so never became insufferably hot, but it was dry enough that there were no mosquitos or flies. But there were cockroaches. They did not often appear in the houses, but the drains were infested with them, and the Sanitation

Department would periodically send round a truck which sprayed pesticide to kill them. It was not a pretty sight when they raised the manhole covers. They were big, tough creatures – simply whacking one with a fly-swatter was not enough. They would just shake their heads and carry on. It was said you were only truly acclimatised when you could crush a cockroach with your bare feet. One of the Philippino cleaners went to the clinic one day complaining of a blocked ear, and the doctor pulled out a cockroach – but just a small one.

Occasionally, particularly in spring, we would experience the *ghibli*, (aka *sirocco, khamsin*) a hot, dry wind from the Sahara drawn northwards by low pressure belts in the Mediterranean. They were usually of short duration, but the sky would become as dark as night, and everything - roads, buildings, cars would be left with a covering of tacky brown mud. The dust often extended over large areas of Europe, where it fell as 'blood rain', and there are dramatic images on the web showing great plumes of sand and dust blowing from the Sahara across the Mediterranean. The *ghibli* made such an impression on the Italians that they named a World War II aircraft for it, and the name has been used for several different models of Maserati sports car.

In the autumn of 1988 Irina suffered a slipped disc and was in such pain that she had to sleep flat on the hard floor. She went to London for treatment and rented a furnished apartment in Redcliffe Square while recuperating (our flat in Warwick Square was rented at the time). Fortunately, the treatment was successful and I was able to go and collect her, and we came back to Libya together. We took the opportunity to visit Leptis Magna, the largest and most impressive city of the 'tripolis', and one of the best preserved Roman cities in North Africa.

Leptis was originally a Punic settlement but was seized by Rome after the destruction of Carthage in 146 BC. The town flourished, particularly from the production and sale of olive oil, and grew in importance, reaching its zenith in the second century AD when the population exceeded 100,000. After Alexandria, it was the most important city in Africa. The Roman emperor Septimius Severus was born there, and embellished it with fine public buildings and monuments, and for a brief period the Roman Empire was governed from Leptis. It is said that Septimius spoke Latin with a Libyan accent, and that with his family he spoke Punic. The coastal towns were always at risk of attack from the Berber tribes of the interior, particularly the Garamantes who inhabited the valleys and oases of the

Fezzan. The Romans established a frontier zone with a defensive line of forts called the *limes Tripolitanus*, and Septimius Severus conducted a five-year campaign against them which extended the *limes* far to the south. The fortunes of the city declined after his death (he died in York during a campaign to establish a similar secure zone along the northern frontier of Britain). Trade declined, the population decreased, Berber raids became more frequent and in 365 AD the central Mediterranean was hit by an earthquake (which originated in Crete) and an enormous tsunami which killed thousands, and carried boats three kilometres inland. It swept over much of the Nile Delta, destroyed the port of Apollonia and devastated Leptis. The city never recovered. Seventy years later the depopulated city fell to the Vandals who destroyed the city walls leaving it open to further Berber attacks. Leptis enjoyed a brief revival under the Byzantines but with the arrival of the Moslems in 647 AD the city was abandoned, and lost under a thick covering of sand.

Today Leptis is a World Heritage site with some wonderful Roman remains. There is a magnificent, opulent bathing complex from the time of Hadrian covering almost three acres, embellished with marble floors, mosaic pavements, black granite columns, cold, tepid and hot rooms, and communal marble latrines complete with running water. The four-sided Triumphal Arch built for Septimius Severus (in a hurry apparently, and not quite complete), is decorated with all the panoply of imperial Rome, winged Victories, weapons, palm branches, panels showing Septimius and his sons, and captive barbarians. But the most impressive ensemble is the Severan Forum (modelled on the Forum in Rome) and the Basilica, originally a roofed hall with marble floors, galleries, colonnades and the most exquisitely carved pilasters showing animals, the Labours of Hercules and Dionysus enveloped in twisting, swirling vine tendrils. The Romans knew how to live in luxury.

Across the main street was the market, another elegant construction of arcaded shops around the rectangular enclosure and two *tholei* in the middle (circular stone kiosks with several openings, each with a marble slab for the display of produce). There is an interesting set of standard measures set into a wall where shoppers could check that they were not being short-changed. Nearby is the open-air theatre with a semi-circular tiered auditorium with the Mediterranean Sea as a dramatic backdrop, looking as though waiting for the next performance. The harbour, from which so much merchandise came and went over the centuries, is now completely silted up and filled with mud and reeds. The two breakwaters built so that ships could berth near to the warehouses are now largely submerged, but on the end of one of them is the base of a substantial lighthouse.

Just inside the city wall to the west is an unprepossessing cluster of four or five linked buildings. They are low, vaulted and domed concrete structures looking almost like Second World War pillboxes, but are in fact another Roman bathing complex called the Hunting Baths, since they contain well preserved murals of hunting scenes, showing hunters with spears killing what appear to be leopards and lions. It is not hard to imagine a group of tired, sweaty hunters stopping off here to clean up before heading home – like a communal shower after a rugby game. Several houses in Leptis have a bas-relief panel near the door showing the curious device of an open eye (in one case with a scorpion attached to it) being sprayed by a phallus. This, according to Pliny, is a *medicus invidiae*, or an apotropaic against the evil eye, a curse transmitted by a malevolent glance. Well, if that's what it takes…

The other two cities of the 'tripolis', Oea and Sabratha, were much less important than Leptis. Oea has been almost totally obliterated by the modern city of Tripoli, and virtually nothing remains apart from the Arch of Marcus Aurelius, some fragmentary traces of a marine esplanade, a harbour wall and the foundations of a lighthouse. Sabratha was much smaller and more provincial that Leptis with a population of maybe 25,000. It had a rectangular forum surrounded by a shady arcade dominated by a huge temple reminiscent of the Pantheon in Rome. The city had numerous temples, civic buildings, public baths, an amphitheatre and a large and thriving port. As a result of coastal erosion the harbour is now totally submerged, and many of the port buildings have been destroyed. In the Sabratha Museum there were outstanding mosaics from the basilica and from the baths. The mosaic portrait of Neptune, with its cherry-red cheeks and lips, dark green foliage for the beard and hair and the startled look in the eyes, is quite hypnotic. But the *piece de resistance* at Sabratha is the theatre. Painstakingly (and perhaps excessively) rebuilt by Italian archaeologists in the 1930s, the theatre has been used to stage plays and opera, including a performance of *Oedipus Rex* given for Mussolini in 1937, and more recently during the Sabratha Festival of 1965. The rebuilt three-storey *scaenae frons* is 25 metres high with over 100 columns. The Michelin Guide would definitely give it a three-star rating – worth a journey.

Letter 44
Gaddafi was the principal supplier of arms to the IRA in Ireland and numerous shipments were sent in the period 1973 to the late 80s, which included Soviet-made heavy machine guns, RPG launchers, flame throwers, AK 47s, surface to air missiles and the Czech-made explosive Semtex. Several shipments were intercepted (the *Claudia* in 1973, the *Casamara* in 1985, the *Villa* in 1986 and the *Eksund* in 1987), but many more got through. By 1987 the IRA was better equipped than at any time in its history, due mainly to the shipments received from Libya. In response, a US Presidential decree in June 1986 ordered the remaining US companies to close down their operations in Libya, and made it unlawful for US citizens to work in Libya. The National Oil Corporation placed the US oil concessions in Libya into a form of escrow for a period of three years (subsequently extended several times) and appointed Libyan proxy companies to take over operatorship of the US interests.

And then, just before Christmas 1988, Pan Am flight 103, flying from London to New York was brought down, killing all 259 people on board and 11 people on the ground in the town of Lockerbie in Scotland. Ironically the plane was late departing. If it had left on time it would have come down in the North Atlantic. The following month the Americans shot down two Libyan MIGs over the Gulf of Sirt. A year later a French plane exploded whilst flying over Niger. Both of these incidents were found to be the result of terrorist bombs, and following intensive investigations the finger of suspicion fell squarely on Libya. Indictments were submitted in French, British and US courts against Libyan personnel for both incidents, and in 1992 the United Nations imposed sanctions against Libya until such time as the suspects were handed over for trial. The sanctions froze Libyan assets overseas, banned arms sales to Libya, and prohibited all airline traffic to and from Libya. On the first day of sanctions a Libyan passenger plane attempted to fly a commercial service to Rome, but was intercepted by Italian fighter jets and forced to turn back. The sanctions were further tightened by prohibiting the export to Libya of oilfield equipment, and it has been estimated that the sanctions as a whole cost the Libyan economy something in the region of $30 billion. Henceforth we were not able to fly in and out of Libya. Usually we took the company plane to Tripoli from where an SOC bus took us to Djerba in Tunisia (often with long delays at the border). From there we flew to Tunis, and onward to Frankfurt or Zurich or Amsterdam. For a while the Libyans introduced a catamaran service from Tripoli to Malta, an uncomfortable trip of five or six hours. The vessel was hired from a

Russian company and was rather small, and not ideal for operating on the open sea. The net result was that where it used to take one day to travel from Brega to London, it now took two.

Several times we had a stopover in Tunis and were able to visit some of the more interesting places. The historic medina in the centre of the city is a maze of narrow streets and alleys with souks selling silks and satins, perfumes and tourist tat. Not so long ago there was a slave market here where Christians, including women and children captured by the Barbary corsairs, were put up for sale. The Bardo Museum is the largest museum in Africa after the Museum of Egyptian Antiquities in Cairo. Housed in a former palace, it has an incomparable display of Roman mosaics from Carthage, Hadrumetum, Ruspina, Utica and other Roman cities. There is a scenic train ride which ambles out on a causeway across the Lake of Tunis to La Goulette (the port of Tunis), Carthage and Sidi Bou Said, a fashionable place popular with artists and tourists, and noted for its striking blue and white painted buildings and attractive sea views.

Carthage is a rather curious place, much of it covered by the urban spread of Tunis, with the President's Palace covering a key part of the ruins. It was of course the centre of the Carthaginian Empire, the legendary home of Dido and Aeneas, which exceeded Rome in splendour at its peak. Near Salammbo station is the Tophet, a macabre place where children were sacrificed to Tanit and Baal for over 600 years, so long in fact that their funerary urns filled the allotted space and then had to be laid one on top of another. Carthaginian power depended on maritime trade and they established control over much of the western Mediterranean. Two adjacent ports were built, a rectangular basin for commercial shipping and a splendid colonnaded circular *cothon* for naval ships. The *cothon* had berths for 220 warships around the perimeter with slipways where they could be winched up for maintenance, and store rooms above for oars and sails and cordage. There was a central island with a watchtower from which the admiral could observe the entire harbour, and look out to sea. Both harbours are still in use, 2,200 years later. The clash between Carthage and Rome reached its climax with Hannibal's epic march over the Alps and victories at Trasimeno and Cannae, but the tide of war turned and twelve years later Carthage itself was invaded, and Hannibal defeated at the battle of Zama. Carthaginian power was broken, and there were repeated calls in Rome for its destruction. The Roman senator Cato is said to have ended every speech with the mantra *'Cartago delenda est'*, and his call was answered in 146 BC when Scipio razed the city to the ground, and sent the entire population into slavery. Subsequently the Romans returned and re-established Carthage as a Roman city complete with all the usual attributes of forum, amphitheatre, basilica, circus for chariot racing and an enormous bathing complex built in about 150 AD. Relics of all these structures can still be seen amidst the sprawl of modern Tunis.

Twice a year we had a formal two-day meeting with a delegation from the Libyan National Oil Corporation, which came to Brega on their circuit of visits to all four of the state companies. The meetings reviewed our progress during the previous six months and outlined our proposals for the future. They were taken very seriously by Taleb (the Exploration Manager) as we had to demonstrate to NOC the competence of our exploration staff and convince them to allocate funds to SOC for our future seismic and drilling programmes. In effect it became a competition in which all the state companies competed against each other to obtain funding from NOC. We might request funding for five wells but if NOC preferred the plans of Agoco or Waha we might end up with funding for only two. Enormous effort was put into these presentations. Staff would come over from our UK office two weeks before the meeting, and there would be interminable dry-runs and practice sessions. It was important to show off the capabilities of our Libyan staff, so they would all give short presentations on their work – some quite competent, but others cringingly bad however many dry-runs they had done. There was always a dinner for the participants, which would have been an excellent opportunity to chat with the NOC delegates or for them to make speeches, but it never happened. I suspect Taleb felt that the opportunity to chat socially with the NOC guys would risk someone making a thoughtless or embarrassing remark – a risk he was not prepared to take. The meetings were always followed by a post-mortem session between Taleb and the senior staff, which often became quite acrimonious. An office can be a stressful place.

In our third summer we took an extended vacation and visited Hong Kong, Australia and the Maldives. Hong Kong was still a British colony at the time but approaching the end of the 99-year lease granted in 1897. This was a very different version of the orient than Singapore. Hong Kong was all bustle, bright lights and the aroma of countless restaurants. In fact Hong Kong was reckoned to have more restaurants per square mile than anywhere else on earth. We took a trip from Hong Kong Island to Kowloon and the New Territories and then to Guangzhou (formerly known as Canton) from where Sun Yat Sen organised the overthrow of the Chinese emperor in 1912. The narrow strip of land between Hong

Kong and the Chinese border seemed to be occupied almost exclusively by ducks, but once over the border you have the overwhelming feeling of arriving in a country with an enormous population (at the last count around 1.37 billion) most of whom seemed to have bicycles, and of the obvious improvements which had taken place since the opening up of the Chinese economy in 1978. We spent a day in Guangzhou and returned to Hong Kong by train pulled by a big steam locomotive with a prominent red star and on arrival, inspired by the ducks of no-man's-land, we decided to order an authentic Peking Duck. This came as a succession of courses where they seem to utilize every part of the duck. First comes the crispy skin served with hoisin sauce and pancakes, then finely chopped and seasoned duck in a lettuce leaf, then a stir-fried duck and spring onions, then duck with noodles, and finally a soup made from the giblets and bones, served with Chinese cabbage. I wouldn't be surprised if they grind up the bones and use them as food supplement for the next generation of ducks, reminiscent of the unofficial Yorkshire national anthem *Ilkla Moor bah't 'at*.

Next day we took a jet-foil across the Pearl River to Macau, the Las Vegas of Asia, a city under Portuguese control for 450 years, and the last European colony in Asia. It is reputed to be the most densely populated place on earth. Of course, we had a flutter in the casinos and went to see the ruined Church of St Paul, a Jesuit church, once one of the great Christian showpieces in Asia, but destroyed by fire in the nineteenth century. The façade is known as a *biblia pauperum*, a sermon in stone for the benefit of the illiterate, epitomising the Nativity and Passion of Christ, the Assumption of the Virgin, the saints and the Triumph of Christianity. The Jesuits were suppressed by the Portuguese in 1762 and the church was turned into barracks. Today only three percent of the population of Macau is Christian.

From there we flew to Sydney and were very quickly seduced by the climate, the prosperity and the egalitarian nature of Australia. We stayed in a hotel overlooking Sydney Harbour, and in the evening went for a drink in the bar with Irina dressed in a smart new blouse and knee-length designer shorts, only to be refused entry. Shorts, we were told, were not allowed – and this was Australia! We went to see a performance at the famous Sydney Opera House, one of the twentieth century's most iconic buildings. The exterior is formed of six interleaved shells covered in white tiles which create a totally unique profile. The interior, by contrast, was rather drab with large areas of stark bare concrete, which gives the impression that they ran out of money before the interior could be completed.

From Sydney we flew to Melbourne, supposedly the most 'English' of Australian cities. In the Fitzroy Gardens, we found Captain Cook's cottage, transplanted from England brick by brick in 1933 and re-erected in Melbourne. Cook was an English naval captain who made three voyages of discovery charting new territories, particularly in the Pacific, and was the first European to set eyes on the east coast of Australia (although curiously he missed the splendid Sydney Harbour). Just before we left Brega one of our friends had shown me an old book with a photograph of Cook's cottage at Great Ayton in Yorkshire where the cottage had originally stood. It was a curious coincidence to come face to face with the same cottage on the other side of the world.

Whilst in Melbourne we drove out to Ballarat the focus of the first Australian gold rush in 1851, and the place where the world's second largest gold nugget (weighing 69 kilos) was found in 1858. An open-air museum has been created at Sovereign Hill in Ballarat to show life in a mining town of the 1860s, but it gives a highly-romanticised impression of a jolly miner panning for gold, carousing in the saloon and heading home to his little shack for supper, and conveys nothing of squalor, hardship, danger and insecurity of life at that time.

I was interested to check out employment possibilities in Australia and compare them with Libya so I visited the School of Science at Royal Melbourne Institute of Technology and the James Cook University in Townsville in Queensland. At RMIT it seemed to me that geology had been subsumed within the much broader discipline of environmental sciences, and had lost its status, but JCU was more appealing. They were looking to appoint a professor of Economic Geology and they were well ahead of their time in developing techniques for the extraction of coal bed methane from the Bowen and Surat coal basins in Queensland. I was tempted, but it would have meant a radical change of direction in my career, and I decided that I was not yet ready for such a change.

We wanted to swim in the ocean at Townsville but the beaches were closed because of the presence of Portuguese men o' war, jellyfish-like creatures which can deliver a nasty, occasionally fatal, sting from long tentacles trailing beneath the float bladder. So we tried to join a boat trip to the Great Barrier Reef, but the sea was rough and the boats were not operating. We did not want to miss the opportunity of seeing the reef so we took a flight on a small Buccaneer sea-plane out to Lodestone Reef (and over a newly installed floating hotel, with a tennis court of all things). It was impressive to see the reef from the air extending as far as the eye could see, but it would have been even better to swim or dive among the corals and reef fish on the reef itself.

From Australia we flew to Male in the Maldives. The Maldives are a collection of over a thousand coral islands in the Indian Ocean extending for a distance of 1,000 kilometres north to south with an average elevation of only 1.5 metres above sea level. Sea level is currently rising at about 3mm per year, so the Maldives could disappear within 500 years (or even quicker if Donald Trump has his way), except that in 1842 Charles Darwin suggested that reef growth keeps pace with sea level changes. We stayed on the island of Baros within the lagoon of the Kaafu Atoll. Baros is a tiny island, only 250 metres in diameter, but the Kaafu Atoll as a whole is about 30 kilometres across. The island was uninhabited until it was developed as a tourist resort and by the time of our visit had accommodation for about 20 visitors. It had its own mini lagoon and 'house reef' surrounding the island. We were able to dive in both the mini lagoon and in the open water beyond the house reef. We saw sharks and lobsters and colourful reef fish and explored caves and grottos, but the undoubted highlight was a troop of enormous eagle rays with long tails gliding silently by like ghosts in the night.

To me the Kaafu Atoll was a very close analogue of the Nasser (Zelten) structure which had proved to be one of the largest oilfields in Libya. It was a similar size, it was surrounded by a ring of islands with a lagoon in the middle, and beyond the outer ring of islands deep-water conditions were present. The Kaafu Atoll had developed on a pronounced submarine ridge, and the same was true at Nasser. The only difference was that Nasser had a cap of porous coral and algal limestones, which were similar to those on the Baros house reef, but thicker and much more widespread. Nasser produced most of its oil from the porous cap, but we had proved by our recent drilling that the outer ring of islands was also oil bearing at Nasser, and that the centre of the atoll was occupied by a lagoon. The comparison was almost perfect.

Shortly after our return to Libya Bob McCrossan the Geology Superintendent resigned. His offshore atlas was complete, he had received many accolades for the quality of the publication and he really did not have much interest in the day-to-day operations of the department. I was offered his job, which unlike Britoil was a management position with plenty of technical involvement, so I accepted. I got a new car and a secretary and it brought me into much closer contact with Taleb, the Geophysics Superintendent and other department heads in Petroleum and Reservoir Engineering. It also gave me responsibility for about twenty geologists, mostly expats but with five or six Libyans who were mostly young trainees. The geologists were assigned to specific areas, or to Regional Studies, and they worked alongside their geophysical counterparts. Their job was to generate prospects for drilling, which involved delineating the structure, assessing which reservoir units were likely to be present, determining whether there was an oil or gas 'source kitchen' nearby from which hydrocarbons could have been generated, how and when the structure might have been charged with oil or gas, and calculating the expected volume of reserves. It was my job to supervise this work and ensure that the prospect recommendations were technically sound, since each well cost anything between $2.5 million and $6 million to drill. The Regional Studies Group had the job of assessing the potential of areas in which SOC was not currently involved.

In most countries well and seismic data are traded on a like-for-like basis so a company can build up a large collection of data for use in its regional assessments. Esso had done this extensively, but SOC either could not or would not do the same, so we suffered from a lack of data on most of the newer wells drilled by other companies. Similarly it is common practice in most countries for companies to farm-in and farm-out of blocks, so that the risk can be spread between several companies, and companies can concentrate their attention on areas which they regard as most favourable. This did not happen with SOC. At no time did SOC join forces with any other company and conduct joint venture operations. These were serious limitations which were largely dictated by political considerations, principally the conviction that Libyan national companies should not be subject to any form of control from foreign interests, but it put great strain on the company as it had to bear the full cost of all seismic, drilling, appraisal, development and production activities from its own limited budget. Coupled with the sanctions imposed by the US and Britain and an oil price of less than $20 a barrel (except during the brief Gulf war), it created a rather constrained working environment.

Letter 45

I began my new job as Geology Superintendent in June 1989. My relations with Taleb were initially cordial. I had actively promoted the concept of the atoll model for the Nasser field which resulted in two small discoveries, and then we drilled some satellite structures around the Nasser field which were also successful, but Taleb was very conservative and averse to taking risks. He had originally been employed by Esso and his attitude was that since they had been so successful their methods must be the best. But circumstances had greatly changed. Esso had drilled up to twenty wildcat wells a year (compared with SOC's three or four) and had been willing to pour money into evaluating their concessions, take risks, and accept a 1 in 4 or 5 success rate. SOC would have none of this, they expected every well to be a discovery. Interestingly though, after the Gaddafi take-over, Esso's exploration activities declined rapidly and, apart from three offshore wells, they drilled not a single exploration well in the last eight years of their Libyan venture. But that was for purely political reasons. I tried to interest Taleb in more effective ways of presenting information to NOC, and of introducing more rigorous ways of calculating potential reserves, but he was not interested. Esso's methods were sacrosanct.

It also has to be said that Esso had a highly professional and well-motivated staff, whereas at SOC Taleb insisted on doing all the recruitment himself, which was sometimes successful, but often not, and the resulting rapid turnover of staff was distinctly alarming and very inefficient. He tended to be highly critical of people's work and rarely, if ever, complimented them on a useful contribution. I tried to persuade him that the carrot was more effective than the stick but I think he felt that would be seen as 'soft' and lead to a loss of his authority. But I was to find that he had two other even worse characteristics. He was Machiavellian, and would squirrel away information about someone until he found an opportunity to use it against them, and he was paranoidally suspicious, not just of expats but Libyans as well, so much so that key reports like Bob McCrossan's offshore study and data on field reserves and success rates were kept locked away, and effectively unavailable. One of the Libyan geologists requested a day off because his mother was ill. Taleb's comment was 'How many mothers does he have?' A Canadian geologist had some medical condition which required him to keep up his fluid intake and he would bring a bottle of flavoured water to the office. When Taleb saw him he would insist of sniffing the bottle to ensure that it did not contain alcohol. I once found him going through my trash can to check what I was throwing away. Nor did he have a good relationship with his Libyan contemporary Faraj ben Salah who had been earmarked by Esso as a suitable Exploration Manager after their departure. Faraj was a nice man, courteous, capable and willing to listen, but somehow he had been outmanoeuvred, and Taleb had been appointed Manager instead with Faraj sidelined into a minor role. The nearest analogy I could find for Taleb's management style was Iago, or maybe Joseph Stalin.

One of my responsibilities as Exploration Superintendent was to conduct staff appraisals as I had done at Britoil. We used the Esso system (what else), which assessed performance according to several different criteria and ended up awarding a rating between 1 and 5 where one was exceptional and five hopeless. The median rating of 3 was 'satisfactory performance'. In a diverse group of mixed ability you expect to find a bell-shaped distribution curve in which 3 would be the most common value. This is what I found among the geologists at SOC when I conducted my first appraisal. There were one or two very capable guys, one or two who hadn't a clue, but most were somewhere in the middle. But when I compared notes with other department heads I found that they consistently rated their personnel much higher than I did. My 'good' was their 'outstanding'; my 'satisfactory' was their 'above average'. My Libyan geologists objected strongly, claiming that they were better than 'satisfactory', but I told them I had followed the guidelines and that the ratings were fair. I only found out later that someone rated as satisfactory would not qualify for promotion, nor for overseas training, nor would they receive a merit increase to their salary. Indeed someone rated as 4 would receive a warning letter. Taleb of course had not explained any of this to me and had deliberately let me blunder into a situation where I was regarded as hostile to the Libyan geologists for holding back their career development, not to mention their meagre pay. And Taleb just looked on and smiled. As my boss, Taleb was obliged to conduct my appraisal, but he had little interest in the task, and on one occasion he simply told me to write my own.

We made some good friends in Brega. There was a large contingent of Yugoslavs. The Medical Department was staffed mainly by Serbians and we became very friendly with a doctor called Jovan who was a GP at the Clinic and his girl-friend Lilijana, who he later married. Jovan was a serious chap and a fitness fanatic, and had been a 400 metre hurdler in the Moscow Olympics. He didn't drink alcohol or smoke, and he was not a party person. He preferred discussing sport or politics with his friends.

Lilijana was a physiotherapist from a village near Novi Sad in Vojvodina where her family owned a farm, and she had worked in Tripoli before moving to Brega. At one point, she had been invited to treat Gaddafi, an invitation she had wisely declined. There was an x-ray machine at Tripoli airport but it was often out of service. Customs would conduct a desultory examination of passengers' baggage, mostly looking for alcohol (which they would then sell on the black market), but quite often they couldn't be bothered, and on at least two occasions Lilijana smiled at them sweetly and brought in a suckling pig for our Christmas dinner in her hand luggage. Imagine what that would have looked like on the x-ray machine.

We also became good friends with a couple from Croatia, Branco and Smilja. They were on family status with their two young children, and were extremely popular since Branco was very extrovert and welcoming, and their house became a meeting place for all the Yugos in the camp, particularly those on bachelor status. It was rare to go to their house and not find three or four visitors. They were noted for their barbecues, particularly of fish caught by the Yugo divers from Marine Department. Branco was fond of saying that a fish has to swim three times, once in the sea, once in oil and once in wine. Like most of the Croatians, Branco worked in Technical Department. He was an excellent swimmer and water polo player and would often play tennis with us. During the early days, as far as I could tell, the Serbs and the Croats mixed quite amicable together and we knew of several mixed marriages between them. We became friends with one such couple, Lana and Den. Lana was from Serbia and Den from Croatia. They were a nice couple, fond of classical music and serious debate, and their grown-up children who visited from time to time were similarly highbrow. I recall one exasperating dinner when the three teenagers each tried to outdo the others with clever repartee. Maybe I am just a low-brow, but I found it rather conceited. Instead of trying to cultivate their garden Lana and Den adopted a Japanese solution, which was curiously effective. They had a few shady trees and some large rocks, and they raked the bare earth between them into intricate swirls and patterns which made for a very pleasant area in which to relax. One year Irina and a Polish friend, dressed in underwear and suspenders, delivered him a surprise kiss-o-gram on his birthday, to find him already in his pyjamas. He was acutely embarrassed.

When Yugoslavia fell apart in 1990 and Slovenia and Croatia attempted to secede from the federation, Serbia invaded Croatia and a bitter and protracted civil war developed. Nationalism became rampant, and in

Brega former friends quickly became enemies. It was sad to see how rapidly this breakdown of trust took place. It became so bad that Croatians were reluctant to take their children to the company clinic, because it was staffed largely by Serbians.

Irina went to visit her parents in Moscow in September 1989 and while she was away Panther died. It is hard to explain the affection you can have for a dog. But Byron knew, and wrote a poem to commemorate his faithful Newfoundland called Bosun:

> *Ye, who behold perchance this simple urn,*
> *Pass on – it honours none you wish to mourn.*
> *To mark a friend's remains these stones arise;*
> *I never knew but one -- and here he lies.*

Panther was a delight, loving, protective, placid, companionable. She was very smart and could understand both English and Russian. At the mention of *khochesh gulyats* she would immediately become animated and start heading for the door, but when told to 'stay and guard the house', she would give you a sad and accusing look, and when we started packing our bags she knew that we were going away. She was our constant companion, in Scotland, in Yorkshire, in the woods, on the hills, on the beach, in the rain, in the snow, she didn't care, it was all good to her. It was impossible to keep her out of water. You could say 'No Panther, it's too cold', but she would give you a look which said 'Just watch me', and in she would go. It was different in Libya of course, but once she had acclimatized, she swam in the sea almost every day. She had been fearless of the sea when we lived in Scotland, until one day at St Andrews she was tumbled by some big breakers, after which she became much more savvy. If we swam with her she seemed to be concerned for our safety and would paddle around us with a look which implied 'your place is on land, not out here'. Labradors are very fond of their food, but she never stole anything, until one day temptation proved too great. We were preparing for a party and someone gave us a plate of cheese straws which we covered and put on a trolley while we went out. When we came back exactly half of the straws were gone, and she had a very guilty look on her face. She died after suffering a series of strokes. I was with her when she died, and I was heartbroken.

1989 was the year of revolution. Communism in eastern Europe had been showing signs of decay for years, but it was the reforms introduced by

Gorbachev in Russia in the late 80s which opened the floodgates, and once begun there was no turning back. In Poland *Solidarity* overwhelmingly won a partially free election and immediately abandoned communism. Hungary followed and opened its borders, allowing a mass exodus to the west. The East to West Berlin border was thrown open on 10 November 1989, which marked the beginning of the end for communism throughout eastern Europe. Within two years a total transformation had taken place. The Soviet Union itself, the birthplace of communism, was dissolved on Christmas Day 1991 to form the Russian Federation with all the former Soviet republics emerging as independent states. To Irina and millions of others it was a transformation they thought would never happen. In Russia the devolution of power did not run smoothly. The privatization of state assets was a scandal, cynically orchestrated by Boris Yeltsin and a clique of associates like Anatoly Chubais and Yegor Gaidar, in which a voucher scheme was manipulated to create instant fortunes for a small group of individuals, who became known as the oligarchs. Some of them possessed insider information, having worked for the former Ministry of Trade or Ministry of Finance. By manipulating the enormous difference between the old Soviet commodity prices and the current worldwide prices they were able to acquire large state enterprises for derisory amounts, and their wealth was invested not in Russia but in the west. The Russian public was shamefully duped and exploited for the benefit of people like Boris Berezovsky, Mikhail Khordokovsky and Vladimir Potanin, and later by Ramon Abramovich, Mikhail Fridman and Mikhail Prokhorov. The oligarchs were eventually confronted by Vladimir Putin, but again a cynical deal was struck in which they were allowed to keep their wealth in return for their political support. *Plus ça change, plus c'est la même chose.*

Letter 46

One of my geological colleagues, Bob Maxwell, had resigned from SOC and moved to Cairo to work for an American oil company. He invited us to visit and offered to show us around. We stayed at the Mena House Hotel at Giza near the Pyramids, and went to the sound and light show which, although a bit kitsch, was nevertheless impressive. The sheer size and antiquity of the pyramids are overwhelming. Imagine the labour involved in creating a structure 280 cubits high (147 metres) and 440 cubits (230 metres) along the base, with individual blocks weighing two and a half tons, with the technology of 4,500 years ago. Incidentally the stones are Eocene limestones, full of little discus-shaped fossils called *Nummulites*. Next day Bob and his wife drove us to Memphis and then to

Saqqara to see the famous step pyramid which is even older than those at Giza. Next to it is the ruined pyramid of the fifth dynasty pharaoh Unas. The burial chamber, which still contains the empty sarcophagus, can be entered along a long underground passage, and on the walls in hieroglyphics is inscribed the earliest known version of the Book of the Dead. It was a hot day, I was dehydrated, there was not much air, and I began to feel faint. I swayed and blacked out momentarily, but Bob managed to grab me and steered me to a convenient ledge. Underground, in a tomb, next to an open sarcophagus, and looking at the Book of the Dead, it was eerie; it seemed like the curse of the pharaoh, and I felt a touch of kinship with Lord Carnarvon.

Bob's wife, Sallee took us to the Cairo Museum, an old-fashioned, cluttered, dusty old place, but which contains the Tutankhamun treasures discovered by Howard Carter (working for Lord Carnarvon) in 1922. This is something you absolutely must see. The treasures are simply fabulous. Lord Carnarvon died shortly after the tomb was opened from a mosquito bite which turned septic, hence the legend of the curse of the pharaoh. Then we went to see the Coptic church of St Sergius and St Bacchus in Old Cairo dating back to the 4th century, on the spot where tradition has it that the Holy Family rested after their flight into Egypt to escape Herod's Massacre of the Innocents. The entrance to the crypt is in a very unprepossessing back street down a flight of steps and several metres below sea level, so that when the level of the Nile is high the crypt is flooded.

The next day we flew to Aswan and had lunch on the terrace of the famous Cataract Hotel overlooking the Nile, a magical place with a bustling landing stage and feluccas on the river, exactly as portrayed in the movie *Death on the Nile*. From there we flew the 300 kilometres to Abu Simbel close to the border with Sudan where Ramesses II built two large temples carved into the living rock to celebrate his victory over the Hittites at the battle of Kadesh. The larger of the temples has four seated effigies of Ramesses, 21 metres high, wearing the dual crown of Upper and Lower Egypt which dominate the façade. Next to the legs of the pharaoh are a number of diminutive figures no higher than his knees which represent his favourite queen, his mother and his eldest children. That gives you some idea of their relative status. However, the smaller temple, dedicated to Ramesses' queen Nefertari, has effigies of both Ramesses and his wife which are of similar size, which presumably indicates that he held her in very high regard. When the Aswan High Dam was built by the Soviets between 1960 and 1970 to create Lake Nasser, it was realised that the Abu Simbel temples would be submerged by the lake, so the entire site was moved 65 metres higher and 200 metres further back from the river, and incorporated into an artificial hill. The entire complex is breathtaking.

Ramesses II was the greatest of the pharaohs. He reigned for 66 years and lived to the age of 90. He was the obsessive builder of dozens of temples and statues glorifying his achievements, and was delightfully portrayed in Tim Rice and Andrew Lloyd-Webber's musical *Joseph and his Amazing Technicolor Dreamcoat:*

> *Pharaoh he was a powerful man.*
> *With the ancient world in the palm of his hand.*
> *To all intents and purposes he,*
> *Was Egypt with a capital E.*
> *Whatever he did, he was showered with praise.*
> *If he cracked a joke, then you chortled for days.*
> *No one had rights or a vote but the king,*
> *In fact you might say he was very right wing.*
> *When Pharaoh's around, then you get down on the ground.*
> *If you ever find yourself near Ramesses, get down on your knees.*

From Abu Simbel we flew to Luxor (the ancient Thebes) to see the tombs in the royal necropolis on the west bank of the Nile. The most striking monument is the Temple of Hatshepsut, looking incongruously like some modern hotel set against a backdrop of red and orange cliffs. In fact, it is a careful reconstruction by archaeologists from the Polish Academy of Sciences of the mortuary temple of Hatshepsut, one of only seven known female pharaohs. But of course, the main attraction is the tomb of Tutankhamun in the Valley of the Kings. The story of its discovery is well known: how the entrance was buried under rubble left by ancient workmen excavating the tomb of Ramesses VI, and that this was Howard Carter's last throw of the dice, as Lord Carnarvon had decided to call it a day and withdraw funding. The tomb is now rather bare since the contents have been removed to the Cairo Museum, but it still gives a *frisson* of excitement to stand in the place where Carter broke through the inner door and peered through the hole to see the fabulous treasure inside. When Tutankhamun's mummy was examined a dagger was found in the folds of the cloth, which had a blade of fine quality hammered nickel-rich iron which subsequent examination has shown was obtained from a meteorite.

The composition is almost identical to that of a meteorite known to have fallen in the Kharga Oasis in the Western Desert in antiquity. He also had a scarab necklace made of silica glass which may have been come from a fulgurite – fused silica sand caused by a lightning strike. Geology left its mark even then.

Tutankhamun was the son of Akhenaten, who introduced a radically new monotheistic religion worshipping the Sun god Aten, and established a new capital at Amarna, 350 kilometres downriver from Thebes. The cult did not long survive his death, and the old polytheistic religion was quickly re-established. Tutankhamun (living image of Amun) was originally named Tutankhaten (living image of Aten), but abandoned Amarna and returned to Thebes. Curious then that the golden throne found in Tutankhamun's tomb shows the pharaoh and his queen, backed by the sunburst image of Aten. Was this perhaps formerly Akhenaten's throne? Akhenaten's wife was the beautiful Nefertiti, of whom there is a wonderful painted limestone bust in the Egyptian Museum in Berlin. Akhenaten and Nefertiti were both buried at Amarna but their bodies were subsequently brought to Thebes. Their location long remained a mystery until recent DNA analysis showed a mummy stored in tomb KV55 was that of Akhenaten but the mummy of Nefertiti has not yet been identified.

Of course, the largest tomb in the Valley of the Kings is that of Ramesses II, and his mummy was initially placed there, but then secretly moved and hidden in an insignificant tomb in which it lay for 3,000 years until rediscovered in 1881. It is now in the Cairo Museum. Examination of the body showed that he was rather short (1.7 metres, but he has probably shrunk over the years), he had red hair, an aquiline nose and showed evidence of arthritis (not surprising at the age of 90). The Ramesses tomb (given the designation KV5) was built to accommodate his numerous wives and children (he is known to have had at least 52 sons), and at the last count (work is still continuing) contained 150 chambers and corridors. In 1974 his mummy was showing signs of deterioration and was sent to Paris for conservation. According to *National Geographic* he was supplied with an Egyptian passport listing his occupation as 'king, deceased', and he was received at Le Bourget with full military honours.

A large (naturally) mortuary temple called the Ramesseum was built to commemorate the great leader, with two granite busts of the pharaoh guarding the entrance. Napoleon made an unsuccessful attempt to remove one of them to France in 1798, but failed. In 1816 the British consul arranged for the bust, weighing over 7 tons, to be removed and shipped to London, and the news aroused widespread interest in England, inspiring Shelley to write his poem *Ozymandias* (the Greek name for Ramesses II). It is a fanciful but powerful poem which captures the mood of the time concerning ancient Egypt, and is worth quoting in full:

> *I met a traveller from an antique land*
> *Who said: Two vast and trunkless legs of stone*
> *Stand in the desert. Near them, on the sand,*
> *Half sunk, a shattered visage lies, whose frown,*
> *And wrinkled lip, and sneer of cold command,*
> *Tell that its sculptor well those passions read*
> *Which yet survive, stamped on these lifeless things,*
> *The hand that mocked them and the heart that fed:*
>
> *And on the pedestal these words appear:*
> *'My name is Ozymandias, king of kings:*
> *Look on my works, ye Mighty, and despair!'*
> *Nothing beside remains. Round the decay*
> *Of that colossal wreck, boundless and bare*
> *The lone and level sands stretch far away.*

The bust is now in the British Museum, and the other one still lies where it has always been, at the entrance to the Ramesseum.

Ramesses' favourite wife Nefertari was buried in a sumptuous tomb in the Valley of the Queens decorated with some of the finest wall paintings in the necropolis, which show her face in exquisite detail (she had almond-eyes, delicate eyebrows and rouged cheeks). The ceiling is covered with hundreds of stars against a blue background. Unfortunately, the tomb was robbed in antiquity and her mummy has never been found. Access to the tomb is now limited due to humidity and bacteria introduced by the large number of visitors.

On the Nile floodplain nearby, and standing in splendid isolation, are two enormous but badly damaged seated effigies of Pharaoh Amenhotep III (the father of Akhenaten) called the Colossi of Memnon. These originally guarded the entrance to a mortuary temple, a complex of buildings which covered an area of 35 hectares of which little now remains. The complex was destroyed by successive inundations of the river and the lack of firm foundations. The effigies were damaged in antiquity either by an earthquake or by vandals, but according to the geographer Strabo the northern statue produced a twanging or cracking sound early in the morning, giving rise to all kinds of legends, although it was probably caused by expansion of the rock as it warmed up after the chill of the night. Rather crude repairs were carried out on the colossi by Septimius Severus in about 200 AD, after which the speaking statue was heard no more.

We ended our Egyptian trip at Luxor (ancient Thebes, known to the Greeks as Thebes of the Hundred Gates) on the east bank of the river. The golden age of Thebes was from about 2000 to 1200 BC, and it was during this period that most of the Karnak temple buildings were constructed. The entrance to the complex is along an avenue lined with ram-headed sphinxes, and inside consists of temples, courtyards, halls and massive pylons (monumental gateways), and includes the famous hypostyle hall with its enormous columns, 25 metres high and 9 metres in diameter, covered with bas reliefs and hieroglyphics. The hall would have been lit by clerestory openings in the roof but the roof has long since disappeared, and in 1899 some of the columns fell, like a row of dominoes, but they have since been restored and stabilized. The columns are huge and closely spaced and some still bear traces of colour, and the effect is overwhelming.

On our last day, we had lunch on the terrace of the Luxor Hilton, to find that the onion soup was full of maggots. When we complained, the waiter explained that the maggots were not in the soup but in the onions! We flew back to Cairo where Irina stayed for a further week with Bob and Sallee, and I returned to Benghazi. I took a taxi back to Brega, a two-and-a-half-hour drive, and I offered the driver an appropriate tip, but he refused to take it. They will not be beholden to a foreigner.

Slowly at SOC we became bolder with our exploration strategy and began to drill some genuine exploration wildcats. We developed several new concepts which all looked to have potential. The first was the Waha play. The Waha is a formation of late Cretaceous age, and you have to imagine a gradually rising sea level advancing over an irregular land surface. The high points stood out as islands, and the areas between were covered by a blanket of sand and silt. The Waha Formation is the sandy component, which of course is absent on the exposed islands, but present around the islands as beach and nearshore deposits. In deeper water it passes into impermeable shales. This sequence had subsequently been buried under several thousand feet of overburden. This play was a difficult target because too high on the structure and the sand was absent, but too low and it passed into shale. We drilled four or five Waha prospects which all proved to be oil bearing, with typical reserves of around 50 million barrels each. It was a considerable success.

Our second success was in western Libya where there are a number of stratigraphic traps in Devonian sandstones several of which are oil bearing in Algeria. We had not explored for stratigraphic traps before, but we had an Indian geophysicist, Harish Joshi, who produced an interpretation showing that one of these traps extended into Libyan territory. We drilled it and made a very significant discovery, subsequently named Al Wafa, containing 4 trillion cubic feet of gas, and forever afterwards Taleb held up Harish as a shining example of someone who was willing to think 'outside the box'. It was a unique compliment.

Letter 47
We had gone to Libya with the intention of staying for a couple of years, but we found ourselves sucked into the typical expat dilemma of having an interesting job, tax free pay, an agreeable life style and no likelihood of finding a better job back in the UK, so we stayed. I got 30 days vacation a year plus a 10-day boondoggle (an American concept actually meaning useless work, but adapted by Esso to mean compensation for overtime or weekend duty), and of course we also had the Moslem holidays. The annual vacation did not have to be taken all at once so it gave us considerable flexibility in arranging our overseas trips. We always spent part of each vacation in England where we would visit my mother in Halifax and spend time with Susannah and Jonathan. Irina would go to Moscow most summers and she would often stay for an extra week or two before returning to Brega. SOC accepted that children wanted to join their parents during the school vacations and Susannah and Jonathan came to visit almost every year, but they did not encourage other visitors; it was an industrial installation, not a holiday camp, so my mother never got to come and see our house and meet our friends, which was a pity. She actually visited Tripoli on one of her cruises, but only for half a day and I could not get a seat on the company plane to go and meet her. As she wistfully said on a postcard 'So near, but yet so far'. She continued to travel widely and made a number of friends on her cruises, but she must have felt lonely. I selfishly thought that she had been a widow for so long that she had learned how to cope with loneliness, but she was not naturally gregarious and it must have been an effort for her to mix with

strangers at dinner or in the lounges of the cruise ships. But she made the effort, and I can only admire her for that.

These letters are in danger of turning into something of a travelogue, but we had some fantastic trips during those years, and I cannot pass by without mentioning a few of them. One year we went to Kenya and the game parks. We visited Karen Blixen's house near Nairobi (the house in the movie *Out of Africa*), drove to the majestic snow-capped Mount Kilimanjaro, saw a cheetah stalking a gazelle, and were chased by an angry elephant. We spent a night in the famous Treetops Hotel, and crossed the Equator, saw the 'big five', and passed Mount Kenya with its jagged peaks, but the main highlights were the thousands of flamingos at Lake Bogoria, in the Great African Rift Valley, which turned the entire lake pink, and a hot air balloon ride at dawn over the Masai Maru to see the herds of migrating wildebeest and zebras.

Sue continued her studies at Greenwich University, but Jonathan decided to take a gap year before beginning his university career. He became part of a pop group with a few of his friends but whether they were into Metal or Hip hop or Britpop I never discovered. I thought it might be interesting for him to spend some time travelling in Europe, so I gave him some financial assistance and suggested a few places that he might like to visit. And so, equipped with his back-pack and Student's Rail Card, off he went. He travelled for over a month staying in hostels and cheap accommodation and got to see Paris and Amsterdam, Berlin and Vienna, Prague and Zermatt, and Florence where he tried to find his grandfather's grave, but unfortunately he was directed to the English Cemetery instead of the British Commonwealth War Graves Cemetery and so was unsuccessful. Your dad subsequently enrolled at the same university as Sue, to study Environmental Sciences which was very popular subject at that time.

In 1991 the Russian Embassy informed the Foreign Office that the ban on my travel to Russia had been lifted, so I applied for a visa which was granted, and I was able to spend a week visiting Irina's family in Moscow. Her parents had moved to a new flat near the Moskva river at Strogino, and we spent lazy afternoons eating strawberries and drinking tea from a samovar at their dacha, like a scene from Tolstoy. We toured the Kremlin and the Lenin Hills where we had celebrated our marriage fourteen years before, and walked in the cool of the evening by the river at Strogino.

Lilijana and Jovan, our friends from the Medical Department in Brega were married in London that summer and they stayed in our flat in Warwick Square the night before their wedding. Not long afterwards Sue and Alex moved into the flat while they searched for a place of their own, and in December they moved into a flat in St George's Square. The family now had properties in three of the Pimlico garden squares – my aunt Eva in Eccleston Square, ourselves in Warwick Square and Sue and Alex in St George's Square. What a pity we didn't keep them, they would be worth a fortune today.

Back in Brega we moved from our original house to a house closer to the sea with a larger garden, and closer to many of our friends. The garden plots were bare and cheerless places to begin with, but the company provided an unlimited water supply and they had an agricultural project from which soil and compost could be obtained so we gradually established a garden with lawns and borders. Not quite the Chelsea Flower Show, but I planted a vine and bougainvillea and honeysuckle, a lovely red hibiscus and two or three fan palms. I had bamboo with delicate feathery tops against one wall, and covered another wall with Morning Glory but I struggled with my banana plant because the climate was really too dry. One of our Libyan friends had a rather more impressive banana plant and when he saw mine he said 'What's that? I have monkeys in mine'. I built a gazebo and trained the vine to grow over it. We acquired garden furniture and a barbecue and eventually a garden bar. I put a string of lights around the walls and over the bar, and in the end had a perfect sheltered garden for drinks, *al fresco* meals and barbecues. In Arabic the word for garden is the same as that for paradise.

The year was marred by the shocking news that Mike Ridd's wife Song had died of cancer at the tragically early age of forty-two. She was a gentle and accomplished woman with whom we had spent many happy times. We had known that she was ill but we had believed that modern medicine would be able to deal with her condition. They had been married not long before us, and we were infinitely saddened to realise how short a time they had had together. Mike, who was still running his own company, now had the additional responsibility of looking after his young son.

That year we took our main vacation over the Christmas and New Year period when we took a trip to South America. We met a couple from Guernsey, Richard and Karen with whom we developed an instant *rapport*, and we have remained close friends with them ever since. We flew to Iquitos and took a rickety old boat for three hours down the Amazon and overnighted in a primitive wooden lodge on the edge of the jungle. Irina got up in the night and went outside, to be passed by a wild boar heading towards the bar. Then we went to see the Nazca lines in a 4-seater Cessna, which the pilot deliberately threw around and made us all sick, and on Christmas Day we headed to the airport in Lima for the flight to Cuzco. Nothing happened, no one showed up, the plane (an old DC8 with plywood patches) stood empty and forlorn, so we went back to the hotel and had Christmas lunch. We had an excellent Queen's speech from an Indian lady in the group, got drunk, and went out on the town. We went busking in the Plaza de Armas pretending to beg for *intis*, which were worthless at the time because of hyper-inflation. Irina went up to a car halted at a red traffic light and thrust her arm out for a tip – not a smart thing to do given the nervousness of the time. The hotel entrances were guarded by armed police to protect tourists from the *Sendero Luminoso*, a communist militant group which frequently shut down the whole of Lima, by cutting the electricity supply. Next day we returned to the airport to find the same plane in the same spot, but at least there were signs of activity. We flew to Cuzco the former capital of the Inca Empire at an altitude of 3,400 metres. What most impressed us were the Inca walls made of massive, irregular-shaped rocks fitted together so precisely that you could not get a knife blade between them.

From Lima we took the train to Machu Picchu, rightly regarded as one of the wonders of the world. The settlement was constructed on top of a lofty peak and is invisible from the valley below. It was never found by the Spanish conquerors (although it was abandoned not long afterwards) and remained undiscovered until 1911. We overnighted there, and the place was just magical. From Cuzco we took a train over the *altiplano* (which looks much the Scottish Highlands except they have llamas instead of sheep), reaching 4,330 metres at one point, to Puno on Lake Titicaca. We were impressed by the black, glossy hair of the locals and were told by our guide that it was due to 'Peruvian shampoo', male urine collected in barrels and left to mature for several months in the open air. By now several of the group were suffering from altitude sickness, but I had brought some pills which turned out to be very effective. On New Year's Eve we visited a floating reed island on Lake Titicaca where we found a woman and her children mourning the death of their father who had died a day or two before from a stoppage of the bowel. If he had been close to a hospital he would probably have been saved. After crossing the lake on a hydrofoil we celebrated New Year in the rooftop bar of the Sheraton in La Paz, Bolivia, or at least those of us who were not laid low by altitude sickness. La Paz lies in a valley sheltered from the bleak altiplano, but the airport is at an altitude of 4,050 metres, and has a 4 kilometre runway because of the thin air. The planes use so much fuel during take-off that they have to refuel again within an hour.

From there we went to Rio and stayed in a hotel overlooking Copacabana Beach (which we found to be rather dirty), then by cable car to the top of the Sugar Loaf, and later Irina and I did a spectacular hang gliding flight (each with an instructor) from the top of Pedra Bonita near Ipanema to the beach at Sao Conrado 500 metres below, a flight of about 15 minutes of pure delight with magnificent views of the city and the beaches. From Rio we flew to the Iguazu Falls on the border of Argentina and Brazil which we saw both from a helicopter and from the ground. They are magnificent from whichever viewpoint you see them. We called in at one of the many rock shops in the area and I bought a wonderful amethyst geode, half a metre tall and weighing 20 kilos, which I had shipped back to London. It arrived there before we did. We ended up in Buenos Aires, a very European-looking city with broad avenues and modern buildings, although it was less than ten years since the overthrow of the brutal Galtieri military dictatorship. The main things I remember are the size of the Argentinian steaks (don't ask for large), and the much-visited mausoleum of Eva Peron at La Recoleta, the legendary actress who went on to become first lady of Argentina and a champion of women's rights. There was a disturbing corollary to our South American trip. We had bought some emerald jewellery in Bogota and Rio and when Irina was returning to Libya by herself a couple of weeks after me, she was robbed at Amsterdam airport. She was approached by two youths while pushing a baggage trolley. One of them asked her a question, the other grabbed her handbag from the trolley and ran. It contained not only the jewellery but her passport, credit cards, Libyan residents permit, keys – everything. She was able to call Jovan, our Serbian doctor friend who was now living in London, and he bought her a ticket back to London (which the British Consulate in Amsterdam had declined to do). Fortunately, the jewellery was insured, but it took seven weeks to sort out all the other problems.

In 1991 we managed to obtain funding from NOC to drill three offshore wells and three onshore wells to test a variety of exploration concepts. Bob McCrossan's regional study had identified a range of offshore plays, and we selected three of the most promising. But the Libyan offshore is a tricky and tantalising area, with some major discoveries but also many dry holes, and we were unable to find the right combination of source rock, reservoir and trap, and all three wells were dry. Offshore wells are extremely costly and it is critical to know that the rig was located on the right spot before beginning the well. The drill ship had a dynamic positioning system but we decided to check the location with a GPS system which we had acquired. The wellsite geologist took it to the rig but when he came to use it he found that there were no batteries, and there were not on board. Taleb was not amused.

Very little drilling had been done in the deep troughs of the Sirt Basin, but it had long been known that all the oil and gas in the area had been generated from source rocks in the troughs. SOC had successfully developed the Wadi field in one of these troughs, and there were several look-alike plays which merited investigation. I wrote a paper on the subject which was presented (with SOC's blessing) at an international conference held in Tripoli in 1993. Our seismic data revealed a structure which looked very similar to the Wadi field, but it was very deep. We had a rig capable of drilling to 15,000 feet, but when we reached this depth we were still several hundred feet short of the objective so the prospect was left unevaluated. Our second well was a deeper pool test on a small oil and gas field, but the deep reservoir was tight, and the third well, again based on seismic data was drilled on what appeared to be a submarine fan, but the interpretation was wrong and no fan was present. Such is the fickle nature of oil exploration; you win some and you lose some, but one big discovery makes it all worthwhile.

Letter 48
President Reagan's presidential decree of 1986 made it illegal for US citizens to work in Libya, but a small number of US personnel defied the ban and continued to work for SOC, pretending that they had business connections in Malta. The Libyans assisted them by not stamping visas in their passports. In order to avoid a regular pattern, they would vary their routes back to the US. One of our American friends opted to travel home via Canada and he presented himself at border control in Buffalo. He was searched and unfortunately, they found his SOC security badge. He was arrested, tried for defying the decree, and given a short jail sentence. It didn't deter him however; in a couple of months he was back in Brega.

After his gap year your dad enrolled to study Environmental Sciences at Thames Polytechnic (which became the University of Greenwich the following year). He spent a few months in our flat in Warwick Square, but mostly he lived at various places in the docklands area of east London, a once notorious district, the haunt of Jack the Ripper, Peter the Painter and the Blackshirt fascists of the 1930s, but now well on the road to gentrification. Susannah graduated with a good degree in Estate Management the following year, and shortly afterwards began work in the Valuation Office of Her Majesty's Revenue and Customs, responsible for valuing properties for Council Tax purposes, and so became a civil servant. Alex started his career in a similar manner but soon moved to a company called Richard Ellis, a private-sector real estate management company based in London. About this time the flat next door to our apartment in Warwick Square was put on the market and we agonised whether to buy it or not. The asking price was high and Sue and Alex suggested we should make a lower offer, a proposal with which we concurred. In the event the flat was sold for the asking price and we missed out on a golden opportunity. We could have combined the two flats and created a spacious three-bedroom apartment which would have been worth more than the sum of its parts. It was a spectacularly wrong decision.

Following that debacle we took off on our next exotic vacation to India and Nepal. We did the Golden Triangle of Delhi, Agra and Jaipur, where the highlights (apart from the cows walking around the city streets) were the Red Fort in Delhi and Lutyens' grandiose vision of New Delhi as the imperial capital of Britain's empire in the east. In Jaipur we went to see the famous Temple of the Winds and the astronomical garden of the Jantar Mantar, dating from around the same time as the construction of the Royal Observatory at Greenwich in London. We rode to the Amber Citadel on the back of an elephant, and stopped off at Akhbar the Great's abandoned Mughal city of Fatehpur Sikri. But, of course, the highlight was the Taj Mahal in Agra, and all I can say is that it lives up to every expectation: the approach, the setting, the ivory-coloured marble, the proportions, the delicacy, and the inlays of semi-precious stones are stunning, but it is the perfection of the ensemble as a whole that leaves you with a sense of wonder. You feel that you could sit there and gaze on it for hours, which I suppose is what the architect intended.

From there we had a complete change of tone by flying south to Khajuraho, a complex of Hindu temples famous for their uninhibited depictions of erotica. People these days tend to snigger at these *risqué* carvings, but to the Hindus of the time it was a celebration of sex without any suggestion of prurience. At Khajuraho airport, as we were leaving for Varanasi, Irina was stopped by security for taking an orange onto the plane. 'Why', we asked. 'You could wrap it is a tissue and pretend it was a grenade', was the reply.

Varanasi on the Ganges is the holiest of the Hindu cities, a city to which Hindus come to perform ritual ablutions in the holy river, and to die. There are special hotels for people to spend their final days, whose bodies are then cremated on funeral pyres on the *ghats* (terraces lining the river). If you are rich you have sandalwood for your pyre, if not it can be any old firewood. Nevertheless Hindus believe that if your ashes are cast into the Ganges at Varanasi your soul will take the direct line to heaven rather than being reincarnated as a goat or a locust. We were taken there at sunrise when a blood red sun rose over the river, several fires were burning, others were in preparation. Ashes were being swept up and cast into the river, people were bathing in the polluted water, others washing clothes. Sometimes, if the supply of wood runs out, partially burnt bodies are thrown into the river. To the Hindus it may be sacred, but to us it was a vision of hell.

We concluded our trip with a visit to Kathmandu in Nepal and a flight to see Mount Everest, along with the magical names of Lhotse, Nuptse, Cho Oyu, Makalu, Ama Dablam and the Khumbu Glacier. It put me in mind of the epic expeditions which had thrilled me as a schoolboy, and of Kipling's refrain:

> *Something hidden. Go and find it.*
> *Go and look behind the Ranges*
> *Something lost behind the Ranges.*
> *Lost and waiting for you. Go!*

In the summer vacation before Jonathan's final year at university he came to visit us in Brega, and because of the sanctions he had been given a plane ticket to Malta and then by catamaran to Tripoli from where he could take the company plane to Brega. He brought with him about £100 as spending money, some of which he spent on a haircut and food at the airport before he left. In Malta there was no one to meet him, so he took a taxi to the port only to find the catamaran had been cancelled and there was not another one for two days. By now he was down to his last few pounds, hungry and with nowhere to stay. He managed to call Irina who gave him the number of a friend of ours in Malta, who came to his rescue.

From being down and out on the Valletta waterfront he found himself picked up in a fancy BMW, taken to Patsy's elegant apartment, given a nice room and taken out for dinner. He did show up in Brega eventually, but I think he would not have objected to a few more days in Malta.

The turnover of staff in the Exploration Department at SOC was alarming. Taleb fired some of our most experienced geologists on a whim – our best wellsite geologist, our best regional geologist, our best log expert and several of our most experienced concession geologists were all dismissed for trivial reasons. It was very counter-productive as it took many months for their replacements to acquire the skills needed to continue the work. It was almost as though Taleb was afraid that they would acquire more knowledge than him, and I began to feel that my own job was not secure.

I had developed an interest in remote sensing during my time at Britoil where I had got to know a lecturer at the Open University called Steve Drury who specialised in that field. During one of our trips to London I had met him and we discussed whether it might be a useful technique to use in Libya. He said he would look into the matter, and I said I would check with Taleb whether we might commission a pilot project. Steve was an enthusiast and didn't hang around, but to my horror I found that he had conducted a preliminary study, prepared a report, and without my knowledge or approval had submitted it to Sirte Oil along with an invoice. This was not at all what we had agreed. We had no budget for this work, and it had not been authorised. Taleb was incensed, and it marked a decisive turning point in our relationship. From being regarded a safe pair of hands, he now saw me as a dangerous maverick who schemed behind his back. My fate was sealed by a further incident when one of our Weybridge-based geologists came over for a brief visit. I was on weekend duty when I was told that he had been refused entry at Tripoli airport and sent back to the UK. When I told Taleb he blamed me of course and said that it had been my responsibility to contact SOC's Tripoli office and arrange for someone to meet him at the airport, a procedure of which I was totally unaware. Shortly afterwards Taleb went on a recruiting trip to Canada and found an experienced Turkish-Canadian, Yener Arikan, who he hired for SOC with the understanding that he would in due course take over as Geology Superintendent. At the same time he hired Lionel, a sidekick of Yener, who had provided technical support to him in their previous company in Calgary. Taleb was not sure whether this would induce me to quit (which with the benefit of hindsight I should have done), so he brought along the chairman of the company Mansur

Benniran, and together they told me how they valued my services, blah, blah, blah, but that it was time for a change, and they offered me the alternative of a move to the Geological Laboratory. On reflection, and after discussing the offer with Irina, I decided that I would accept until I could find a suitable job elsewhere. The lab was nominally under the direction of Faraj ben Salah, an amiable, long-serving Libyan geologist, but he always deferred to the expertise of the expat specialists, and rarely interfered in the day to day running of the division.

I cannot say that the arrival of Yener and Lionel produced any positive results. The following year Mr Benniran, the forceful chairman of the company, was killed in a car crash which deprived the company of its powerful and influential leader. Yener may have been an effective manager in Canada, but he knew nothing about Libya and it was left to his colleague Lionel to oversee the technical work. I remember when they proposed a new well they did not even know how the well numbering system worked. About that time some major discoveries had been made by another company in western Libya in a very interesting formation deposited at the end of the Ordovician, during a brief glacial period (at that time Libya was located close to the South Pole). The reservoir was limited in extent and was patchy in distribution, but the quality could be excellent and some of the recent oil discoveries were very large. SOC had a concession just to the north of these discoveries and Yener oversaw the drilling of four wildcat wells to test this area, but unfortunately the good quality reservoir ended close to the concession boundary, and all the wells were dry. Yener fell sick after a year and returned to Canada where he died shortly afterwards. Lionel did not hang around, and was followed by the one remaining regional geologist, so Taleb had to go recruiting yet again.

That summer Irina's brother Yuri, Alla and their two children came to visit us in London, the first time they had all come together. We drove them to Halifax and visited the local attractions, the Piece Hall, Hardcastle Crags, the Bronte Parsonage in Haworth, the Worth Valley Railway, the Castle Museum in York, and on the way back we took them to Burghley House near Stamford, England's finest Elizabethan house, and later to Beaulieu Abbey in the New Forest where there are feral ponies roaming free in the forest. I left them with Irina and returned to Libya via Tunis and Djerba from where I took the SOC bus to Tripoli. On a country road about 20 kilometres from Tripoli airport a tyre blew out, the driver lost control, the bus came off the road, just missed a tree, ploughed through a mound of gravel and ended up buried in a sand dune. Some people were injured by flying glass, and the bus was wrecked, but we were lucky and no one was seriously injured.

Letter 49
Our final 'exotic' vacation was to China. We flew to Beijing and went to see Tiananmen Square where Mao Zedong proclaimed the establishment of the People's Republic of China in 1949. Tiananmen translates as Gate of Heavenly Peace, a bitter irony in view of the insurrection which took place there just five years before our visit, resulting in the deaths of hundreds of students, workers, civilians and soldiers. The square was (and still is) overlooked by a massive portrait of Mao dominating the entrance to the Imperial City. In the middle of the Imperial City is the Forbidden City, the Chinese Imperial Palace, for 500 years the home of the Ming and Qing emperors and their *entourages*. There are hundreds of buildings, mostly dating from the fifteenth century, which established a style of architecture whose influence spread over much of the Far East. The buildings are intricately carved and decorated but sparsely furnished. The last emperor was forced to abdicate in 1912 but was allowed to live in the palace until 1924 after which the Forbidden City became a vast museum. It has been occupied by foreign powers, survived the Japanese occupation, the Chinese Civil War, the Communist seizure of power and the Cultural Revolution. Many of its treasures have been looted, and a large number were taken to Taiwan by the Nationalists when they were forced out of mainland China in 1948.

From there we went to see the Great Wall of China at Badaling, where the wall was rebuilt in the sixteenth century to protect a pass on the approach to Beijing. From the pass, the wall climbs dizzyingly up into the mountains. We made it as far as watchtower number 7, and it occurred to me that Hadrian's Wall in northern England must once have looked rather similar. The Badaling section was built in the fifteenth century but has been extensively restored in recent times. It is nevertheless an amazing spectacle.

Next, we flew to Nanjing on the Yangtze, the former capital of China, where we visited the shrine of Sun Yat-sen, the Father of Modern China, revered by nationalists and communists alike. Nanjing however has a darker claim to fame, since it was the scene of the horrific 'rape of Nanjing' by the Japanese army in 1937/8. Following the fall of Shanghai, Chiang Kai-shek ordered the Chinese army to evacuate Nanjing in order to preserve the army and entice the Japanese deeper into China where they could be harried (like Kutuzov at Moscow in 1811/2). This left Nanjing undefended and the Japanese took full advantage, killing, raping and looting as they went. The detailed accounts are about as horrific as any I have ever encountered. A Japanese newspaper reported a contest between two officers who wagered on who could kill 100 people in the shortest time using only a sword. About 200,000 people were killed altogether, and the massacre has soured relations between Japan and China ever since. Today at the Yasukuni shine to the war dead in Japan there is a panel in the museum denying the Nanjing massacre, but stating that 'Chinese soldiers in plain clothes were dealt with severely'. In Nanjing, they have long sold coloured and polished pebbles from the Yangtze river called Yuhua stones. Nowadays each one is claimed to represent the soul of a victim of the massacre. It is surely a paradox that a nation of such artistic sensibility, capable of producing a Sesshu landscape, a Hokusai painting or the Ryoan-ji rock garden, could descend to such bestiality. But of course, the phenomenon is not confined to Japan.

From Nanjing we took the train to Wuxi and then by boat on the Grand Canal from Wuxi to Suzhou (formerly Soochow). Not at all like the Grand Canal in Venice, the Chinese version is a major commercial highway, crowded with industrial boats and barges, 100 metres wide and six metres deep, all dug out by hand – and this is just one branch of an enormous network extending over 1,800 kilometres. On the boat, our guide offered us deer-penis wine, which she claimed was highly therapeutic and especially good for healing injuries. I am willing to try most things, but this I politely declined. From there we took the train to Shanghai, another city with more than its share of tragedy. When we were there however the city was booming. It is one of the most populous cities in the world teeming with people, bicycles and cars, where crossing the road is a perilous undertaking. We found T-shirts for sale with the legend 'I crossed the road in Shanghai and survived'.

It owes it importance to its location at the mouth of the Yangtze River and it developed rapidly in the nineteenth century when it was shamelessly exploited by the western powers, particularly the British, who were determined to open up China to international trade. Britain used the cynical tactic of importing opium into the country, and when the Chinese tried to stop the trade Britain used gunboat diplomacy to intimidate them. China was forced to open up treaty ports to the western powers, of which Shanghai became the most important. An International Settlement was established in which Britain, the United States and France were granted over 20 square kilometres of prime land in downtown Shanghai to establish an independent, self-administered trading enclave, an arrangement which persisted from 1863 to the Second World War. During this period, the foreign powers transformed the enclave into a European city, and The Bund on the waterfront of the Whangpoo (Huangpu) River, was lined with palatial banking houses (Hong Kong and Shanghai Bank, Jardine Matheson, Chartered Bank), hotels and clubs. It has an uncanny resemblance to the waterfront in Liverpool. There is a small park, the Whangpoo Park, at the northern end of The Bund which, during colonial times, had a notice on the gate stating 'No dogs or Chinamen allowed'. This privileged life style came to an abrupt end in December 1941 when the Japanese occupied the International Settlement (following occupation of the rest of the city in 1937), dramatically portrayed in the film *Empire of the Sun*, and the city remained under Japanese control until 1945. Life was little better after the Communists gained control of the city in 1949 and wholesale purges were carried out against the 'capitalist roaders' and 'imperial lackeys' of the old regime. The communists of course are still in charge, but since the death of Mao more liberal policies have been adopted, and Shanghai has again emerged as a powerhouse of trade and commerce.

Xi'an is an ancient city in the centre of China, and a former capital of the Chinese Empire. It was the world's largest metropolis during the Tang Dynasty, located at the eastern end of a network of trade routes, collectively known as the Silk Road, reflected in the presence of a large Moslem population and the largest mosque in China. Xi'an became famous in 1974 with the discovery of the Terracotta Army by local farmers digging a water well. Subsequent excavation revealed thousands of life-size terracotta soldiers standing in formation, surrounding the burial mound (still unopened) of the Emperor Qin Shi Huang who lived from 259 to 210 BC. The figures were assembled from mass produced body parts, but the faces were finished individually and originally were brightly painted. The army contains foot soldiers, spearmen, archers, horsemen and charioteers, many wearing padded armour, but almost all of the weapons – lances, swords and cross-bows have long since rotted or rusted away. There are also terracotta horses, wooden chariots and other military hardware. Several pits have been opened up, but many more await excavation. The pit which is shown to tourists contains about 6,000 warriors. Most were badly broken when discovered but have since been painstakingly restored. It is a truly remarkable sight.

Our next stop was Guilin in southern China, famous for its spectacular karst landscape of tall, limestone pinnacles standing up from a flat plain like a stage-set for some fantastical opera. We took a morning boat trip on the Li River when the shadows and the reflections on the river added a surreal touch, made even more exotic by the presence of Chinese fishermen in coolie hats, using captive cormorants to dive for fish. The birds have a string around their necks, just tight enough to prevent them swallowing the fish, but occasionally they would get lucky by catching a small one. There are some impressive cave systems nearby with artfully illuminated stalactites and stalagmites, which provided a refuge for Chinese refugees fleeing the Japanese during the war. One of these caves, called the Reed Flute Cave (which sounds like something by Tchaikovsky) was found to contain graffiti left by visitors 1,200 years ago.

We ended our trip at Bali in Indonesia. This is an area of intense geological activity driven by the collision of the Australian Plate against the Eurasian Plate, and marked by a line of volcanoes extending from Sumatra to Java and Timor and then through the Moluccas to the Philippines. Krakatoa is located between Sumatra and Java, and Tambora, the volcano which caused 'the year without a summer' when it exploded in 1815, is located 150 kilometres east of Bali. We took a trip to see Mount Batur, one of Bali's active volcanoes. The present volcanic cone is located within a giant caldera, and the extensive lava field from the 1968 eruption is clearly visible from the volcano's rim. Mount Agung, a few kilometres to the south-east, erupted violently in 1964, killing over 1,700 people, and is still highly active. But for the rest, after three weeks on the road, we just relaxed at the hotel and enjoyed the local hospitality.

My move to the Geological Laboratory gave me the job of coordinating the activities of the geochemical, petrographic, micropaleo and sedimentological specialists. Cores and samples arrived from each well, which were processed and analysed by our staff to provide information on the age, reservoir characteristics and oil generating potential of the rocks penetrated. These results were then passed to the concession geologists to use in their interpretation work. In addition, I decided to update some of the out-of-date exploration databases which had lain neglected ever since the departure of Esso in 1982. SOC did not even have inventories of all the wells drilled on their concessions or of the reports in their archives, let alone an up to date well data base, which is the *sine qua non* for a concession geologist. Taleb, in his inimical way, decided that the task of creating the well data base should be given to the Computer Division of Technical Department who knew nothing about geology, so that plan came to nothing. Taleb however was coming under pressure from NOC to relinquish SOC acreage, so that they could offer it to foreign companies, and he asked me to compile a report with recommendations on which areas could be safely relinquished. By safely I mean unattractive areas with little oil or gas potential. By this time I had an extensive knowledge of Libyan petroleum geology, and the history of oil and gas exploration in the country, and Taleb came to realise that I was rather good at handling large data sets. He asked me to prepare reports on remaining SOC oil and gas reserves, success rates, drilling statistics and finding costs which he used in his meetings with NOC, and in presentations at various conferences - without any acknowledgement of course. Certainly, whilst assigned to the Lab I acquired a great deal of knowledge which was to prove useful in later years.

These were lean times for SOC. The oil price remained stubbornly low, and sanctions began to have a major impact. Some state-owned acreage was offered to foreign companies – a practice unthinkable four or five years earlier. Not only that, but SOC was forced by NOC to offer a number of field development projects to foreign companies, as we lacked either the resources or the skills to undertake them ourselves. One of these was the Mabruk field, discovered by Esso 35 years before. Mabruk is a difficult field, beset with problems. The reservoir is thin, extensively faulted, under-pressured and with poor reservoir quality, but it is large, with over a billion barrels of oil in place. Esso drilled 26 wells to appraise the field and SOC drilled a further three, but in the end judged the field uneconomic. The French company Total showed an interest in developing the field and sent a number of geologists to our lab to examine cores, logs and completion reports, and I spent considerable time with them discussing the exploration history of the field and the problems which had been encountered. On the basis of these findings Total submitted a development plan to NOC based on the relatively new technique at that time of horizontal drilling. Their plan was accepted, and the field transferred to Total who over time drilled more than 70 horizontal wells which converted a sleeping giant into a significant producing field. Having seen the success of this project NOC then awarded the field development contract for our Al Wafa discovery to Agip. SOC was beginning to lose some of its prime assets.

Our exploration budget was greatly reduced, and the company closed down its office in Weybridge, offering the staff either a transfer to Brega or redundancy, and staff who quit were not replaced. One of my very good friends, a German geologist called Peter Brenner, who had notably started one of his presentation to NOC with a quotation from Goethe, was terminated for medical reasons. He had developed cancer and was admitted to the Royal Marsden Hospital in London where he died shortly afterwards. He was a charming, cultured man and his death was a great loss to all of us.

By this time Jonathan had completed his bachelor's degree and had embarked on a Masters course at Royal Holloway College. The college, which formed part of the University of London, was based at Egham in Surrey, 25 miles from London, in a splendid building built by the Victorian millionaire Thomas Holloway, who had made his fortune from patent medicines, pills and potions, first produced in his mother's kitchen. Following his environmental interests, Jonathan conducted a study of pollution caused by fine-grained effluent discharged in waste water from an aggregate processing plant on Southampton Water, which he completed the following year.

Late in 1994 my mother went on a cruise around the eastern Mediterranean to Turkey, Egypt and Jordan, which was rather marred by poor weather. We spent Christmas in Brega but came to London for Irina's birthday in January, after which she went to Moscow, and I returned to Brega. A week later I had a call from Susannah to say that my mother had had a fall at home, and could not immediately get up. I called her and she sounded confused. She was examined by a doctor, but characteristically resisted any suggestion of going to hospital. Susannah and Alex went to see her and found her to be functioning but rather confused, and a couple of days later Irina visited her *en route* back from Moscow and found the same thing. I called her daily and she alternated between muddled and more or less normal. I am astonished now that I did not realise it for what it was. She had had a stroke, but no one mentioned the word, and I persuaded myself that it was a mild attack and that she was showing signs of recovery.

On one of the Eid long weekends a group of us decided to take a trip to Ghat in south-west Libya to see the prehistoric cave paintings (*gravures rupestres* as they call them in Algeria). It was organised by one of our exploration geologists who managed to hire a 19-seater Twin Otter from the Tripoli Flying Club to fly us down there. It is a 1,000-kilometre trip from Brega to Ghat and we made a refuelling stop at Sebha. From there the route passes over the Acacus mountains before reaching Ghat, which is close to the border with Algeria. As we flew over the mountains it became clear that the pilot was lost, and because of the sanctions Ghat airport was unmanned. Bob Weimer, one of our geologists, had done field work in the area and said 'He just needs to fly over the next ridge and he will see the airfield', so we told the pilot, who was using a small road map to fly by. He said he had been warned by the Algerians about encroaching into their air space and he was proposing to turn back. We persuaded him to fly over the ridge, and sure enough Ghat was exactly where Bob said it would be. We stayed in a small hotel called the Tibesti and in the lobby was a wonderful framed picture from the Libyan Tourist Office showing a copy of a cave painting of a Garamantan two-horse chariot where the horses are shown with nine legs between them, with the caption 'Libya: the driving force of civilization'.

Ghat is an oasis town of 20,000 people in a valley called the Wadi Tanezuft. It lies on an important trade route between black Africa and the Mediterranean and had long been a Tuareg stronghold. In ancient times it formed part of the Garamantan kingdom which dominated the region from about 500 BC to 500 AD. They had an ingenious system of irrigation based on wells and underground tunnels known as *foggaras* (a system still used on the Arabian peninsula), and they had a flourishing agricultural economy. According to Herodotus, and supported by cave painting evidence, their cattle had prominent forward-directed horns which meant that they had to walk backwards to graze. The Garamantes challenged Roman control of the Mediterranean hinterland (making very effective use of their two and four horse chariots), and the Romans devoted much effort to establishing a defensive zone to keep them out. Several punitive campaigns were waged against the tribe culminating in an all-out war led by Septimius Severus which destroyed their capital at Garama in 202 AD and most of their other settlements. The Garamantan civilization declined after about 600 AD perhaps due to deteriorating climatic conditions. During the Italian pacification of Libya, Ghat was occupied by the Italian army and a large fort was constructed on a hill overlooking the town. The walled old town lies at the foot of the castle hill and is almost completely abandoned. It consists of a maze of mud brick dwellings and narrow shaded alleys clogged with sand, a desolate town square and an old mosque. Just as we found in Algeria, the former inhabitants have now moved *en masse* into modern houses and high-rise apartment blocks, where they have electricity, plumbing and running water.

We hired local guides to drive us in Land Cruisers to the cave paintings in the Acacus Tadrart (Tadrart is the Berber word for mountains). To reach them we had to make our way up a 300 metre escarpment through a steep, rocky, sand-filled defile, emerging at the top to see a wonderful panorama of sand-filled valleys and mountains stretching as far as the eye could see. It is quite a white-knuckle experience to drive up the lee side of a 30 metre dune, reach the crest and suddenly find nothing beneath the front wheels as it goes over the top. Once in among the dunes it is very easy to become disoriented; it all looks the same. There was a story of a seismic crew operating in an area nearby where there are 60 metre linear dunes up to twenty kilometres long. Two of the surveyors climbed the lee slope of a dune and jumped over the crest, just for fun. But the sand was soft and the scarp slope steep, and they found that they could not climb back to the crest. If they had not had colleagues who went to their rescue, they could easily have died out there.

The cave paintings are remarkable. For the most part they are found beneath rocky overhangs where they have been protected from the elements. They are principally done in red ochre mixed with animal fat, and span a period of perhaps 8,000 years, and show a range of different styles. There is the Round Head style of bloated bodies with round heads and featureless faces, a style showing hunters with spears, bows and arrows, chariots and dogs, there are ibex and elephants and giraffes, depictions of rituals, sports and dancing. Then there is a pastoral style showing Garamantan cattle and horses, and an example in which the camel makes its first appearance. Finally we saw some cryptic symbols, which we thought might be an early version of a Berber alphabet. At the time of our visit there was some evidence of mindless vandalism, and sadly this has increased as the area has become better known and more frequently visited. But we saw no-one during our visit. We had the entire place to ourselves, and were left to reflect on how radically the climate has changed since the pictures were created. Today there is not an elephant or giraffe within 1,500 kilometres of where we stood.

The next day we prepared to depart, but were told there would be a delay. Time passed. We received a message that the plane needed refuelling and because of the Eid holiday that was proving difficult. They considered bringing fuel from Sebha, but that was 570 kilometres away. We finally left after lunch. They would not tell us how they got the fuel, but there was another plane on the tarmac when we arrived, and we suspected that they had called one of their buddies from the Tripoli Flying Club to fly down with jerry cans of fuel on board. Totally illegal of course, but I would not be surprised if it were true.

Letter 50
Immediately after the spring NOC meeting I flew back to the UK. My mother had been able to travel to London and was staying with Eva at Eccleston Square, but she didn't look good, and her left foot was dragging as she walked. I drove her back to Halifax and arranged for a home help to come and attend to her each day, and I submitted a request to Halifax Social Services for a place in sheltered accommodation, only to be told there was a long waiting list. She soldiered on, saying that she could cope, and I thought with the home help she would at least be spared some of the more onerous tasks, but I had no sooner arrived back in Brega than she told me she had got rid of the home help because she was too intrusive.

One of our good friends from Brega, Franco Caparrotti, lived in Rome and he invited us to visit his family for a few days. I had a friendly bet with him about the superiority of French wine over Italian and we agreed to put it to the test. I brought a nice chateau-bottled claret, and he could have bought a similar quality Italian red, but he didn't, so my claret won the bet, and we celebrated by drinking it. The next day Franco drove us to see the sights of Rome, the city wall, the *'Quo Vadis'* church where according to legend St Peter, who was fleeing Rome, encountered Christ who asked him 'Where are you going?', whereupon St Peter turned back and went to his martyrdom. Franco then took us to the Catacombs of St Sebastiano, and the very impressive Appian Way which runs straight as an arrow for 60 kilometres, its paved surface of interlocking flat stones just as good as it was 2,000 years ago. This is where 6,000 slaves of the Spartacus Revolt were crucified, as portrayed in the movie *Spartacus*. He drove us to Tivoli in the hills east of Rome to see the Emperor Hadrian's country retreat, and the Villa d'Este with its celebrated fountains. On our arrival back at his apartment we found a state of panic. Franco's daughters were sitting outside with a packed suitcase. There had been an earthquake during our absence.

We spent another couple of days exploring the famous sites of Rome – Colosseum, St Peter's, the Sistine Chapel, and the treasure house of the Vatican Museum with its celebrated double helix staircase (like a DNA molecule), the amazing map gallery showing the world as it was known in 1580, and the splendid library (but lacking presumably the works of Voltaire, Rousseau, Kant, Hulme, Bacon, Milton, Victor Hugo and all the others on the *Index Librorum Prohibitorum* – curiously Charles Darwin's works were never on the list. From there we walked on the Lungotevere to the foot of the Aventine Hill, to the place where legend – and Macaulay - has it that Horatius and his dauntless companions held the bridge against the might of the Etruscan army 'in the brave days of old'. Winston Churchill was no great shakes at school, but he was able to recite from memory the entire 1,200 lines of Macaulay's poem.

We then headed south. I had been impressed by Hadrian's Wall in England, and by Timgad and Tipasa, Leptis and Sabratha in North Africa, but these remains are nothing compared to Pompeii. The city was famously destroyed by an eruption of Mount Vesuvius in 79 AD, witnessed and described by Pliny the Younger whose uncle, called – guess what? - Pliny the Elder, army commander and inquisitive naturalist, died in the eruption trying to rescue a friend. He crossed the bay by boat but was overwhelmed by the climactic finale of the eruption, and was never seen again. The city was covered with ash and debris and remained hidden for almost 1,700 years and was not seriously investigated until 1748. Since then the city has been thoroughly excavated and has revealed some remarkable finds. Bodies buried in the hot ashfall were totally incinerated, but their shape was preserved as cavities in the ash. These were filled with plaster to produce a cast of the bodies in their death throes, quite a macabre thing to see. The town was home to about 11,000 people. The streets were well paved, and even had stepping stones in case of flooding - or worse. There were food shops containing carbonised traces of wheat, millet, olives, dates, onions, garlic and grapes, wine shops with the day's takings still on a shelf, several brothels, as well as the usual forum, temples, bath-houses, theatres and sports grounds. Almost all of the buildings were left roofless after the eruption, but a few with stone vaulted roofs remained intact. Some of the houses have been restored and many of the gardens have been replanted and give the impression that the occupants will be back shortly. There are some well-preserved frescoes and graffiti, some of them so explicitly sexual that they were locked away in the Secret Museum in Naples for over a hundred years.

From Pompeii it is a short drive to Sorrento and the Amalfi coast, and then a brief boat ride to Capri, the villa San Michele and the Blue Grotto. We overnighted at a small hotel in Sorrento and were seduced, as many before us, by the beauty of the place. On the return to Rome we passed Monte Cassino, a major obstacle which held up the Allied advance in 1944 for four months and cost the lives of 25,000 Allied troops, including 4,000 Poles, half of their entire force. In the victory parade in London in 1946, the Poles were not invited to attend - in case it offended the Russians.

We drove to Florence and visited my dad's grave in the military cemetery by the banks of the Arno, and as usual I found it a very emotional experience. We thought Florence a bit tacky, which is a pity because it is a wonderful city – but don't go on a Monday because everything is closed, apart from the shops. I took Irina to see the leaning tower in Pisa, the *Torre Pendente*, which by the time of our visit was perilously close to collapse and closed to visitors. During the war it was used as an observation post by the German army and the Allies actually considered shelling it! Urgent remedial work was carried out in 2000, directed by the Professor of Soil Mechanics at Imperial College in London, and the tower was eventually straightened by 51 centimetres, back to its state in 1838 and supposedly made stable for the next 300 years.

We ended our trip in Venice, which always impresses – not only for its canals, but for its unique history as a maritime republic which existed for a thousand years. It controlled at various times much of Lombardy, the eastern Adriatic coastline, Crete, Cyprus, Rhodes, the Greek islands and even part of the coastline of Crimea, and its outlook was always east rather than west. Venice's wealth was based on trade throughout the Mediterranean and Black Sea, and further afield - Marco Polo was a Venetian, and much of the trade of the Silk Road ended up in Venice. Like many other European states its independence ended with the Napoleonic wars. But Venice is more than a historical gem, it is a sybaritic place of opulence, self-indulgence and decadence. It almost offends my puritanical soul to lounge in the Piazza San Marco with a glass of wine and simply watch the world go by listening to the competing orchestras of the red jackets and the green jackets. Did you know, by the way, that the prows of gondolas are full of symbolism: curved like the Grand Canal, with six prongs representing the six districts of Venice and terminated at the top by curious device representing the doge's hat? No? Well not to worry, it's what Irina would call Don's useless information. But Venice has a problem: it is sinking by two to three millimetres a year. It has always been subject to flooding but recently this has become much worse due to excessive extraction of groundwater and rising sea level.

Nowadays it is common to see St Mark's Square flooded, and raised walkways have had to be put in place for pedestrians. A flood barrier is currently under construction which will operate like the Thames Barrier in London, which should protect the city from particularly severe storms and high tides, but for how long? Venice can be a rather spooky place. Away from the main thoroughfares there is a maze of old, ill-lit, narrow alleys and tiny squares, and I remember stumbling on a deserted spot one night called something like *Calle dei Morti*, which seemed strangely appropriate.

In August Susannah and Alex were married at St Saviour's Church in Pimlico, close to where they lived. We came over from Brega and my mother came from Halifax, but she was poorly prepared for the wedding. Happily Irina gave her a nice blue suit, and did her hair, and in the end made her look very presentable. The wedding went well. Your dad was an usher, Irina's niece came over from Moscow, and I proudly walked Susannah down the aisle and gave her away. She looked wonderful and I felt very proud of her. We had photographs taken in the garden where I spoke quite amicably to Wendy's mother who I hadn't seen for twenty years. I discovered that, like me, Alex's parents had both studied at Durham University which gave us a topic of common interest. Alex hired an old London bus to take guests to the reception, which continued until late, and Irina rather overindulged on the champagne, but on such an occasion, why not?

In Brega I had experienced a few occasions on the tennis court where I had felt dizzy, but I put it down to dehydration and the heat, but I went to see my GP in London who dismissed it as trivial. She was wrong, but it took another few years to find out how wrong she was. A couple of days after the wedding I took my mother and aunt Eva to see Jonathan's college at Egham. He met us and showed us his quarters and walked us round the very impressive quadrangle and college buildings. He had almost completed his Masters course and was preparing his thesis. He submitted it a couple of months later and was awarded his degree in November. Jonathan had a real commitment to Environmental Sciences and often used to criticise the oil industry for the pollution it caused. He was right of course, but the fact is the world cannot function without the oil industry, particularly for aviation fuel and petrochemicals. At that time there was simply no alternative.

Since then 'fracking' has become a contentious issue. Fracking involves fracturing tight rocks containing oil or gas by the injection, under very high pressure of a fracking fluid containing sand or other proppants. When the rock has fractured the proppants hold the fractures open and allow the oil or gas to be produced. Millions of frac jobs have been conducted worldwide, and in some areas have totally transformed oil and gas production, but this comes at a price: potential pollution of water supplies, ground subsidence, and the triggering of earthquakes. The technique has been particularly fiercely opposed by environmentalists in the UK, and the government forced a company called Cuadrilla to shut down its operations in Lancashire when a number of small earth tremors were linked to its fracking operations. The Scottish Parliament banned fracking in 2017, but is happy to import oil produced by fracking from Pennsylvania. Is that hypocrisy or just nimbyism?

The following month my mother went on a cruise by herself which I found very reassuring, but I think it proved too much for her and she never went again. At Christmas Jonathan drove her to Scunthorpe for Boxing Day and then she went to stay in our flat at Warwick Square. It seemed that she was still capable of getting around, and I deluded myself into thinking that with the help of her neighbours Steve, who offered to cook for her, and Margaret who helped with her pills, and with a community service called Meals on Wheels, she could get by. She was also provided with an emergency call button with which she could summon help in the event of an emergency.

I took her on a winter break to Madeira, which I thought would cheer her up, but it turned out to be a nightmare. I hired a car and drove her to see the sights, but she was not responsive and became so ill that I had to take her to hospital in Funchal. It was a cold and unfriendly place, the staff could not speak English, and she hated it. They did a brain scan and it was only then that I saw how much damage had been caused by her stroke the previous year. Back in Halifax we tried to persuade her to go into a respite home for a week or two, but she could not bear the idea. What she wanted was for me to leave Libya to look after her, and that is what I should have done. But to my eternal shame I did not. I could not see how we could live without an income, how I could abandon everything to look after her full time, where could we live – Warwick Square only had one bedroom, Halifax was out of the question. What was Irina supposed to do? The questions seemed insoluble. I arranged for a carer to visit her three or four times a week, and this time she was a bit more appreciative, and I started to look for employment in the UK, but I waited in vain for a vacancy to become available in a protected housing scheme, and to cap it all just at this time her oldest friend died, which made her even more depressed. She struggled along sad and lonely, and I felt a terrible burden of guilt. When I was a child she had sacrificed everything for me. Now when she was old and needed me I could not, or would not, make the same sacrifice for her. That was the second time I had been weighed in the balance and found wanting. What the good of all my education and position, if I could not be a caring son?

> *Yet each man kills the thing he loves*
> *By each let this be heard*
> *Some do it with a bitter look*
> *Some with a flattering word.*

In summer, we came home via Paris where we overnighted in St Denis with our Iranian friend from Brega and his wife, and then travelled on the recently opened Eurostar train to London. I drove to Halifax and found my mother rather calmer, and I spent a week with her during which we went to some of her favourite places in the Yorkshire Dales, Skipton, Kilnsey, Buckden, a picnic by the river at Yockenthwaite, and then over the top to Hawes and Ribblehead and back through Settle. It was like a page from the past in happier days. My mother was always tearful when I left, ever since I was at college, but I remember on one occasion after a weekend visit I forgot my keys and had to return to find her quite calm and composed, and she said 'You see I am alright, I am not upset'. The pain was bravely hidden, but it was there inside.

Letter 51

Irina and I had planned a trip to Germany and we knew we could never get my mother to go there, so we arranged for her to come to London and stay in our flat where Eva could keep an eye on her and she would be able to see Jonathan and Susannah and Alex.

Even in those trying times our trip to Germany was memorable. We flew to Berlin and stayed on the Kurfürstendamm. We hiked from the Zoo to the Victory Monument, through the Tiergarten (with its Russian T34 tanks still in place) to the Reichstag. The Berlin Wall of course was gone, but the no-man's-land which ran behind it was still bare. It was a strange feeling to walk through the Brandenburg Gate and reflect on all the events that had happened there. At the time of our visit the area was a major building site. The Reichstag was being restored and the famous Adlon Hotel which had survived the wartime bombing, but had been burnt down by drunken Russian soldiers at the end of the war, was being completely rebuilt. We walked down Wilhelmstrasse to see the site of Hitler's Chancellery and the infamous bunker where he met his *Götterdämmerung*, but there was nothing to be seen. Much of the street had been razed and there was a housing estate and children's playground on the former site of the bunker.

We walked on *Unter den Linden* to the Opera Square where the burning of the books had taken place in 1933. As Heine remarked a hundred years earlier 'Where they burn books, they will, in the end burn people'. How very prophetic. Whilst East Berlin was under Communist control little had been done to restore the public buildings. The Gendarmenmarkt was a typical example, a splendid square badly damaged during the war, with no significant reconstruction until the fall of the Communist regime. Schinkel's elegant *Neue Wacht* had several incarnations, firstly as a guardhouse for a royal palace, then a First World War memorial, then a shrine to Nazi glory, and in the GDR era a memorial to the victims of Fascism. Now it houses a poignant *pieta* sculpture by Kathe Kollwitz placed beneath an open oculus symbolically exposed to the sun and the rain. Further east was the site of the stolid City Palace of the Hohenzollern kings and German emperors, built in a style known as 'Protestant Gothic'. In 1918 the establishment of the short-lived German Socialist Republic was proclaimed from one of its balconies by Karl Liebknecht. The palace was badly damaged during the Second World War and demolished by the Communist authorities in 1950 – except for the Karl Liebknecht balcony. In its place was built the ugly, out-of-character Palace of the Republic, to house the East German parliament. Astonishingly in 2007 the decision was taken to rebuild the Royal Palace more or less in its original form, and the Palace of the Republic was torn down. *Plus ça change*....... We completed our perambulation at the rather undistinguished Alexanderplatz and the old Nicholas quarter, having walked at least ten kilometres.

The next day we took the train to Potsdam, past the sinister Wannsee, and went to see Frederick the Great's summer palace at *Sans Souci*. As an admirer of Frederick, I was pleased to see that his 'palace' was a modest villa of only ten principal rooms overlooking a terraced vineyard and a large park with numerous temples and follies. It is a single storey construction in the rococo style in which Frederick was able to indulge his passion for music and entertain guests such as Voltaire, Bach, Kant and

Casanova. Most of the furniture and fittings were looted after the war, but eventually most of Frederick's library and some of his pictures were returned to Potsdam and his body was brought back and reburied at Sans Souci in 1991. Following the Seven Years' War Frederick built the much grander *Neues Palais*, 'just for show'. It is capped by a Prussian royal crown held aloft by the three graces - his enemies Empress Elizabeth, Catherine the Great and Maria Theresa.

We completed our short visit to Berlin with a visit to Charlottenburg Palace which originally contained the fabulous Amber Room, its walls lined with panels of amber, which was gifted to Peter the Great in 1716. It has a bizarre history. It was installed in the Catherine Palace at Tsarskoye Selo near St Petersburg, but during the Second World War the amber panels were removed by the Nazis and sent to Königsberg Castle in East Prussia where they were put on display. At the end of the war it was planned to send them to a secret location in the Alps, but the Red Army arrived before they could be dismantled. The Soviets shelled the city and destroyed the castle, most probably with the amber still inside. Traces of hinges and fittings were found amongst the burnt debris. In a curious reversal of fortune the so-called King Priam's Treasure, discovered by Heinrich Schliemann at the site of Troy in 1873, housed in the Royal Museum in Berlin, was taken to Moscow by Soviet agents after the war. There is a replica display at Charlottenburg Palace with the poignant caption 'Original in the Pushkin Museum Moscow'. Across the road from the palace is the Egyptian Museum which contains the famous painted stone bust of Queen Nefertiti discovered at Amarna on the Nile Delta just before the First World War.

We ended our tour back on Museum Island at the Pergamon Museum, the most visited museum in Germany. It contains the reconstructed remains of three stupendous architectural ensembles, the monumental Altar from Pergamon on the Aegean coast of Turkey, the Ishtar Gate from Babylon and the Market Gate from Miletus near Bodrum, so huge that a special museum had to be built to house them.

The following day we took a train from Berlin to Prague which required a long taxi drive through former East Berlin along Karl Marx Allee, a monumental example of socialist urban architecture which would not look out of place in Moscow. We arrived eventually at Berlin-Lichtenberg station (used in the movie *The Bourne Ultimatum*), a terminus for trains travelling to the east. It is possible to travel to Novosibirsk and Kazakhstan from Lichtenberg, but Prague was far enough for us.

I had last seen Prague thirty years before, but since the collapse of communism the city had become much more prosperous and welcoming. We did the tourist circuit to Wenceslaus Square, the Old Town, Jewish Quarter, Charles Bridge, Hradcany Castle and the Cathedral, but one of the most appealing features was the myriad of intimate evening concerts performed by just a small group of musicians in any available assembly room, church, bar or private house.

On we went by train to Vienna, then drove to Schönbrunn, Gmunden, Hallstatt (and visited its salt mines and underground lakes), to Salzburg where we heard *Eine Kleine Nachtmusik* in the Archbishop's Palace, where Mozart had entertained the Prince Archbishop – before they fell out.

I have long been fascinated by Hitler: how a man with basic education, a corporal during the First World War, with no connections to the elite of German society, could become Chancellor of a highly cultured and civilised nation, and then proceed to dominate and intimidate it. It is a conundrum to which I have no answer. Virtually no trace of his 'thousand year Reich' is preserved in Berlin, but I thought there might be more to see at his mountain retreat in Bavaria. We drove to Berchtesgaden and then up to the village of Obersalzberg which Hitler first visited in 1923, and where he wrote part of *Mein Kampf*. In 1928 he rented a chalet called Haus Wachenfeld which he subsequently purchased and rebuilt as a large villa which he renamed the Berghof. In later years the entire village was commandeered for the Nazi hierarchy, and prominent Nazis like Göring and Bormann built properties there, all closely guarded by the SS. The Berghof faced north-west and had breathtaking views to the Kneifelspitze and the Untersberg range. Behind the village to the south rose the summit of the Kehlstein (Eagle's Nest) at an altitude of 1,850 metres, and in 1937/8 the Nazis constructed a 6½ kilometre road to the foot of the peak, whence an elevator rose through the mountain to the summit where they built a stone chalet, the Kehlsteinhaus, presented to Hitler as a 50th birthday gift. It has fantastic views as far as Salzburg, the Untersberg Range, the Berchtesgaden valley and westwards to the Konigssee and the sharp peak of the Watzmann.

Hitler used the Berghof to entertain his cronies and visiting distinguished guests such as Lloyd George (bemused by Hitler), Mussolini (overawed by Hitler), Neville Chamberlain (duped by Hitler), Lord Halifax (intimidated by Hitler), the Duke of Windsor (reputedly an admirer of Hitler) and Unity Mitford (an ardent admirer of Hitler). Hitler left the Berghof for the last time in July 1944 and the village and the house were heavily bombed on 25th April 1945, just five days before Hitler's suicide. The remains were set on fire by the retreating SS, and the ruins of the house were cleared away in 1952.

Little remained of Nazi Obersalzburg by the time of our visit. The site of the Berghof was an overgrown wilderness with only the remains of the garage still visible, but we took a bus to the Kehlsteinhaus, which deposits visitors at the portal of a tunnel cut 120 metres into the mountain to the original 1938 elevator, decorated with brass panels and a clock salvaged from a U-boat. This was the only authentic relic of the Nazi era we found during our entire trip. It was spooky to ride the elevator once used by the man responsible for the deaths of 60 million people. The Kehlsteinhaus is now a cafeteria catering for visitors curious, inquisitive or just plain casual. We set off to walk around the summit which is very rocky and precipitous with Irina wearing totally unsuitable shoes; she accused me of trying to get rid of her!

We ended our trip with a visit to Ludwig's fairy-tale castles in Bavaria which I had last seen as a student, thirty years before, but the weather was not kind, and the crowds were oppressive. We flew back from Munich, but before leaving I went to Dachau – by myself since Irina avoids such horrors. Dachau was the first concentration camp built by the Nazis and housed a wide variety of German dissidents, along with communists, intellectuals, scientists and resistance fighters from many countries, even including some minor European royalty. There are 32,000 recorded deaths in the camp but many more went unrecorded, and thousands more died on the death marches at the end of the war and from a typhus epidemic which swept through the camp. It was a sombre experience to walk beneath the cynical slogan over the entrance *'Arbeit Macht Frei'*, but all of the original barrack blocks have been demolished, and only two have been rebuilt to show what conditions were like, but the place has been sanitized, there are manicured lawns and trees and painted fences. It has none of the horror which I was to find when I visited Auschwitz several years later.

From there it was all downhill, or so it seemed at the time. Not long after our German trip our house in Brega was burgled. The villas were fitted with metal louvred sun screens operated from inside. Someone had obviously seen us heading off to tennis and broke in through a small side window, forced open the louvres and climbed in. We suspect they pushed a child through since the gap was quite small, who then opened the door for the burglar. They stole a several watches, Irina's costume jewellery and a few other things of no great value. But it was not so much the value of the stolen property, as the thought that someone has been in our home, rifling through our drawers and personal effects. We reported it to Industrial Security, but of course they did nothing.

At Christmas I thought it would be nice for my mother to come to Djerba in Tunisia where we could spend Christmas together. She couldn't travel alone so your dad very kindly came along as her escort. She rallied a bit and enjoyed the sun, and the pianist in the hotel lounge, but she was forgetful and vague, and extremely upset when the time came to leave. By this stage she was becoming desperate, she really could not cope by herself. One on occasion I asked her what she was having for lunch and she answered with a wail 'a baked potato'. I blushed in shame and impotence. On another occasion when she was complaining about how many years I had been overseas, she put the phone down on me. She had never done that before. It was awful.

Fortunately, not long afterwards I met Paul Bathurst, a partner in a small UK consultancy called Exploration Geosciences which was looking for a senior geologist to join their expanding business. He seemed to think my background was appropriate, and frankly I would have accepted anything. This was the first UK employment opportunity that had come up in a long while, and we agreed to have further talks. Meanwhile we took a trip to Benghazi one weekend to buy fish for a barbecue. As usual we called in at the Tibesti Hotel where we had a buffet breakfast of scrambled eggs, fruit and coffee. A few days later I developed a fever, temperature and headache which felt like a bad hangover, but a blood test showed a serious liver disfunction. I had contracted hepatitis, and the SOC doctor was obliged to get me out of the camp as quickly as possible as the disease is highly contagious in its early stages.

They sent me by ambulance to an old Italian-built isolation hospital in Benghazi. It is hard to convey the appalling condition of the place. In front of the main entrance was a stagnant pool of muddy water. I was put into a ward with about fifteen others, all Libyans I think, suffering from various infectious diseases. They were sitting around smoking, listening to Arabic music, chatting with their wives and children and cousins and aunts, and eating pungent food – and no doubt passing infection around. There was a bed with a stained mattress – nothing else. Irina had been warned, so she had brought sheets and a pillow. Only the most basic and unpalatable food was provided and you had to rely on friends to bring something more suitable. There were no showers and only one bathroom for all of the patients in the unit, but the bath was full of cloths and pails and mops used to clean the floors. The mattresses were cleaned periodically by taking them out into the yard and hosing them with a high-pressure fire-hose, and then leaving them in the sun to dry. Fortunately, Irina made a big fuss and I was transferred to a small side ward with just one patient, a Sudanese guy with malaria, but at midnight they wheeled in an old Libyan man who was very sick indeed, noisy and vomiting, and I suspect he did not survive the night; certainly he was gone by the next morning. Irina stayed in the Tibesti Hotel and visited each day, usually accompanied by one or other of our Libyan friends. Eventually, thanks to their efforts, I was put into a two-man room with a young New Zealand engineer working for Occidental called Warren Wright who also had malaria, caught not in Libya but on an earlier assignment to Nigeria. He had frequent malarial attacks when he would lie shivering and sweating as the fever raged. All I could do was to help him keep hydrated. They took a blood sample from me for analysis but only sent them to the lab twice a week so it took several days for the results to arrive. Then they put me on a drip, but missed the vein, so my arm swelled like a balloon.

I was in there for six days, the blood tests confirmed hepatitis but they could not determine which type. At SOC we had a very capable Libyan geologist, Ali el Sogher who lived in Benghazi, and when he heard about my plight he got his mother to prepare rice and vegetables and salad which he brought to the hospital for me. I was very touched by his kindness and consideration. Irina and Walid, another of our Brega friends, spoke to a Libyan woman doctor, trained in the UK, who advised them to get me out of there asap. She said if I stayed I was likely to contract other nasty diseases. In the end, since I was no longer infectious they agreed to release me back to Brega. By now my skin had taken on a rather deathly pallor and my eyes were yellow, like a werewolf. There is no medical cure for hepatitis. It is a matter of nurturing the liver back to health and that takes time.

I was given medical leave of absence and Irina accompanied me back to the UK and I was quickly admitted to the Cromwell Hospital in London. What a contrast; I had a private room with clean linen and even spare pillows in the cupboard! My mother came to London and I had frequent visits from her and Irina, Eva, Susannah and our friend Lilijana, who was now married and living in London. I was treated with steroids which were supposed to speed up recovery, and told not to drink alcohol for six months. I was released after ten days and then had two months convalescence, during which I had further discussions with Exploration Geosciences, and they offered me a job which I gratefully accepted. I told them I would have to go back to Brega to resign and work my one month's notice in order to qualify for my end of contract bonus, and so it was agreed.

Jonathan was finding it difficult to find employment in the field of Environmental Sciences. It was a popular subject and there were far more graduates than available jobs, so he took temporary employment as a driver with DHL in London, followed by a couple of months backpacking with a friend in Thailand. I felt that he would have a much better chance of employment if he gained some practical experience, and I helped him to obtain a temporary job with a company called Welsh Gold. This involved living on site in a caravan at one of only two gold mines in North Wales (which traditionally made the wedding rings for the British royal family). The job was without pay but they covered his travel and subsistence expenses. He was involved in analysing the effluent from the mine and assessing its effect on the local environment. I spent some time with my mother in Halifax and we drove over to see Jonathan in his caravan in the remote Welsh mountains. It was a tough life, and a complete change from DHL, but it made his CV look more attractive to potential employers. A short time afterwards he found a job with an Indian company which was involved in obtaining and marketing semi-precious stones from Africa. He put his affairs in order and headed off to his first real job in Windhoek, Namibia.

Our friends from Aberdeen, Tony and Laleh were now living in Oxshott in Surrey, but they also had a holiday home in Boulouris close to St Raphael on the Cote d'Azur, perched beneath the rocky peaks of the Massif d'Esterel, which they kindly invited us to use. It was a real tonic for an invalid, and I returned much rested and refreshed. Shortly afterwards my mother celebrated her 80th birthday. She came to London and stayed with Eva for a few days. We got her a nice birthday cake, and had a family party in the gardens of Warwick Square. We also spent time looking for a larger flat in London, or for a small flat for my mother, but without success. We returned to Brega and I made a final trip to the desert with Bob Newport, our lab photographer, to take photographs at Zelten and Meghil for the cover of my final geological report, after which I submitted my resignation. Taleb had the gall to say that he had been offered a job at NOC and was hoping that I would go with him as his technical advisor! We had gone to Libya for two years, and had stayed for ten. In retrospect, I should have left several years earlier.

Letter 52
And so 'to fresh woods and pastures new'. We moved into our small flat in Warwick Square, and at last I was able to call my mother without the frustration of the Brega telephone exchange, and I drove to Halifax to see her most weekends. I took her to buy the things she needed, and drove her to some of her favourite places in the countryside, but these trips were tinged with melancholy. It was as though she was saying goodbye to places she had known in happier times – times that would not come again. At Christmas, she came to London and I took her to a carol concert at the Albert Hall. It proved to be her last concert.

I started work with Exploration Geosciences almost immediately. The company was run by two enterprising and capable geologists, Mark Groves-Gidney and Paul Bathurst who had established a successful business as petroleum consultants, and they had a network of contacts who they could call upon to provide specialist services in geophysics, geochemistry, petroleum engineering and micropalaeo. Mark's wife Gavrielle was an independent-minded American geophysicist, who wrote fiction in her spare time. She had recently published a novel called *The Last Assignment* based on her experiences in Pakistan, which impressed me no end. They operated from a delightful place on a farm in rural Kent, adjacent to an old oast house, a kiln for drying hops, one of the traditional crops of Kent. They are typically circular in shape with a conical top. Mark and Paul were very good at designing projects for oil companies which were useful, but for which the company did not have sufficient staff to tackle themselves, and they also produced a weekly newsletter on North Sea oil activity to which companies could subscribe.

Fortunately, I was able to do much of the work from home or from the Geological Society in London, which has an excellent library, and I only needed to travel down to Goudhurst, where Mark and Paul were based, once or twice a month. The projects were very varied. I began with a review of the potential of the Browse Basin offshore north-western Australia based on well logs, and then I provided geological input to a geophysical study of offshore Angola. This led to a much larger project providing geological data for many offshore basins around Africa in which I was able to draw upon the experience I had gained with Esso in Morocco, Senegal and Portuguese Guinea (or Guinea Bissau as it was now called).

It was an interesting illustration of how science advances. In 1955 the Head of Research at one of Esso's principal subsidiaries brashly claimed that 'continental drift is just European bullshit'. By the time I was an undergraduate, continental drift was generally accepted, but the mechanics of how it operated were still unclear. It had been observed that sediments in the Atlantic Ocean were thin and contained no fossils older than about 180 million years. Then it was found that most oceans showed evidence of a volcanically active mid-ocean ridge at which new oceanic crust was continually being generated, and the situation was finally explained in 1963 when it was demonstrated that magnetic minerals in the newly formed oceanic crust retain the polarity existing at the time of their formation, and that there have been repeated reversals of polarity (during which a compass needle would point to the South Pole), and these magnetic stripes are detectable on the sea floor. When plotted on a map they show a mirror image on each side of the mid-ocean ridge with the oldest crust furthest from the ridge. Since the end of the Cretaceous there have been at least 170 reversals in the Atlantic Ocean (the last one about 780,000 years ago), allowing the evolution of the sea floor to be charted in detail. Incidentally, the marine magnetometer was first developed for detecting the presence of submarines during the Cold War. The Atlantic Ocean has been opening at a rate of about 20 millimetres per year since the early Jurassic, which explains why the sea-floor sediments are thin and contain no fossils older than 180 million years. This finally provided an explanation for the movement of large masses of the earth's lithosphere, which by 1967 came to be known as plate tectonics.

So, by the time I became involved in the Round Africa project, our knowledge of the West African margin had advanced enormously since my Esso involvement thirty years before. Then there had been great optimism for finding major oil and gas reserves in the world's oceans. Now it was clear that the potential was limited to the shallow shelf seas and continental margins. Not only that, but it was now evident that the evolution of the Atlantic Ocean was reflected in the sediments preserved on the West African margin. There were rocks and structures formed before the opening of the ocean, those formed during the break-apart, and those formed after the break-apart. These concepts were of great importance in assessing the oil and gas potential of the area, and in assisting the interpretation of the seismic data to which my work was linked.

At the end of January, I went to visit my mother. I took her for a drive to the Pennine hills and valleys west of Halifax and returned past the house at Scar Head where my dad had lived as a boy, and then to Norland where they had courted in the 1930s. I thought it might lift her spirits to remember the days of her youth. Three days later at two o'clock in the morning I was awoken by a call from the police to say my mother was in hospital with a broken hip. She had fallen down the stairs, but fortunately was wearing her emergency bleeper which enabled her to call for help. The emergency services managed to gain entry and had taken her to hospital. We drove to Halifax and found her in the Royal Infirmary looking very frail, but quite calm. She said she had been going downstairs to fill a hot water bottle when she had tripped or had a blackout, and found herself at the bottom of the stairs and unable to get up. The following day as we drove to Halifax, she had a hip operation, and Irina and I went to see Social Services, who said they could provide home care and equipment when she came out of hospital. We also went to see some sheltered accommodation but were told that there were no units currently available. Susannah arrived the next day and we all spent the afternoon in the hospital with my mother before heading back to London, where I was scheduled to give a presentation to Mark and Paul the following day. No sooner had I arrived home than we had another call to say that my mother had taken a turn for the worse, and had an irregular heartbeat. Her condition was serious but not critical. Back I went again and found her better than I expected, still uncomfortable from the operation, but her heartbeat had stabilized and she had been cheered up by visits from two of her neighbours. I told her about the sheltered accommodation idea which she liked, so I submitted a request hoping that a one of the units would soon become available. By the next day she was sitting in a chair, but after ten days she was transferred from the Infirmary, which was a nice, bright, friendly hospital with views across the town to Beacon Hill, to a rehabilitation unit at the General Hospital in Halifax, a former hospital for bed-ridden patients of the Halifax Workhouse, a dark, grim Victorian building with no view at all. She was not happy, and it did not help her recovery process.

My mother's last year was very difficult and stressful. We persuaded her, with difficulty, that she should go into a respite home for a week or two on release from hospital. I think she suspected we were trying to make this a permanent arrangement. Despite having a nice room of her own, and sympathetic care, she hated it, and could not wait to get back home. I set up a bed for her downstairs and arranged for a carer to visit each day. Social Services provided meals and equipment to help her get about, and her next-door neighbour, Steve, kept an eye on her, but she was depressed and confused, to the extent that sometimes she did not know whether it was morning or evening. Each weekend I went up Halifax to be with her, and with no progress on the sheltered accommodation front in Halifax I applied in London, only to discover that as a non-London resident she was not eligible. We searched for a small flat in London that we could rent for her, or a larger place for us in which she could have a room, and we found two places, a nice townhouse by the river in Isleworth and a house in Surbiton with a wonderful garden. We submitted offers on both, but both our offers were 'gazumped'. I juggled these activities with work on the Round Africa project – at this stage on Cote d'Ivoire, the Niger Delta and Equatorial Guinea, which involved further travel to the office in Kent, and to Devon where our geochemistry consultant was based. It was an exhausting schedule.

In March I had a call from a former colleague at Britoil, now working as a consultant, who asked if I could give him a briefing on Libya. He had a yacht in Brighton Marina on the south coast and suggested it might be nice to have the meeting on board. We had our meeting but as it progressed I began to feel nauseous and threw up a couple of times. I must be the only person to become sea-sick on a boat moored in a marina. Actually, it wasn't seasickness, it was food poisoning, but I'm not sure they believed me. They said next time we had better meet in London.

My mother slowly became more mobile and I was able to take her to the shops and on small excursions around Halifax. After the ceaseless activity of recent months, I felt in need of a break and we had a long-standing invitation from several of our old Brega friends to visit them in Canada. I felt that with the help we had arranged my mother could be left for a couple of weeks, and so we flew off to Calgary, and had a pleasant time looking up our old Brega acquaintances. We visited Kent and Lois in their comfortable house at Springbank Hill, with views over the Elbow River valley west of Calgary. I checked Google maps recently, and the house is gone. Where it formerly stood is now a vacant lot.

They drove us to Banff in the Rockies where they had an apartment. The town is dominated by the enormous Banff Springs Hotel with over 750 rooms, built by the Canadian Pacific Railway in Scottish baronial style, in the days when rival railway companies vied with each other to create the ultimate in luxury and extravagance. We drove along the valley of the Bow River to Moraine Lake where the water is turquoise in colour due to the rock-flour from the nearby glacier. The setting, in the so-called valley of the ten peaks, is stunning. At nearby Lake Louise, there is another gigantic railway hotel built to lure wealthy passengers to vacation in the mountains. Despite the opulence we found it rather obtrusive in such a setting. Just north of Lake Louise the railway leaves the Bow Valley and crosses the continental divide at Kicking Horse Pass into British Columbia. The gradient to the west is very steep, so in 1909 they built the amazing spiral tunnels which make two complete loops through two separate mountains and reduce the gradient of the ascent. Following our excursion, Kent and Lois organised a Brega reunion barbecue to which they invited a dozen of our former Brega friends (half the ex-pat staff of Sirte's Exploration Department were Canadians).

We were fortunate in having good friends in Calgary, Edmonton and Vancouver, and for the next part of our trip we joined Mariola and Arthur, a couple we had known well in Brega. They took us to the Columbia Icefield and the Athabasca Glacier and to Jasper (where we stayed in a chalet once occupied by Marilyn Monroe). We white-water rafted on the Athabasca River, and rode the Sky Tram, and visited Maligne Lake, dammed at its northern end by a text-book example of a terminal glacial moraine. From there we drove to Edmonton (Deadmonton to Calgarians), where they lived, stopping off *en route* for a quick visit to the Miette Hot Springs. We saw all kinds of wild life – black bears, moose, caribou, bison, bald eagles. We covered 500 miles through some of the most spectacular scenery in Canada, and were much impressed – and grateful. Next day we took the scenic train trip through the Rocky Mountains from Edmonton to Vancouver. From Jasper, the route follows the valley of the Miette River to the Yellowhead Pass, then down the valley of the Fraser River and southwards to Kamloops and Vancouver. The train has an observation car, restaurant, club car and sleeping accommodation and takes 22 hours to complete the 725-mile journey. It was an interesting trip but the weather was poor, the views obscured by cloud and the most scenic part of the trip was passed through during the night. I had long had an ambition to travel the Trans-Siberian Railway, but after our Canadian experience, perhaps not. However, we met a guy on the train, travelling alone from Halifax, Nova Scotia to Vancouver, a distance of 3,800 miles, and he told us he did it every year.

They say there are two types of Canadians, those who live in Vancouver and those who wish they lived in Vancouver, and it is true that the location is impressive, although the US border is inconveniently close. The ferry from Sturdie's Bay on Vancouver Island to Tsawwasen in Vancouver actually travels through US territorial waters, and we had friends with a property on the peninsula south of Vancouver, which is actually in the US state of Washington. Vancouver is commonly known as Hongcouver because of the large number of Chinese immigrants in the city (500,000 ethic Chinese in Greater Vancouver from a total population of 2,500,000). The first Chinese appeared at the time of the gold rush of 1858, and stayed to work on the Canadian Pacific Railway, and large numbers arrived from Hong Kong, Taiwan and mainland China in the twentieth century. The setting of the city is beautiful - a large natural harbour on one side, False Creek in the middle, and the Fraser River to the south, all set against a backdrop of mountains. Our friends Tom and Dorota lived in West Vancouver, curiously in a house, like that in Calgary, which no longer exists. The entire neighbourhood was bought by a property developer and they were made an offer they couldn't refuse. They showed us the Lion's Gate Bridge, Stanley Park, Granville Island and Gastown, and the next day we took the ferry to Vancouver Island where we were met in great style by Dave Wilkie, a petroleum engineer from Brega, in his enormous 1959 Lincoln Continental. Dave drove us to Mill Bay to visit Bob McCrossan, my former boss in Brega, who had retired with his wife to a delightful house with its own private dock on Cowichan Bay. He had installed a 'geological' bathroom with granites and marbles he had selected himself. He spent his time sailing, wind surfing and monitoring the stock market - just as he had done in Libya.

Dave then took us to see his house at Cordova Bay, close to Victoria, which his wife Vesta ran as a very smart B&B. Each guest room was elegantly furnished with the curtains and covers and cushions all made by Vesta herself. Victoria is very English-looking city with red double-decker buses, a Gothic-revival Anglican cathedral, public gardens, the Empress Hotel, and a statue of Queen Victoria wearing a crown and full regalia. I suspect it is a deliberate attempt to highlight the differences between Canada and the neighbouring USA - although it is interesting to note that 80% of Canadians live within 50 miles of the US border. I left Irina with Mariola for a further couple of weeks. My route home from Vancouver to London followed the great circle over hundreds of miles of trackless barren tundra, over the Northwest Passage where Sir John Franklin's expedition came to grief in the 1840s (and where the wrecks of his ships were finally located in 2014 and 2016), over Baffin Island (the world's fifth largest island, but with a population of only 11,000), and the Greenland ice cap, where the thickness of the ice reaches almost 3,000 metres in places. If all this ice were to melt sea level would rise by five or six metres, inundating most of the world's coastal cities. I arrived home knowing that we faced an uncertain future.

Letter 53
I found my mother better than expected. She was able to do a few simple tasks with the aid of her helper, but she was not able to negotiate the stairs without assistance. I collected her from Halifax and brought her to London where Irina and Eva and Susannah were on hand to keep an eye on her. Each year Mark and Paul organised an 'outing' for their clients and associates, and that year they decided to go to the Le Mans 24-hour car race in northern France, and I was invited to join them. They hired a minibus and we headed to Dover, loaded with cases of beer and booze. We took the midnight ferry to Calais, and drove through the night to arrive at Le Mans in the early morning. There is a long tradition of Brits attending Le Mans, dating from before the War when Bentley's were dominant, and from the 1950s when Jaguars and Aston Martins won six out of eight races. Those were the days – Bentley is now owned by Volkswagen, Jaguar by Tata Motors of India and Aston Martin by a consortium of Italian, German and US investors. The race is divided into categories, depending on engine size. The winner is the car to cover the greatest distance in 24 hours (usually around 5,000 km). Each car is driven in turn by two or three drivers, and it is not unusual for a car to come in for a lengthy pit stop for major repairs. The circuit contains the famous Mulsanne Straight where the cars clock speeds of around 330 kph, and there have been several horrific accidents, one in 1955 in which a car catapulted into the crowd killing 83 spectators. The Le Mans 'tradition' involves camping near the race track, cooking barbecues and consuming large quantities of booze, while paying occasional visits to the track to check on the race. It is particularly impressive at night with the headlights piercing the night sky, the roar of the engines, and the brake discs and

exhausts glowing red hot. That year it was a battle between Toyota and Porsche. Toyota led for long periods until the suspension broke on their last remaining car an hour before the end, and Porsche's took both first and second place. Of the 47 cars which started, only 23 completed the race. It was a unique experience.

Life continued much as before. I continued my work on West Africa, extending southwards from Nigeria to Gabon, Angola and Namibia, and we continued to look for more suitable accommodation in London, in Richmond (where our friends Lilijana and Jovan were now living), and even further afield, but still without success. I commuted regularly to Halifax and my mother was still capable of fixing breakfast, and of accompanying me to the shops, and I often took her to London. On one occasion, I drove her and my aunt Eva to Windsor and we went to Victoria Barracks to see where Eva and Dick had lived forty years before. The gate was guarded by soldiers of the Welsh Guards (the guards' regiments are distinguishable by the pattern of buttons on their tunics), armed with impressive-looking assault rifles (this was during a period when the IRA was still active), but when we explained our mission, they let us in. We found it barely recognisable. The layout was the same, but all the buildings were new. The former parade ground was now a car park, the flat where my aunt and uncle had lived, forever etched on my mind as the place where we had received the shocking news of my cousin's polio, the Sergeant's Mess where my mother had proudly played the piano to entertain the troops, and the Aladdin's Cave of my uncle's quartermaster's store, were all gone. *Sic transit gloria mundi.*

For my mother's 81st birthday I decided to take her to a leisurely five-day trip to Wales, and intended to visit Llandrindod Wells where my dad had been commissioned as an officer during the war, but the weather was dreadful. We got as far as Oxford and decided there was no point in continuing. Instead we took some shorter trips to Burnham Beeches, Leith Hill and Epping Forest. We made a nice birthday reception for her at our flat with Eva, Margaret, George and Susannah. It proved to be her last hurrah.

Your dad arrived back from Namibia with a badly infected wound on his back which showed no signs of healing, and eventually we had to take him to hospital to have it properly treated and dressed. His company had changed their focus from Namibia to Tanzania, where there was a lucrative trade in tanzanites, a lovely cornflower-blue gemstone found only in Tanzania. They set up a small office in Arusha where they acquired rough tanzanites from small mining operations or from individual dealers, and sent them to India to be cut. As he was travelling to Arusha, Al Qaeda carried out truck-bomb attacks on US embassies in Nairobi and Dar es Salaam in which over 200 people were killed - not a good omen for the start of his new assignment.

In August that year I went for a routine medical check which showed an abnormal ECG result which required further investigation with an angiogram. The other patients in the recovery room were discharged one by one, but when the cardiologist came to me he said 'I'm afraid with you there is a problem'. I had partial blockages in all four coronary arteries. He said I needed surgery, and when I asked if this could be deferred until later in the year, he said 'Certainly, but without an operation your chances of surviving until this time next year are 50:50'. That grabbed my attention. Two days later I drove to Halifax, and whilst preparing lunch I collapsed – out like a light. One moment I was talking to my mother, the next I was coming round on the floor. I had no idea how long I had been unconscious. The curse of Unas all over again, I thought, but it was more likely the effects of the angiogram anaesthetic. I dared not drive back to London, so Irina had to come to Halifax by train and drive me back.

A week or so later my mother collapsed in the garden and was taken to hospital again. She was confused and depressed, and I suspect she had had a stroke. I desperately contacted housing associations in both Halifax and London but nothing was available. We found a nice large flat in Warwick Square with a spare room that my mother could have had, and we immediately submitted a bid, only to find that the flat had been let the previous day. I called the owner who said she would have accepted our offer but the letting agreement had been signed and was legally binding. I was frazzled – trying to take care of my mother, driving to Halifax, looking for accommodation, trying to do my job, and preparing for a major operation. Our friend Mariola came over from Canada to stay with us for a week or so. Her son had been diagnosed with cancer which at first seemed to be responding to treatment, but now she received news to say that the cancer had spread. If there is a tide in the affairs of men, ours was definitely on the ebb, but we had not yet reached low tide.

I went for my heart operation at the London Bridge Hospital – having first made my will and put my affairs in order. I called Jonathan to let him know, and he said that he was near the top of Mount Meru, a 15,000-foot volcano in the Arusha National Park in Tanzania. Lucky for some! A heart by-pass operation is complex. Replacement veins are obtained from the legs and chest. The chest and ribs are opened up and the heart is stopped, using a pump as a temporary replacement. The blocked arteries are by-passed with the replacement veins which are carefully grafted in place. The heart is (hopefully) re-started, the coronary arteries checked for leaks, and the chest sown up (I still have wire loops in place which were used to join my rib cage together). The operation took about four hours, and I was then taken to the Intensive Care Ward for 24 hours. They got me up and walking as soon as possible and assured me that 'the stairs are your friend', and after a week I was allowed home. A week later I was walking the streets, and within a month I was able to climb to the top of Box Hill in Surrey, courtesy of our friends Tony and Laleh, who generously put their house in Oxshott at our disposal for my recuperation. But as I was improving my mother was failing. She fell several times and could not get up, and had to be readmitted to hospital. She was moved to a Rehabilitation Unit, which at least was a nice, bright, modern hospital, but of course she couldn't wait to get out. In the midst of this we received news from Canada that Mariola's son Raph now had a tumour on his spine, which was inoperable. He died, aged just twenty, a week before Christmas.

I slowly eased back into work (now on Morocco, where I had worked thirty years before), and I must say Mark and Paul were very understanding about my personal problems. We were invited to spend a restful weekend with Mark and Gavrielle (and their two sheepdogs) at their home in Kent, but then, just before Christmas, my mother fell again on a polished floor in hospital. They did an x-ray and reported nothing broken, but they were wrong. She had fractured her hip. She was in constant pain and could barely stagger about on her Zimmer frame. I took her home for a weekend, and then again for Christmas, but it was a disaster. I simply could not cope, and on Boxing Day I had to call an ambulance to take her back to hospital. They carried her out on a wheelchair, and I knew she would never return to the house where she had lived for sixty years. After several weeks, they finally realised that her hip was fractured, and had to do another painful operation. There is something worse than death, and that is not having anything to live for, and by now she had reached that stage. Eventually I arranged for her to be brought to a nursing home in London, despite the fact that I had promised I would never do that. She lingered on for a few weeks, sometimes thinking she was in our flat, and sometimes realising where she was. She stopped taking her medication and in March had a massive stroke. She was taken to St Thomas's Hospital in London, but never regained consciousness and died a week or so later.

> *Fear no more the heat o' the sun,*
> *Nor the furious winter's rages;*
> *Thou thy worldly task hast done,*
> *Home art gone and ta'en thy wages.*

The funeral was held at St Gabriel's Church at the end of our square with most of the family in attendance plus some of our closest friends. We had a reading of Christina Rossetti's *Does the Road Wind up-hill all the way*, and some of her favourite hymns. We buried her at Putney Vale cemetery on a bright crisp day, just as the first signs of Spring were beginning to appear.

I felt an awful sense of guilt. I should have done more. I could have done more. I stayed overseas when she begged me to come home. I was her only child; she had sacrificed everything for me when I was young, I did nothing for her when she was old. Firstly, I had failed my children, now I had failed my mother. I felt that I had reached rock bottom.

Looking back now, twenty years later, I can't avoid the conclusion that she had a tragic life. She was a good woman who had more than her share of adversity. To lose your husband at 27, with a young son to bring up, to live alone for forty years after your son left home, and still keep your sanity shows great fortitude and courage. She lost heart towards the end when she could no longer cope and when it was obvious that I was not going to look after her full-time, and her despair must have been awful. It will haunt me for the rest of my life.

Letter 54
But, as they say, life goes on. My final large project for Exploration Geosciences was on Alaska, an area totally new to me, but Mark and Paul had worked on the Beaufort Sea and the North Slope previously. We had a large amount of data from wells drilled in the area, and regional seismic data and our task was to try and identify undrilled areas with oil and gas potential. The problem was that some of the most attractive areas were off-limits, as the US government had designated part of the North Slope as a National Petroleum Reserve (originally a strategic oil reserve for the Navy), and another part as the Arctic National Wildlife Refuge, with yet

other areas reserved as Native Lands. Despite this, the area had produced the largest onshore oil discovery in the United States when the Prudhoe Bay oilfield was discovered by Arco in 1968, which at its peak produced 1.5 million barrels per day. The western part of the field was operated by BP, and when Arco was acquired by BP the entire field came under BP operatorship. An 800-mile pipeline was laid over the Brooks Range reaching an elevation of 4,600 feet, across environmentally sensitive permafrost in central Alaska, and through the Thompson Pass to Valdez, a former gold-rush port on the south coast of Alaska. It was near here that the *Exxon Valdez* ran aground in 1989 and spilt half a million barrels of crude oil into the pristine waters of Prince William Sound.

We identified a number of under-explored areas with good potential and others which had undeveloped potential, but frustratingly (for us) the evidence suggested that the best potential was out of reach on the NPRA and ANWR. We thought that the powerful Republican oil lobby, might obtain some relaxation of drilling restrictions, but it was not to be, and the reserves in these areas remain untouched.

We presented our findings at a three-day meeting to potential clients at Kelowna in the Okanagan Valley of British Columbia. It was a good choice of location – away from the bustle of Calgary, and in a delightful setting beside a lake, and surrounded by vineyards. The meeting went well and several companies expressed an interest in our report. At the meeting, I met a French-Canadian Exploration Manager called Jean-Claude Beauvilain, who quizzed me about opportunities in Libya and asked if I could give a brief presentation to his company in Calgary, which I did following the meeting. I was also able to look up some of my old Brega friends in Calgary. Rigel, the company for which Jean-Claude worked, also had an office in London and he asked me to give further presentations on Libyan potential during the following weeks (at which I found Mike Ridd, my old friend from BNOC, who was acting as a technical advisor to Rigel). An economic recession in the Far East led to a collapse of the oil price at the end of 1998 to only $11.48 per barrel, and Exploration Geosciences felt the pinch. I suggested that they might consider producing a promotional report on Libya similar to that on Alaska. They considered this proposal but decided against it as the area was outside their area of expertise and because of the political problems in Libya. It was a mistake. The situation in Libya was about to change for the better. The Lockerbie suspects were handed over for trial in April 1999, as a result of which the UN sanctions were lifted and Britain re-established diplomatic relations.

In order to stimulate interest the Libyan National Oil Corporation offered large tracts of land for licensing to foreign companies, and many companies showed an interest.

I continued working for Exploration Geosciences through the summer on small projects on Yemen and Mozambique, but it was clear that they did not have sufficient work to keep me employed. We agreed that I would continue to provide consultancy services as required, and in October I became an independent consultant – the first time in my life I had worked for myself rather than for a company. It was the start of a new career.

Jonathan and Susannah were on the move too. Jonathan was relocated from Tanzania to the Virgin Islands which was a major outlet for the semi-precious stone company he worked for, and he became more and more involved in establishing computer systems for marketing the stones, and after a short time he was moved to New York where he continued working on web-site design. He rented an apartment in a quiet street in Astoria, within easy reach of Manhattan, which he furnished with a rather Spartan simplicity. Susannah completed her qualification and became a Chartered Surveyor and she and Alex moved out of their small flat in St George's Square and bought a house in Wimbledon. Alex by this time had switched from the public sector to a private estate management company called Richard Ellis which managed both commercial and residential properties across London and beyond.

Our flat in Warwick Square was too small to live there permanently so our quest for a larger place continued. Fortuitously our Serbian friends Lilijana and Jovan, now with two young sons, decided to move from their flat in Richmond to a larger house. We had always thought that Richmond was too far from the centre of London (Irina has always been a city girl, and was used to shopping on King's Road), but we decided that we could buy their flat in Richmond, and let our flat in Warwick Square, which is what we did. We quickly discovered that Richmond is a delightful place, with a large park stocked with deer, a beautiful, Arcadian riverside, and that ten miles is actually not that far from London.

We spent the summer clearing my mother's house in Halifax and preparing the flat for our move to Richmond. There can be no sadder task than sorting through the cherished possessions of a loved one. It felt like an intrusion, a desecration, almost. The jewellery that she had proudly worn on her cruises, outdated dresses from long ago, an ill-chosen green coat which never suited her, letters from my dad sent during the war, photographs of a happier Eden, childish souvenirs from my schooldays, tea and coffee in the cupboard, her yellow gardening gloves, a purse of coins she would never use, her favourite china cup – all now useless and unwanted. Irina and I went through her worldly possessions. We gave much of it to Steve, my mother's next-door neighbour who had helped her during her final years. He was a habitué of car boot sales, and had once sold a pair of Russian felt boots I had given him, claiming that they had been worn by a Russian soldier at Stalingrad. The rest of my mother's things we sent to charity shops in the town. It was a sad end to a sad life. I put the house for sale and it was sold just before Christmas, ending my association with a much-loved house that my mother had occupied for sixty years, and with the town of my birth.

Curiously amidst all this turmoil, I received a letter from the Foreign Office informing me that the Russian Embassy had lifted the ban on my travel to Russia and I was now free to travel there. Yeltsin was in power at the time and the country was in a state of flux. Irina had recently returned from visiting her family, so we decided to defer using the new authorisation until the following year. It was a mistake. By the time I decided to go, Putin was in power and the door closed again. What else do you expect from a former KGB officer? I was obviously still a danger to the state. Curious, when you think of some of the people they allow in.

Nevertheless, we felt in need of a break, so we took a short trip to Andalusia in Spain and stayed in *pardores*, historic buildings converted into modern hotels – surely one of the most inspired tourist initiatives in Europe. I had been impressed by Arabic architecture and decorative art in North Africa, but to my mind the zenith was reached during the years of the Moorish occupation of Spain, in Seville, Cordoba and Granada. This was the golden age which saw the building of the Alcazar in Seville, the Mezquita or Great Mosque in Cordoba and the Alhambra in Granada, and they are stunning. There are other attractions too, in Seville the eighteenth-century Plaza de Toros and opulent Royal Tobacco Factory were made famous by Bizet's fiery opera *Carmen* (there is a statue of her too). In Cordoba the Moors built the wonderful *mezquita* (mosque) with its hypostyle hall, 850 pillars of jasper, onyx and granite (salvaged from earlier Roman buildings) with the *voussoirs* painted in red and white. This was brutally vandalised after the *reconquista* by having a heavy, dull Christian cathedral erected in the middle of it. When Charles V saw the result, he commented that 'they have destroyed something unique in all the world and replaced it with something you can find in any city'. When Moslems requested permission to use the mosque again in recent times they were refused by both the Spanish authorities and the Vatican. In Istanbul, of course the situation is reversed, and the famous Byzantine basilica of Hagia Sophia became a mosque. In Algiers the Ketchaoua mosque became a cathedral for 120 years and then, following independence, became a mosque again.

From Cordoba we drove up the valley of the Guadalquivir through Villa del Rio and Andujar, locations which were to prove of great significance a few years later, to Ubeda where our *parador* was a former palace, and then to Granada. We spent a whole day at the Alhambra ('red' in Arabic, which is entirely appropriate, since the entire edifice is built of red sandstone). It is just breath-taking. It was here in 1492, immediately after the *Reconquista*, that Christopher Columbus obtained approval (and funding) from Ferdinand and Isabella to embark on the first of his famous voyages. In addition to the incredibly intricate carving, like lacework in stone, there are wonderful gardens, one called *Jardines del Paraiso*, a tautological expression in Arabic, as garden and paradise are one and the same. Adjacent to the fortress is the Generalife, a summer palace with a wonderful 'Persian' garden, a design known from Spain to India, where there is another fine example at the entrance of the Taj Mahal.

We spent a night at Malaga where we were able to watch a bull fight from our hotel balcony on the hill behind the town, and next day drove into the Betic mountains to find the *Caminito del Rey*, a precarious balcony traversing a vertical cliff over a vertiginous gorge. It had been built for King Alfonso XIII in the 1930s and clearly not maintained since then. The handrail was bent and rusty, boards were missing from the walkway and the whole thing sagged alarmingly. There were danger notices posted at the entrance, and after a few gingerly steps I retreated – it was a potential death trap. But I see that since then it has been restored and reopened. I'd better go back and try again.

We drove on to Ronda, a town spanning a spectacular gorge called *El Tajo* or The Cut. It is a setting very reminiscent of Constantine in Algeria, and the bridges over the gorge are equally impressive. Ronda is also the place where the rules of modern bull-fighting were established in the late eighteenth and early nineteenth centuries, and there is an impressive *plaza de toros* in the town, which of course attracted Ernest Hemingway. We ended our trip at Jerez de la Frontera, named from the time when it stood on the frontier between Arab and Christian Spain, and sampled some of the local brew, known as sack during Shakespeare's time (maybe because Sir Francis Drake sacked the port of Cadiz in 1587) but now known as sherry. Sir Francis captured almost 3,000 barrels of the stuff, which he brought back to England, where it has been popular ever since.

Later that year we visited Cuba. I don't want this to become a travelogue, but I can't resist a few comments about the trip. This was a time when Fidel Castro was still firmly in control. There was no contact between the US and Cuba, politically, economically or culturally. For forty years Cuba's main ally had been the Soviet Union (and to a lesser extent China), and Cuban sugar, tobacco and agricultural products were traded in exchange for large scale construction assistance and military aid. This worked fine until the collapse of the Soviet Union in 1991, but then Soviet troops were withdrawn from Cuba, economic aid dried up, and the country entered a period of harsh austerity. By the time of our visit Cuba had established links with some of its South American neighbours, and had been forced to accept Spanish investment in order to rebuild Cuba's moribund tourist trade. We found Havana to be very run-down, but the people extremely friendly. Nowhere else could you find Cadillacs, Chevrolets and Studebakers from the 1950s, patched and battered, but still operating as taxis; nowhere else would you find families offering a room to paying guests in order to provide extra income. Isn't it ironic that nowadays, when relations between the US and Cuba have thawed a bit, Americans are buying up the old Cuban cars and taking them back to the States, where they are now collector's items? On walls throughout the city there were painted slogans invoking *Hasta la Vittoria Siempre*, and images of Che Guevara and José Marti, but very few of Fidel. We visited a park where there are relics of the 1958 revolution in which the corrupt Batista government was overthrown by 82 revolutionaries who unsportingly arrived during the New Year's Eve celebrations. You can see the motor launch *Granma* (built to accommodate only 12 people) which brought Fidel, his brother Raul, Che Guevara and their intrepid band from Mexico. In addition there are two light tanks, a red delivery truck labelled 'Fast Delivery S.A.' used in the assault on the Presidential Palace, and two small planes, looking like something from the Spanish Civil War. Is that all you need to start a revolution?

The other big hero in Cuba is Ernest Hemingway, who is revered because he defied the US ban and continued to live in Cuba until 1960. There is room 511 in the Hotel Ambos Mundas, which he occupied from 1932 to

1939, his house, *Finca Vigia* (Lookout Farm) on a hillside outside Havana, where he wrote *Old Man and the Sea* and a large part of *For Whom the Bell Tolls,* and his favourite bar at *Floridita's*, serving of course his favourite daiquiris to tourists like us. Hemingway had an adventurous life. He was a journalist, novelist and big-game hunter. He was an ambulance driver in Italy in World War I, and a journalist in the Spanish Civil War and the Second World War, where he was present at the Normandy landings and the liberation of Paris. After the war, on one of his game-hunting trips in Uganda, he was involved in two successive plane crashes when the first plane collided with a telegraph pole, and the second, taking him to hospital, burst into flames on take-off. He arrived at the hospital in Entebbe to find his obituary in several local papers. In the end, seven years later, suffering from heart disease, depression and paranoia, Hemingway took his own life at his home in Idaho.

In our hotel there was a porter who spoke Russian, and Irina got into conversation with him. He had been a student in Moscow and had qualified as an engineer. When she asked him why he was working as a hotel porter, he said he could make more money from tips than he could earn working as an engineer in Cuba.

Letter 55
We saw in the new millennium with our friends Karen and Richard at their flat near Vauxhall Bridge from where we had a grand-stand view of the firework displays and celebrations. The roads and bridges below were completely choked with a sea of revellers, and on the following day the street cleaners removed 22 tons of champagne bottles – and that was just in Westminster. But the celebrations were overshadowed by an awful tragedy which struck just a few days later.

In the aftermath of the collapse of communism in eastern Europe many state assets were put up for sale very cheaply and Branco, our great friend from Brega, had discovered that a state-owned leisure company in Croatia, where he lived, was selling off its fleet of Bavaria and Adria boats. He persuaded his family and friends to invest in the venture, and they began buying boats and setting up a boat hiring business. He decided that once the business was up and running he would quit his job in Brega and devote himself full time to the company. At the beginning of January, he came to the London Boat Show and stayed with us in Richmond. From there he returned to Brega for what would probably have been the last time. On the news that night we heard that a plane belonging to Sirte Oil

Company had crashed into the sea approaching Brega, and that several of the passengers and crew were reported killed or missing. Had Branco travelled back that day? Had he been on that plane? Even if he had, he was an excellent swimmer. Surely, he would have survived? We called Smilja, Branco's wife in Zagreb, and the phone was answered by a friend. Was he on that flight? 'Yes'. Was he OK? 'No'. What do you mean, 'No'? 'He was killed in the crash'. It was a brief, stunning conversation. We had been laughing and joking with him 48 hours earlier, and talking about his plans for the future. We eventually learned what had happened. The aircraft was a 40-seater turboprop plane built by Short Brothers in Northern Ireland. It was normally used for flying personnel from Brega to the oilfields, but in view of the large number of incoming staff after the New Year break it had been diverted to help transport them from Tripoli to Brega. When used for flying to the south the pilots had not needed to use the de-icing equipment, but flying for an hour over the sea in winter was a different matter. However, the pilots had not used the de-icers, the fuel lines had iced up, and when coming into land the melting ice had killed both engines. The plane broke up when it hit the sea, the tail section smashed into the front part, and everyone in the first eight rows was killed, 22 in all, including Branco and several others we knew well. We went to the funeral in Zagreb which was attended by several hundred people and it was an extremely emotional experience. Not an auspicious start to the new millennium.

We went to visit Jonathan in New York that Spring at his apartment in Queens, and he took us to Staten Island and Wall Street and to the top of the World Trade Center. We visited the Guggenheim and the Metropolitan Museum and took your dad to a Wagner and Mozart concert at the Lincoln Center. I don't think he was much impressed, but he enjoyed the late-night supper afterwards. We very briefly met your mom one evening, and we thought her very pretty with her petite figure and sleek Burmese hair, and wanted to get to know her better, but she was very shy and would not join us for dinner. We took a helicopter flight over Manhattan and did the usual tourist attractions, including climbing up to the corona of the Statue of Liberty, and Jon introduced us to Katz's Deli on East Houston, which I discovered is pronounced Howston in New York. The Texans would not be impressed. Just down the street is Red Square, a reddish apartment block topped with a large statue of Lenin, salvaged from a dacha near Moscow. The building has a clock with the numerals in random order, presumably to suggest the confusion which reigned in Russia following the collapse of communism. Jonathan visited us in

Richmond a couple of times that year. It was a rare event, so we killed the fatted calf, but could not help wondering whether he would ever return to England.

My contact with Jean-Claude and Rigel looked promising, but it was not a full-time job, and in casting around for other possibilities it occurred to me that this might be an opportune time to write a book on the petroleum geology of Libya. There were a number of reasons for this. There was no existing book on the subject, and I felt that I had sufficient knowledge to undertake the task. But more than that, the country had large proven oil reserves and future potential. UN sanctions on Libya had been lifted and diplomatic relations restored with Britain and several other countries (but not the USA), and after thirteen years of austerity the Libyan government was desperate to encourage foreign investment. In 2000 NOC appropriated inactive acreage from state companies and added this to the portfolio of acreage available to foreign operators. Over 100 exploration blocks were made available, making this the largest offering of acreage in over 30 years. This was a real window of opportunity. I contacted Elsevier, a well-known publishing house of technical books, and submitted a proposal. I was successful in convincing them that this was a worthwhile project and signed a contract in January 2000. Now began the serious task of collecting data and planning how the book would be organised.

In September Rigel Energy was acquired by Talisman, a large integrated oil company based in Calgary, and Jean-Claude was taken on as Team Leader of their Middle East/North Africa New Ventures Group. I was invited to work with them as a Consultant. I made several presentations to their management, and they asked me to prepare reports giving an overview of the oil industry in Libya, the present state of exploration, licensing and operating conditions, the role of NOC and the Libyan national companies, the major oil producing basins, recent discoveries, field development and farm-in opportunities, offshore exploration and so on. The attraction to companies like Talisman was that US companies were prevented from participating, as Libya was still regarded as a rogue state which supported terrorism. So, I worked the two projects in parallel, producing reports and presentations for Talisman, and gradually assembling the book chapter by chapter.

With Talisman we slowly homed-in on certain areas and concepts and I gradually gained the confidence of Talisman management as an expert on

Libyan petroleum geology – my long years of work for SOC were finally paying off. We examined all the available open acreage, field development opportunities and farm-in possibilities. In September, I accompanied them on a visit to NOC in Tripoli to outline our ideas and to discuss a possible bid, and we were well received by NOC. Following the meeting we selected a short-list of blocks which we felt offered the best potential (and they were excellent blocks, some of which yielded discoveries in future years). By the following spring we had made our block selection and arranged a final meeting with NOC. But just at this point, out of the blue, I received a call from your dad inviting us to his wedding with your mom in New York in three weeks' time! And at exactly the time I had committed to be in Tripoli. I asked him if he could change the date, but he said no, it was all arranged. Of course, I wanted to be at the wedding, but this presented me with a serious dilemma. I had agreed to accompany Jean-Claude and the others to Tripoli weeks before. It was the culmination of two years' work and I felt if I pulled out now it would reflect badly on my commitment to the project and probably mean the end of my consultancy with Talisman, and perhaps with other companies too. And so, as your mom and dad were married in New York, I was four and a half thousand miles away in Tripoli. Another error of judgement? Probably. I suspect your mother felt that I did not approve of the marriage. If so, she was wrong. I think she brought exactly the stability that Jonathan needed after his peripatetic wanderings, and provided a safe and secure environment in which to raise a family.

Following the plane crash in Brega, in which Branco died, a class action had been raised by a firm of London lawyers attempting to secure compensation on behalf of the victims' families. I suggested to Smilja, Branco's wife, that she should join this action, and I helped her to prepare a submission to the lawyers. Smilja invited us to visit her in Croatia and we went there in September 2001. We went to see Branco's grave in Zagreb, and then flew to Zadar on the Adriatic coast to see the fleet of boats which they had bought, and which Branco had intended to manage. By then they had sixty or seventy boats and the business was beginning to take off. Branco's daughter Lana drove us to Vodice where they had a family house, and his son Alan took me sailing in his Elan 19, a popular boat for pottering about in the sheltered coastal waters. Smilja loaned us her car to tour the coastline. We drove to Dubrovnik, crossing a narrow strip of land, the only part of Bosnia to reach the sea, which since Croatia joined the European Union, has become a major source of friction. The corridor dates from the time when the Republic of Ragusa (Dubrovnik) was forced to cede it to the Ottoman Turks. This is a troubled part of the world. Dubrovnik is located close to the border with Montenegro, and as Yugoslavia fell apart in 1991 the city was shelled and besieged for seven months by Serbian and Montenegrin forces. The old town is located on a peninsula, very picturesque, all umber and lime, and surrounded by a formidable defensive wall. Towards the end of our visit we were joined by Zbigy and Mariola who had driven from Poland. We all went off to Split to see Diocletian's Palace, a rambling fortress which from the seventh century was progressively occupied by the locals, who built houses and shops in the courtyards, rooms, walls and basement of the palace. Today it is a jumble of old and new, with parts of the former palace cheek by jowl with shops and cafes and souvenir stalls. We stopped off for dinner in Troghir, a peaceful little coastal town, and drove back leisurely along the coast road, to find Smilja's place in a state of shock. It was 9/11 and we watched in disbelief as the first television pictures came in. The year before we had stood on top of the World Trade Centre. Now we saw it collapse.

We went to New York for your dad's birthday the following year. Your mom and dad had set up home in Forest Hills and Jon arranged for us to meet your mom's dad and sister over dinner at a little place in Greenwich Village. Everyone was agreeable, and polite, and sizing each other up, but your grandpa's English was not very good, and our Burmese was zilch, so the conversation flagged a bit. Your mom's sister, on the other hand, was very well adjusted to the American way of life, and determined to live life to the full. Your mom of course had her Burmese name, Shwe Zin Pe, but that doesn't roll too easily off the tongue, so she was generally known as Jenny. They had planned to take us to a Burmese restaurant for your dad's birthday dinner, but first we went to a little bar on 71st Street for drinks, and in the end decided to eat there. It was a mistake. The food was indifferent; Burmese would have been much better.

I continued work on the book. There was plenty of material available, scores of publications in various scientific journals, proceedings of symposia and conferences, many of which were held in Libya, maps and reports published by the Industrial Research Centre in Tripoli, and of course the experience gained from my ten years in Libya where I had been exposed at various times to regional studies, producing fields, concession evaluations, laboratory analyses, and statistical compilations with of course some knowledge of how NOC operated. It was simply a question of bringing all this material together in a coherent form, producing maps and charts to accompany the text and trying to produce a book that would be useful to companies planning to apply for acreage or field development opportunities in Libya, and to academics and financial advisors. I sub-contracted the drafting of maps and charts to a number of draftsmen that I knew, which was expensive, but I was fortunate in securing funding from Talisman and from a couple of smaller companies I had worked with. I was also able to obtain a very nice colour satellite image of Libya for the cover, on which you could actually see the flares from some of the producing oilfields.

Elsevier had requested that I presented my final draft in page-ready format, that is with the text and illustrations organised exactly as they would appear in the book. I was fortunate to have the services of Cathy Hickey, the draftswoman at Exploration Geosciences, who had the skills necessary to bring it all together. Elsevier gave me complete freedom to assemble the book as I thought most appropriate, but I had to be very careful not to include any confidential information belonging to Sirte Oil Company. I heard subsequently that Taleb had appointed a committee to go through the book line by line to check that I had not done so. I sent copies to NOC but I did not ask for their prior approval, as I knew that I should not get it. In fact, I think I did them a favour. The book gave a very positive picture of Libya as a major petroleum province with lots of remaining opportunities, and as I was to discover later encouraged all kinds of companies to seek an involvement in the country. The book was published in February 2002, and sold like hot cakes – no, not really, but it received some complimentary reviews, and sold about a twelve hundred copies over a period of years.

I wanted to acknowledge Talisman's assistance with the drafting costs by putting their logo on the title page, but after some deliberation they declined the offer. It turned out that they were beginning to get cold feet. Talisman is a Canadian company but they had many US investors who apparently were uneasy about Talisman's involvement in places like Sudan, and as far as Libya was concerned the US still had in place legislation which blacklisted companies which did business with Libya. Despite the fact that NOC had agreed terms with Talisman and had even printed maps showing the blocks ear-marked for Talisman, the deal was never signed. It was an excellent commercial opportunity sacrificed on the altar of political pressure. Not long afterwards the US ban was lifted, and US companies returned to Libya and acquired acreage all across the country.

Susannah meanwhile continued to work for the Valuation Office, part of the omnipresent HMRC (think IRS), still based in Finsbury Square in the City of London, but in 2001 she became pregnant and had her first baby at Kingston Hospital in March 2002. It was a difficult birth and she eventually had to have a caesarean section. James was Sue and Alex's first child and our first grandchild, and we celebrated in the time-honoured fashion. Shortly afterwards my long-time friend and colleague Mike Ridd was married in London to a very talented Japanese lady called Mikiko. She was a professional singer and proved to be a very capable choir master and conductor too. Mike had a riverside flat overlooking the Hurlingham Club in London, plus his Scottish *dacha* near Loch Lomond and he loved nothing more than planting trees, digging ponds, or building a shed. He was a multi-faceted man, and we were happy to see them settle down together. Irina quickly adjusted to life in Richmond. She joined (and became very active) in the local tennis and bridge clubs, and she became a County Master in bridge. She soon established a circle of friends, including Zora, a former professional ice-skater born in Prague, and we also kept close contact with our old friends from Aberdeen, Tony and Laleh, and Lilijana and Jovan from Brega. We got to know 'the matchless vale of Thames' and King Henry's Mound in Richmond Park, from where there is an uninterrupted view of the dome of St Paul's Cathedral ten miles away, which is protected by Act of Parliament. We frequented Kew Gardens and Hampton Court and Windsor Great Park, Strawberry Hill and Syon House, and then of course there was London. Life after Brega was not so bad.

Shortly after the book was published Mark and Paul asked me if I would join them to provide assessments and advice to a rather shadowy company which was looking into the possibility of becoming involved in Libya. They had offices in London and Amsterdam, a British Chief Operations Officer, Russian Vice President and American CEO. We made a number of presentations to them and accompanied them on visits to NOC in Tripoli and to ETAP (Tunisia's national oil company) in Tunis. We pointed out that NOC had very experienced negotiators whose only aim was to obtain the most favourable deal for Libya, but they told us that was not a problem, they had contacts 'at the very top'. How often have I heard that? Each time we reached a decision point they would tell us they had to consult their shareholders. It turned out that the company was owned by ex-Soviet oligarchs who controlled most of the mineral industry in Kazakhstan and they were looking for ways to diversify into oil. The number of shareholders could be counted on one hand. Their aim was to acquire promising acreage or producing fields in Libya, make a quick profit and then move on. Needless to say, it did not work out.

Letter 56
We had an open invitation to visit our friend Zbigy and his family in Poland, so that summer we decided to pay him a visit. Zbigy worked as a Fire and Safety Officer in Brega (and despite all the recent turmoil, he is still there). He has spent most of his working life in Libya and had accumulated sufficient wealth to build a large, modern detached house in his home town of Siedlce, 90 kilometres east of Warsaw and close to the borders with Belarus and Ukraine. This is a troubled corner of Europe and the town has been repeatedly sacked - by Swedes, Russians, Cossacks, Tartars, Czechs, Germans – you name it, they've been there, and its ownership, like pass the parcel, has gone from the Habsburgs to Poland to Russia, and back again. They suffered pogroms under the Czarist government, and ethnic cleansing by the Nazis. The town had a Jewish population of 15,000 before World War II, who were all rounded up and sent to Treblinka, and exterminated there, and the Jewish presence in the town has never been restored. Under the communists the inhabitants merely suffered austerity and a low standard of living. Poland truly has had a tragic history.

Warsaw suffered particularly badly during the war (as portrayed in movies like *Kanal, The Pianist* and *Warsaw '44*). It is still possible to trace the outline of the sealed-off ghetto where 400,000 Jews were crammed into an area of just 300 hectares, most of whom were shipped off to Treblinka in the summer of 1942. The following year, the Nazis tried to clear the remaining residents of the ghetto, but by now the Jews had learnt the fate of the deportees, and they put up ferocious resistance. It took four months and 2,000 troops to finally overcome the revolt. Fifty thousand Jews were killed and the entire ghetto, including the famous Great Synagogue, were razed to the ground. After the war, new buildings were erected, which stand over a metre higher than the rest of the city, since they were built on the rubble of the old ghetto. And of course, the story of the Warsaw Uprising of 1944 is well-known – how the Polish resistance rose up against the Nazis when the Soviet forces appeared on the banks of the Vistula, how they expected the Soviets to drive the Germans out of the city, how Stalin halted Rokossovsky's Belorussian Front for sixty days, allowing the annihilation of the (western-sponsored) Polish Home Army, and how at the end of the war, 85% of the buildings in Warsaw lay in ruins. By the time of our visit the city had been rebuilt, the Old Town, the Royal Castle, the famous Sigismund Column – all except the Great Synagogue. In 1939 there were 3.2 million Jews in Poland, in 1945 100,000, and in 2010 3,200.

Zbigy was amazingly hospitable. He devoted ten days to driving us around Poland, from the Baltic to the Carpathians, and from Warsaw to the border with Belarus. There is a subtle change in the landscape 150 kilometres north of Zbigy's home town. The countryside becomes dotted with lakes and forests, and the towns look more German than Polish. That is because we had crossed into what used to be East Prussia before the war, a land of wealthy, Junker families with huge estates worked by peasants who were little better than serfs. This was Masuria, a historical battleground where Teutons, Poles, Lithuanians, Russians and the odd Ruthenian and Moldavian slugged it out over the centuries. Napoleon passed through here in June of 1812 *en route* to Moscow with 680,000 men, and back again in December with only 120,000. In August 1914, two Russian armies invaded East Prussia, but were unable to link up due to the difficult terrain, and General Samsonov's Army was surrounded at Tannenberg and annihilated. This was regarded as settling the score for the destruction of the Teutonic Knights at the same location 500 years before. In the first five days of the war the Russians lost 70,000 men killed, wounded or missing. But we had come to see something else.

Deep in the forest near a village called Gorlitz, a few kilometres from the town of Rastenburg, Hitler constructed his Eastern Front command centre in 1941, called the Wolf's Lair - the name was designed to create an image. From here he planned Operation Barbarossa, the invasion of Russia in June 1941. The complex was heavily defended and the buildings constructed of reinforced concrete over a metre thick. The location was carefully camouflaged, and in fact was not located until Russian ground troops arrived in January 1945. It was here in July 1944 that an attempt was made to assassinated Hitler by Colonel von Stauffenberg with a bomb placed in a briefcase. The attempt failed and thousands of people were rounded up, many of whom has no connection with the plot. Almost 5,000 were executed or committed suicide, including three field marshals, 19 generals, 26 colonels and two ambassadors. Hitler survived because it was a hot summer day, so the meeting was moved from the *Führerbunker* to a flimsy wooden building, and because someone moved the briefcase which von Stauffenberg had placed close to Hitler's chair. The Nazis tried to blow up the complex at the end of the war, but the buildings were so massively constructed that they largely defied the efforts of the demolition squads.

Since then everything has changed. After the war Poland moved westwards, that is the eastern provinces, east of the so-called Curzon Line, were annexed by the Soviet Union, and the Polish inhabitants expelled, and similarly the eastern provinces of Germany, east of the Oder-Neisse Line, were annexed by Poland and the former German inhabitants expelled. This resulted in the transfer to Poland of large parts of East Prussia, Pomerania, Silesia and Brandenburg. The northern part of East Prussia including Memel, Koningsberg, Tilsit and Insterburg was incorporated into the Soviet Union. All the old German names were replaced by Polish or Russian names, so Memel became Klaipeda, Tilsit became Sovetsk, Konigsberg Kaliningrad and Rastenburg Ketrzyn. I can't resist mention of the Konigsberg puzzle. The city is located between two arms of the river Pregel in which there are two islands. Seven bridges link the two banks and the islands. The puzzle is to find a route in which all the bridges are crossed, but only once. The puzzle was tackled by Leonhard Euler, one of the most eminent mathematicians of the eighteenth century, who proved that it is impossible. Why not give it a try? I bought a map showing East Prussia in 1938. It was not called a tourist map or a historical map but a 'homeland' map which says something about the German attitude to the region. Zbigy told us that Germans are now buying up land in former East Prussia and said it would not be long before it reverted back to Germany, but I think not in my lifetime.

From there we drove to Gdansk, the former Danzig, a city with a complex history. Today the restored waterfront area of the Old Town is very attractive and we had a typical Polish lunch of *jurek* (a sour soup served in a large, hard bread roll from which the inside has been removed) and *zander* (pike-perch). Gdansk is famous for its amber, fossilised resin found along the Baltic coast where it has been washed up by the sea. The most valuable pieces are those containing insects which were trapped in the sticky resin and are perfectly preserved. I bought a small piece, but the larger pieces with multiple insects can be seriously expensive.

We drove back to Siedlce and next day were joined by Mariola our Polish-Canadian friend from Edmonton. Zbigy then drove us in his Mercedes van to Krakow and Zakopane, a winter resort in the foothills of the Tatra Mountains in Galicia. From there it is not far to Oswiecim, the notorious Auschwitz extermination camp which Zbigy took me to see. Irina and Mariola, not surprisingly preferred to spend the day in Krakow.

Auschwitz is a chilling place. Unlike Dachau, there is a tangible air of fear still lingering over the place. The original camp was a former Polish Army barracks converted in 1940 to house Polish political prisoners and intelligentsia, but it was rapidly filled to overflowing with Jews and Russian POWs. The brick-built barrack blocks at Auschwitz I, with their attractive overhanging gables and homely chimneys give no clue to the horrors inside. They are crammed full of closely-spaced bunks from floor to ceiling. There was the cynical inscription *Arbeit Macht Frei* over the gateway, the area where the camp orchestra was forced to play as the slave workers marched out to the nearby factories, the assembly area where prisoners were held for hours in the snow and rain during the frequent roll-calls, and the 'wall of death', a narrow alley between two barrack blocks with a wall at the end where prisoners were shot. There were rooms full of the pathetic possessions of the inmates, who thought they were being rehoused in the east – mountains of suitcases bearing labels of long forgotten trips to Lausanne or Nice or London, great piles of shoes and spectacles and clothing, and boxes of hair and false teeth. Truly shocking. But Auschwitz II (Birkenau) is worse. Built some distance away, this was developed as a forced labour and extermination camp, mostly for Jews, gypsies, Slavs and other 'undesirables'. The barrack buildings here were cheap and flimsy and not intended for long-term occupation. It had accommodation for 200,000 inmates, but most of them did not stay long. A railway line ran into the heart of the camp, and the old, young and physically infirm were sent directly to the gas chambers, which operated on an industrial scale. The fitter prisoners were put to work as slave labourers and when they became unfit they too were despatched to the crematoria. There is no precise record of how many people perished at Auschwitz, but the most likely figure is around one million over a period of about 50 months, or 20,000 every month. I have never felt so overwhelmed by the evidence of man's inhumanity to man, or the fragility of so-called civilisation. It was a relief to get back to the normality of life in Krakow.

On our final day, we went to see a geological wonder – a salt mine at Wieliczka in the Beskid mountains south of Krakow. This area lies within the thrust front of the Carpathian Mountains and contains highly folded and thrust rocks formed during the Alpine orogeny. The salt formed as an evaporitic lake in front of the rising mountains and was eventually overridden by the thrust sheets, forming a complex assemblage of sheared *in situ* salt at the base and irregular broken masses of salt in the overlying section. Salt was very important in medieval Europe and salt mining in Poland was a royal monopoly which contributed one third of the total royal income. The mine at Wielikzka has been worked since the thirteenth century and is still in production today. There are incredibly 300 kilometres of roadways in the mine linking 2,000 chambers on nine different levels. Most of these have long been abandoned, but others can be visited, including a chapel carved out of the salt complete with a salt altar, pulpit, salt chandeliers and bas reliefs on the wall cut in salt, including a representation of *The Last Supper*. There is also a lake, a sanatorium for asthma sufferers, and during the war the Nazis used one of the chambers to assemble aircraft engines using slave labour. Salt was also sent from Wieliczka to the I.G. Farben synthetic rubber factory at Auschwitz (the factory was actually commissioned by the Italian state after the failure of its own synthetic rubber factory). Michelin, I think, would rate it three stars – 'worth a journey'. I was so impressed with the place that I did a little research on the geology of the mine, and wrote an article for a popular geological magazine.

Not content with Poland we took another couple of short breaks that year. We had visited our friends Karen and Richard, who we had met in South America, several times at their house in Guernsey. Guernsey is a rather delightful place with lovely bracing cliff walks, empty unspoilt beaches, fortifications from Napoleonic times and from the Second World War when the Channel Islands were occupied by the Nazis. They built bunkers, gun emplacements, observation towers and an underground hospital in an attempt to make the island impregnable. They need not have bothered, the British had higher priorities than to attempt to liberate a group of well-fortified islands situated close to the French coast. There is also the house of Victor Hugo who lived in St Peter Port during his exile from France, an eccentric house full of tiles, carpets on the ceiling, home assembled furniture, and a glass fronted pavilion on the roof with a view of the French coastline, in which he did his writing (standing at a lectern, as Churchill did at Chartwell). On this occasion Richard took us on his boat called *Res Ipsa Loquitur*, to Sark, an island without cars and of great tranquility, and we walked to a lovely restaurant on Little Sark called *La Sablonniere* where we ate freshly-caught lobsters in the garden. We had arrived at high tide when the sea was calm and pacific, but when we returned to the boat it was low tide and there were rocks and shoals exposed all over the place. Sailing in the Channel Islands is not for the faint-hearted.

Later that year Irina was called for jury duty at the Old Bailey, the Central Criminal Court in London. She was assigned to Court No 1, the so-called hanging court, where the most serious cases are tried. The trial lasted for several weeks and she was sworn to secrecy and told not to discuss the case with anyone outside the court, and I must say she obeyed that instruction. I discovered later that it was a very sad case of child abuse by drug-addicted, young, and totally inadequate parents who had cruelly treated their baby daughter and caused her death by neglect and ill-treatment. They were found guilty and sentenced to long prison terms, and of course there were lots of statements that it must never happen again, but it always does.

Our final trip that year was to Barcelona, primarily to see the astonishing architecture of Antoni Gaudi. Gaudi was a Catalan, and as an architect he was unique. He rarely prepared plans on paper, but preferred to work from models. His imagination knew no bounds. He used shapes and forms from nature, and developed a style which can best be described as 'fairy-tale' or 'gingerbread'. He designed houses for wealthy clients with crazy minaret-like chimneys, churches with impossibly tall parabolic arches, crypts which looked like natural caves, and a park with a stunning serpentine bench covered with colourful tile fragments called *trencadis*, surrounding an esplanade originally intended for theatrical performances. The park also has a hypostyle hall with over 80 Doric columns, with more than a nod to the temple at Karnak. We saw some of his most iconic buildings, the Palau Guell, surely the ultimate in Mudéjar architecture, Casa Batllo, a Hansel and Gretel-like house with balconies resembling bleached animal skulls, and Casa Mila built in sinuous, flowing lines on a corner plot at a major intersection. The *pièce de résistance* for me was the roofscape built on an undulating surface with chimneys resembling warriors, pepper pots and twisted abstract figures, mostly monochrome but a few tiled, and one faced with broken fragments of champagne bottles from some long-forgotten celebration.

His *magnum opus* was the *Sagrada Familia* Basilica (not a cathedral) begun in 1882, and not expected to be complete before 2030. It is designed in a most eclectic style with flamboyant, extravagant decoration. The nave was designed to have columns representing tree trunks spreading their branches towards the roof. The basilica will have 18 spires in ascending order of height, representing the 12 apostles, the Virgin Mary, the four evangelists, and Christ himself. So far eight have been completed. Work was interrupted by the Spanish Civil War (when some damage was caused by Republican anarchists, and Gaudi's models and plans were destroyed), and at various times due to lack of money, but work continues. Gaudi, I think, would not approve of the modern work which has decorative features which can best be described as *kitsch*. It is also a sad commentary on Spanish jealousies that in 2010 the Ministry of Public Works in Madrid approved a plan to build a tunnel beneath Barcelona for the high-speed AVE, to pass beneath the basilica. The local authorities objected and claimed that it would destabilise the building, but to no avail; of all possible routes beneath the city the line was built directly beneath the main façade of *Sagrada Familia*. Towards the end of his life Gaudi became more and more obsessed by the building and took up residence in his workshop, and became something of a recluse, dirty and unkempt. In 1926, he was hit by a tram and taken to a hospital where he was unrecognised and thought to be a beggar, so he received minimal attention. A priest recognised him the following day, but by then it was too late.

Having visited Barcelona, we thought it only right to visit Madrid too. We spent four days there and of course visited the Prado and El Escorial. The Prado is without question one of the world's great art galleries. Its collection of Spanish paintings is unsurpassed – particularly those of Goya and Velasquez. The impact of paintings like Goya's *The Colossus* and *The Executions of 3rd May 1808* is immediate and visceral, and his portrayal of the *Family of Charles IV* is a real 'warts and all' picture which perfectly portrays the domineering Queen Maria Luisa and her feckless husband. My favourite Velazquez is *Las Meninas*, an enigmatic and puzzling picture showing the daughter of Philip IV as a young girl surrounded by her ladies in waiting, dwarfs and dogs, but most curiously from the viewpoint of the sitter, portraying Velazquez himself painting at an easel, with a mirror on the rear wall reflecting a tiny image of the king and queen, presumably the subject of the painting. It is a picture where reality and illusion become blurred. We took the train to El Escorial, an austere palace in an austere setting for an austere king. Philip II was a humourless monarch whose mission in life was to counter the Reformation and dominate western Europe, financed by the wealth flowing in from Spain's possessions in the Americas. El Escorial was more than a palace; it was also a monastery, a royal mausoleum, basilica, convent, school and library. When I tell you that Philip arranged to have a view of the high altar from his bed, you will get some idea of the place. There are many interesting places in Madrid, and maybe it was because we visited in winter, but my overwhelming impression was that, compared to Barcelona, the city was cold, formal, restrained. Maybe it was the spirit of Franco still lingering over the city.

You may recall that in Algeria we had lived in a place called Boumerdès on the coastal plain east of Algiers. By the turn of the century it had grown into a sizeable town, but on the evening of 21st May 2003 the region was stuck by a major earthquake, in which 2,200 people died, 10,000 were injured and 200,000 made homeless. The epicentre was offshore, about 25 kilometres north-east of Boumerdès and damage was caused over a wide area. It was established that the quake was caused by compression of the African Plate against the European Plate, which are converging at a rate of about 6 mm per year. Boumerdès was left in ruins, and the Petroleum Institute where I had once been an external examiner was totally destroyed. Along with Agadir, that makes two cities where I lived for several years, which have been destroyed by earthquakes. I don't thing I shall be taking up residence in San Francisco.

Letter 57
We came to the States in May of 2003 and met up with your mom and dad in Las Vegas for a brief encounter. On the first afternoon Irina went to check the shops, I went to the pool, Jenny rested and Jonathan went to play blackjack. That tells you all you need to know! It was Memorial Day and your dad treated us to a gala buffet at Bellagio's where they served Alaskan crab legs, beef Wellington, elk, ribs, and lots more besides. We made our contribution to the local casino, saw the musical fountains, and meandered along the strip. It is an experience not to be missed – but once is enough!

Las Vegas however was just part of a much longer trip. We started at San Francisco and drove to Sutter's Mill, a replica of the old sawmill on the banks of the South Fork American River between Sacramento and Lake Tahoe where gold was discovered in the mill race in 1848, sparking the California gold rush, which attracted 300,000 people within the first few years. This, incidentally, is another example of the Matthew Effect where the name of the discoverer (James Wilson Marshall) has been largely forgotten, whilst the name of Sutter, the mill owner, has entered the history books. Technically the discovery was made when California was still a province of Mexico. The treaty which transferred it (and most of the south-western states) to the United States did not become effective until six months later. We overnighted at Lake Tahoe. The sky was blue, the water limpid, the wild flowers in bloom and the air balmy. We drove right around the lake and thought it delightful.

We had intended to head south and drive to Yosemite from the west but the two main access highways were still closed by snow, so instead we crossed the Sierra Nevada over the Monitor Pass and the Devil's Gate, which brought us to Mono Lake in a rift valley which forms part of the Basin and Range Province. This is an area of mountain ridges and long flat-bottomed valleys caused by faulting due to crustal extension, but despite being located in a rift valley Mono Lake is still at an elevation of 6,700 feet, and pretty bleak. We overnighted in Lee Vining (population about 300), and ate in a little diner where they gave us bean soup and about two pounds of roast pork. Mono Lake is an enclosed lake with a salinity twice that of the ocean. It has numerous 'tufa towers' projecting twenty feet or so above the surface. The pillars developed underwater but have been exposed by lowering of the water level caused by diversion of some of the feeder streams to augment the water supply for Los Angeles. The lake has a food-chain extending from algae to brine shrimps and alkali flies to birds – and even humans, who historically used the pupae of the flies as food. The flies have developed the ability to swim underwater encased in an air bubble and graze on the algae and to lay their eggs. Eventually, after the lake level had fallen by 45 feet, legal action forced Los Angeles to stop taking water from the area, and the lake level is slowly recovering.

From there we went to Bodie, a ghost town, totally uninhabited now but once one of the largest settlements in California. It is located in the mountains north of Mono Lake at an elevation of 8,400 feet in a desolate, raw, wind-blasted valley and fourteen miles from the nearest highway. Gold was discovered in 1876 and a Wild West boomtown blossomed overnight. Within three years the population had reached 10,000, attracting not only prospectors but pimps and cheats, prostitutes and card sharps. There were frequent gun fights, bar brawls and stage coach robberies, and on the edge of town there is a 'boot hill' cemetery where many of the victims were buried 'with their boots on'. There were two churches in the town, but sixty-five saloons. It was such a lawless place that a group called the 601 vigilantes was formed, which administered summary justice - six feet under, zero trials, one rope. A 10-year old girl whose family was heading for Bodie wrote in her diary 'Goodbye God, we're going to Bodie'. The boom began to fade in the late 1880s as the prospectors moved on, and the principal mining company folded in 1913. The town was gradually abandoned, but is now most effectively preserved in a state of 'arrested decay', which means that no attempt is made to restore the town, but what remains is preserved. There is a clothing store with dresses still in the window, a general store with goods on the shelves, houses with tattered curtains and cheap furniture and pictures on the wall, privies tilted at crazy angles, a bar with a pool table and cues still in their rack, broken waggons in the street, bits of old mining machinery, and a Shell sign riddled with bullet holes. It is a most atmospheric place, and I congratulate whoever thought up the idea of arrested decay. It is infinitely better than restoration.

Highway 120 to Yosemite was still closed by snow so we had to make a long detour via Carson Pass to Sonora, a delightful little town where we overnighted, before arriving the following day at Yosemite. Scenically Yosemite is majestic and owes much of its appeal to geology. It is located on a granite batholith which has subsequently been weathered and sculpted by glaciers during the Ice Age into deep U-shaped valleys and sheer vertical cliffs like El Capitan and the Half Dome. We stayed the night at Yosemite Lodge and delighted in the peace and tranquillity of the place. Next day we visited Sequoia National Park and paid our respects to General Sherman and General Grant (and collected some seeds), the two tallest giant redwoods in the world, said to be 2,500 years old.

From there we drove over the Sierra Nevada to Death Valley. Two days before we had been at Bodie at an elevation of 8,400 feet near the snow line. Now we were 260 feet below sea level with a temperature of 46 degrees Celsius. It was so hot that when we went to see the sand dunes, Irina's sandals began to melt. It vies with Libya for the distinction of having logged the highest air temperature ever recorded. Death Valley is another rift within the Basin and Range province, like that at Mono Lake. It is an endorheic basin with no outlet to the sea and the run-off simply evaporates, depositing a crust of salt on the surface. We stayed at Stove Pipe Wells which at least had a nice pool and cool bar. Just a short distance to the west is the Race Track Playa with its mysterious moving stones. The playa is perfectly flat and surfaced with a thin layer of mud onto which a number of stones have fallen from the surrounding mountains. Many of the stones exhibit a sinuous trail behind them, often hundreds of feet long, looking as though they had been dragged over the surface, but how? By what? The explanation was only revealed by detailed observation, time-lapse photography and experiment. Movement only happens in particular circumstances. Occasionally a thin layer of water collects on the playa, in winter this freezes at night and lifts the stone clear of the mud. In the morning, the ice begins to melt leaving the stones sitting on thin ice above a slick surface of wet mud. In these circumstances, even a light wind can move the stones over the flat surface, sometimes at a rate of several feet per minute, although for relatively short periods. Isn't that fascinating? There will be a quiz later.

After our few days in Las Vegas with Jon and Jenny we pushed on to the Grand Canyon. It is too well known to need a detailed description, except to say that standing on the rim you feel very small. It is a stupendous gash in the earth's surface, a mile deep, and showing a cross-section which spans 1.8 billion years (but with some major gaps), from the Permian limestones on the rim to igneous Basement on the canyon floor. The canyon was formed by uplift of the plateau and rapid downcutting by the Colorado River, particularly during the last Ice Age when the flow rate was much greater than now. It was first mapped by John Wesley Powell, a geologist who spent three months exploring the Green and Colorado rivers with a small party, and suffering so many misfortunes that several of his party abandoned the expedition before the end of the passage through the canyon. Three of the men left just two days before the party reached safety, and were never seen again. The Grand Canyon is a place to go and be humbled. It is nature at its most sublime.

From there we returned to Williams and Kingman and then turned on to the famous Route 66, the road made famous by Steinbeck's *Grapes of Wrath*. Here the route twists and turns through the mountains to Oatman, a wonderful relic of the 1930s where we found the splendid old adobe Oatman Hotel (a survivor from 1902), with its wooden false front and period balcony, where you half expected to see a drunken gun-slinger being thrown out of the bar. Unfortunately, their water supply had failed so we couldn't stay there but they assured us we could find a place 'just down the road'. Just down the road turned out to be 30 miles away.

We eventually left the mountains, crossed the Mojave Desert and made it to the 'promised land' of farms and fruit trees and abundance which the Okies and Arkies had been seeking in the 1930s. We passed Bakersfield in the San Joaquin Valley one of the principal oil cities of California, and passed the hundreds of 'nodding donkeys' of the Lost Hills Oilfield, a field comparable in size with some of the fields in Libya. In 1998 one of the gas wells blew out and caught fire, creating a column of flame three hundred feet high, which took two weeks to extinguish and six months to plug. We overnighted at Paso Robles, a lovely little town surrounded by vineyards, but it lies not far from the San Andreas Fault, and in 2003 was hit by a magnitude 6.6 earthquake which destroyed several of the non-reinforced buildings in the town. Our last stop of note was at Hearst Castle, an OTT construction overlooking the Pacific, built for the newspaper mogul William Randolph Hearst (the *Citizen Kane* of Welles' 1941 movie). Like William Burrell in Scotland, he acquired ornate ceilings, entrances, door cases, fireplaces and such like in Europe and had them reassembled in California. The pool is modelled on the Villa Adriana in Tivoli, and is overlooked by the portico of a Roman temple from Italy, and from a distance the mansion looks like a church, with two towers flanking the entrance copied from a church in Ronda, Spain. You might say he had illusions of grandeur, and we found it rather ostentatious and tasteless. We ended where we had begun, in San Francisco, and flew back home to find the oil price edging inexorably upwards, almost by the day. Within a year it had reached an all-time high, and just kept on going. Things were looking up for the oil business.

With the oil price booming and the sanctions lifted, Libya was desperate to attract foreign investment into the country. The offer of large tracts of acreage to foreign companies in 2000 had bogged down in protracted negotiations with individual companies, and in order to break the log-jam NOC decided in 2004 on a radical new approach. Firstly they standardized the blocks into a grid pattern, much like the North Sea, then they selected 15 very large areas, and invited sealed bids from foreign oil companies. The companies were asked to indicate what equity share they would be willing to accept, what work programme they would commit to, and how much signature bonus they would offer up-front. This was a win-win situation for NOC. At a stroke it introduced keen competition between rival oil companies. They fell over themselves to acquire acreage which a few years earlier could have been obtained for peanuts, and offered ridiculously high signature bonuses, and a willingness to accept ludicrously small equity shares. Some of the blocks attracted bids from seven or eight different consortia. NOC ended up with a pot of money, something like 85% of the equity on most of the blocks, and no obligation to pay anything until oil had been discovered and brought on production. You might say they had the oil companies over a barrel. Thus encouraged, NOC introduced three more offers of acreage during the next few years. It was a great time to be a consultant.

Letter 58
Royalties on technical books are distinctly modest, but it was rather gratifying to receive a cheque from the publisher each year. The book generated a lot of interest and I started to receive enquiries from companies about oil and gas potential and opportunities in Libya, and several of them asked me to give talks to their exploration management. It started modestly enough, but then I thought why not offer a Petroleum Geology of Libya Workshop to interested companies. I already had most of the data, and the workshops could easily be tailored to the specific requirements of each company. All that was needed was for the material to be re-worked into a user-friendly format such as Powerpoint. The first workshop I did was a four-day presentation in Japan, in association with Nick Blake who worked for a data processing company which had co-sponsored my book. This was my first visit to Japan and I found it fascinating. How could a place with a population of 120 million, crammed into a habitable area smaller than Greece, contrive to be so orderly, clean, polite and efficient? In order to make my talks more interesting I invited discussion and questions, but they just smiled enigmatically and asked me to continue. Talk about the inscrutable orient – it was difficult to know whether they found the material useful, interesting or just plain boring. Nick knew Tokyo quite well so in the evening we took the metro to Ginza and marvelled at how the Japanese lined up to board the train, and politely queued to cross the road at traffic intersections. The audience had been very formal during the lectures but on the last evening, the company entertained Nick and I to a banquet where they really let their hair down. They became boisterous and voluble and high-spirited. We sat cross-legged around a low table in a private room. There was a little bell to summon the kimono-clad waitress when they wanted more *sake* (which was quite often) and they insisted that we tried all the regional varieties. It was all quite uninhibited and friendly and I took it as an indication that the course had been a success.

After that things started to take off. I received requests to conduct workshops all over the place, in Calgary, Dallas, Houston, London, Tripoli, Kuwait, many of them with major oil companies. I did about twenty of them in total. BP, Shell and Esso had left Libya for political or economic reasons in the 70s and 80s, but were now keen to return. In March of 2004 Tony Blair, the British Prime Minister, visited Gaddafi in Tripoli and very shortly afterwards Shell was awarded a very large concession to explore for gas. He went again in May 2007 following which BP was awarded several key blocks both onshore and offshore.

I assumed that these companies would have extensive archives of material from their earlier operations in Libya, but over the years they had been lost, misplaced or destroyed, and they had to begin the process of data collection all over again. Other companies were totally new to the country and knew little or nothing about the geology and oil and gas potential. My book and workshops, I like to think, helped companies to decide whether Libya held any interest for them.

During these trips I got to meet some of the key exploration personnel of these companies, and develop a feel for how they conducted their business. I also took the opportunity to visit places I had not seen before. From Calgary I flew to Montreal where Irina joined me, and I remember being bowled over by the Gothic Revival Catholic Basilica of Notre Dame, which is probably the most pleasing church I have ever seen. The use of colour, lighting, artifice and decoration is wonderful. The ceiling is like the night sky dotted with stars, and the ornate, intricate altar within the artfully lit semi-circular choir is breath-taking. Curiously it was built by an Irish-American Protestant. We drove to Quebec which is dominated by the gigantic Chateau Frontenac Hotel, another Canadian Pacific Railway hotel, similar to those we had seen in Banff and Vancouver, and we walked the Plains of Abraham where General Wolfe defeated the French in 1759 in one of the decisive battles of the Seven Years' War, which led to the loss of all French possessions in eastern North America. The French have long memories. When Quebec demanded independence from Canada in the 1960s, they had the active support of General De Gaulle. We left Quebec on my birthday in a snow storm (just like the day I was born), and from there we went to Ottawa, the city chosen by Queen Victoria to be the capital of Canada. The story goes that she was shown two paintings, one of Toronto and the other of Ottawa, and said 'I want this one'. The Americans said it was a very good choice because when they invaded, they would never be able to find it.

We had of course to visit Niagara Falls, and someone had recommended a nice hotel at Niagara on the Lake, a delightful little town about 20 miles to the north. We checked in there, and went to see the falls from the Canadian side. They were swollen by spring floods and very impressive, but the town itself was cheap and tacky and touristy. We did the bit where you get kitted out in waterproofs and walk behind the waterfall, but then headed back to the peace and quiet of Niagara on the Lake. On the way, we called in at a vineyard where we bought some of the famous Niagara ice wine, where the grapes are left on the vine until the winter by which time they are shrivelled and sugar-rich and have to be picked by hand.

During my workshop in Dallas I used my free time to visit the infamous Book Depository from which Lee Harvey Oswald shot President Kennedy in 1963. The setting is still much the same, and it is an eerie feeling to stand on the spot from which the shots were fired. It is one of those pivotal events, like 9/11, where you can recall precisely where you were when the news broke. I did a series of workshops in Houston and arranged to fly back to London via New York, where I was able to spend a couple of days with your dad and mom, and catch up on family affairs, and hear about Jonathan's new job as a Customer Interface Manager at a logistics company called TNT in New York. He had moved away from mining and minerals into IT, and gradually began to climb the corporate ladder. Susannah and family were on the move too. She transferred to a new Valuation Office, covering a more affluent area, not far from the Imperial War Museum in London, and they moved to a large Victorian villa in Wimbledon, which they found by the simple expedient of putting a note through the letter box of any house that they fancied in the area.

At the end of one of my workshops in Houston I went out for dinner, and followed the meal with coffee and a celebratory liqueur. Walking back to the hotel I started to feel ill with angina and tachycardia. I had to stop and sit on a bench, and then could only walk very slowly. Not a good sign. I went to see a cardiologist on my return to London and after a series of tests he suggested an angiogram. He recommended the Royal Brompton, a noted cardiac hospital in London for the procedure, and said that if the angiogram showed any problems they could do an angioplasty at the same time. For an angiogram, the patient is doped with Vallium but remains conscious, and it is a weird experience to watch on the monitor as a probe checks out each of the coronary arteries in your own heart, without feeling a thing. I came round totally confused. I didn't know where I was or how I had got there. The nurse called Irina, and I asked her what happened. Did I have a heart attack? Why was I in hospital? She realised that they had done an angioplasty and that they had given me a general anaesthetic on top of the Vallium. She came over, and still I was spaced out. I could not remember going there, what had happened, or why I was connected up to all kind of monitors. The doctor came to see me, and asked me the day of the week – no idea, the year – I struggled, the Prime Minister – was it Thatcher? Then he asked me about the book I had on the table, and I was able to give him a pretty fair resume. He said no cause for alarm, amnesia is not an unusual occurrence, and would soon clear up. It took 48 hours, during which my memory slowly recovered. The thing that bothered me most, and the last memory to return, was how I had travelled to the hospital. They had inserted three stents, which was a whole lot easier than the by-passes I had had five years before.

The petroleum geology workshops led in turn to consulting opportunities, and I was hired by a medium-sized oil company called Apache in Houston to advise them on Libyan opportunities. Apache had been founded in Minneapolis in 1954 by three friends (the initials of their names forming the first half of the name) as a small diversified company involved in all manner of activities, but in the 1970s they sold off their peripheral interests to concentrate on oil and gas. It developed a policy of 'acquire and exploit' and moved its offices to Houston in 1992. Apache expanded internationally with operations in Australia, Egypt and Canada, and in 2003 bought BP's Forties Field, the largest oil field in the North Sea. The field was well past its prime, but they introduced innovative new techniques, and extended the life of the field by several years. They established a New Ventures Group to evaluate other opportunities, which is why they commissioned me to advise them on Libya. I have worked with many companies on acreage evaluation, but I have never come across a company which was as focussed as Apache. They knew exactly what they wanted, and exactly how to achieve it. I prepared many briefing summaries for Apache's management, and they asked me to attend an NOC presentation in London (where NOC showed illustrations from my book, completely unattributed). They then bought a data package from NOC, and brought in a geophysicist and petroleum engineer to work alongside me at their office in Houston, and within a couple of months we completed a thorough evaluation of the three or four blocks which we had highlighted as being the most prospective. Irina came over for my last two weeks just before Christmas and we visited Galveston and San Antonio, and went to see *A Christmas Carol* at the Alley Theatre, not expecting very much from an American production of Charles Dickens, but we were treated to a fantastic performance, which would have gone down well in London. We were also impressed with the Museum of Fine Arts which has a wonderfully representative collection spanning all periods and all genres. It was here that I first appreciated the skills and inventiveness of Georges Braque, particularly his picture *Fishing Boats*, which shows a harbour crammed with boats viewed from many different angles, abandoning the convention of a single-point perspective. Despite the fact that the US government lifted sanctions on Libya in September 2004,

Apache decided that the licence terms were simply too harsh for the potential rewards, and decided to try their luck elsewhere. Many people dislike Houston, but I found it affluent, cultured and comfortable, and I enjoyed my time there. But the one essential is to have a car; without one, life is virtually impossible.

Letter 59
Between my various reports and meetings with Apache we took a trip by car to France and Belgium to see the battlefields of the First World War, which is something I had wanted to do for a long time. We drove first to Ypres (called Wipers by the British troops), a town which was virtually obliterated during the war. The medieval Cloth Hall has been rebuilt, and the Menin Gate is the site of a nightly commemoration at which the Last Post has been sounded every night since 1928 (with the exception of the Second World War). The present gate was built as a war memorial on which to record the names of 54,000 British troops whose remains were never found or could not be identified. But the memorial was too small to display all the names, so a further 35,000 are commemorated elsewhere. I find it quite shocking that of the 300,000 who died around Ypres, 90,000 were never found or identified. It is hard to escape the conclusion that the Allied commanders were either incompetent, insensitive or just boneheaded. 'Lions led by donkeys' is a phrase that comes to mind.

We drove around the infamous locations of the Messines Ridge, Wijtschate, the Menin Road, Hell Fire Corner, Railway Wood, Polygon Wood and Poelkapelle. It was at Poelkapelle that Fred Greaves my uncle Dick's first cousin once-removed, won his Victoria Cross in October 1917. He was a corporal in the Sherwood Foresters, who had already survived Gallipoli, the Somme and Loos. When his platoon was held up by a machine gun post occupied by 25 German soldiers equipped with four machine guns, he rushed forward with another NCO and bombed it from behind. Five of the occupants surrendered and the rest were either killed or fled. Later during a German counter-attack, all the officers being either dead or wounded, Corporal Greaves organised the platoon, established enfilading fire and held up the enemy advance. He was subsequently offered a commission, but turned it down and ended the war as a sergeant. Back in England he was given a new suit in which to receive his medal, and whilst travelling to the ceremony at Buckingham Palace he was accosted by a woman on the train who presented him with a white feather (a symbol of cowardice) because he was a young man wearing civilian clothes. After the war, he was invited by the king (along with other VC recipients) to a garden party at Buckingham Palace. Eventually we reached Passchendaele, the main objective of the six-month battle, located just eight miles from Ypres. One man had died for every inch-and-a-half of ground gained – and that was just the British, the German losses were even greater. And they called it a victory. We stopped at Tyne Cot Military cemetery, the largest Commonwealth war cemetery in the world, with almost 12,000 graves, many of them simply inscribed 'A soldier of the Great War'.

From there we drove across Belgium to see a battlefield of another age, at Waterloo. Waterloo is remembered in Britain as the final defeat of Napoleon by the Duke of Wellington. A century before a grateful nation had built the Duke of Marlborough an enormous country house called Blenheim Palace, named after his greatest victory; Wellington got a bridge and a railway station, named for his victory. But as Wellington said 'it had been a damned nice (close-run) thing, the nearest run thing you ever saw in your life'. He is also quoted as saying 'Give me night, or give me Blücher'. Blücher and his Prussians, much to Wellington's relief, arrived in the nick of time, and saved the day. There is another nice story too. The Earl of Uxbridge was a cavalry commander, not much liked by Wellington, who led a charge which stopped the advance of a French infantry unit. Late in the afternoon he was with the Duke of Wellington when he was hit by one of the last cannon shots of the day, and his leg was smashed. He exclaimed to Wellington 'By God, sir, I've lost my leg!', to which Wellington replied, 'By God, sir, so you have'.

An enormous mound, the *Butte de Lion*, has been built which provides a panorama of the entire battlefield. From there you can look south over the ground where Napoleon's troops advanced. To the right is the large farm of *Chateau Hougoumont*, occupied by the British and Dutch, which held out all day against furious French assaults, in the middle distance the cross roads of *Quatre Bras* which Wellington relinquished before the battle in order to occupy the higher ground of *Mont St Jean*, and in the centre *La Belle Alliance*, a small inn used by Napoleon as a command post. To the left is the walled farmhouse of *La Haye Sainte*, which the French seized in the late afternoon when the defending troops ran out of ammunition. The French were about to launch an attack from *La Haye Sainte* when the Prussians arrived and turned the tide of battle. By late evening the demoralised French were streaming from the battlefield; Napoleon had met his Waterloo. In the days following the battle local inhabitants and souvenir hunters collected what they could from the numerous corpses, including teeth which were sold to surgeons to make into dentures; they were known as Waterloo teeth.

We drove up the valley of the Meuse through Dinant, Waulsort, Hastière and Givet, all classic geological localities which I had explored with my Belgian friend during my PhD studies, and we overnighted at Reims, where we checked out the Venoge and Esterlin champagne houses (my concession to Irina for all the battlefield tours). We headed northwards to Craonne and the Chemin des Dames where a disastrous, suicidal French offensive was launched by General Nivelle in 1917. The French *poilus* by that time regarded themselves simply as cannon fodder, and went into the attack baa-ing like sheep, which precipitated a mutiny which then quickly spread and threatened the entire war effort. The name Chemin des Dames dates from the time of Louis XV when his daughters and their entourage used the road visit the chateau of their father's mistress. There is a very poignant poem written at the time which begins:

> *In silks and satins the ladies went*
> *Where the breezes sighed and the poplars bent,*
> *Taking the air of a Sunday morn*
> *Midst the red of poppies and the gold of corn.*

and ends:

> *The living cringe from the shrapnel bursts,*
> *The dying moan of their burning thirsts,*
> *Moan and die in the gulping slough,*
> *Where are the butterfly ladies now?*

We concluded our battlefield tour on the Somme, another scene of mindless carnage in the summer of 1916 where the British suffered 57,000 casualties on the first day, of which 19,000 were killed (almost exactly the same number as American casualties during the Battle of the Bulge in 1944, but that was spread over forty days). It was the bloodiest day in British military history. The allies (Britain, France, Russia and Italy) had agreed to launch simultaneous offensives in 1916 in an attempt to overwhelm the Germans. The French, supported by the British were to launch an attack along the Somme valley from Albert toward Bapaume, 20 kilometres to the east, against a heavily defended, but relatively inactive part of the front. The plans were upset when the French had to transfer many divisions to plug the gap at Verdun, the site of by a major

German attack in spring 1916. The British then took on the principal role in the Somme offensive. The Germans (as at Passchendaele) occupied the high ground and had established strong defensive positions. The British and French launched a five-day barrage of continuous shelling which it was believed would obliterate the German defences, after which the British and French would launch their attack. Unfortunately for the allies, the Germans had constructed deep dug-outs in the chalk which offered a measure of protection against the shelling, and as the barrage lifted the Germans emerged and the advancing troops were 'mown down like dry grass before the scythe'. On a front of 25 kilometres minor advances were achieved in the southernmost sector, but no progress at all was made further north. Throughout the summer and autumn the offensive continued, and the Germans made progressive strategic withdrawals, until the fighting ground to a halt in November. 'The Somme' has become synonymous with slaughter. Even now it is quite common for shells, human remains and the detritus of war and to be found by farmers ploughing their fields.

We toured some of these tragic locations, Mametz Wood, Delville Wood, Thiepval, Beaumont Hamel and some of the German defensive positions, on a bright spring day with the trees in bud, and it was hard to imagine that day, ninety years earlier, when the whistles blew and the troops were told to walk, not run, as the enemy had surely been obliterated. And all around were the results of that fateful decision: scores of war cemeteries containing the countless graves of a lost generation of doomed youth. Sassoon's poem *The General* sums up the situation very succinctly:

> *'He's a cheery old card', grunted Harry to Jack*
> *As they slogged up to Arras with rifle and pack.*
> *But he did for them both with his plan of attack.*

Letter 60
We had a busy summer. In my spare time I had started gathering information on lost mansions of London, with the intention of writing a book on the subject. London used to have scores of town houses belonging to prelates, the aristocracy, politicians and the *nouveau riche*. Throughout history those seeking royal favour and patronage have aspired to live close to court, and gradually the owners of medieval mansions in the City of London gravitated westwards. Henry VIII evicted the Catholic dignitaries and nobles from their riverside mansions, and after the Palace of Westminster burnt down in 1512, he oversaw the construction of the

Palace of Whitehall, one of the largest (but most ramshackle) palaces in Europe. In the following century the area between the City and the Palace was developed with squares like Covent Garden, St James's Square and Leicester Square. Then estates owned by aristocrats like Bedford, Cadogan, Portman, Burlington and Grosvenor (not to mention the Crown) were developed as fashionable streets and squares, and by the time of Queen Victoria there were dozens of large, opulent houses, vying with each other in ostentation and wealth. The entire process went into reverse after the First World War, principally as a result of social and political changes, taxation and a gradual erosion of the power and influence of the aristocracy. Many of the aristocratic town houses were sold to developers and demolished to make way for smaller properties, others were converted into clubs or offices or galleries. By the end of the Second World War, there was barely a single aristocratic town house left in London which still performed its original function. I thought it would be rather fun to describe some of the houses which had disappeared: their architects and architecture, contents and paintings, the foibles and eccentricities of their owners, and some of the memorable events that had occurred there. I assembled information on about seventy houses and began writing what I hoped would be a popular and entertaining account. But I kept having to interrupt this task as paid employment came along, and ten years later it is still unfinished.

On the 7th of July Irina was away with a group of tennis friends in France, when the television news reported an explosion at Aldgate Tube station, which was attributed to an electrical fault. It turned out to be the first of three terrorist bombs on the Underground, and another which blew the roof off a double decker bus. 52 people were killed and 700 injured. The four attackers were all British-born Moslem extremists. It was a scenario which was to become all too familiar in the years ahead, and not only in Britain.

It was against this background that your mom and dad came to visit us. It was Jenny's first (and only) visit to London so we showed her the sights, and introduced her to aunt Eva, and visited Sue and Alex in Wimbledon. Then they headed off to see your gran, before taking the Eurostar train to Paris for a day or two. No sooner had they left than we had a visit from our Polish friend Zbigy and his daughter, so we did the tourist trail all over again. Zbigy's daughter Anya, was quite an accomplished pianist so we took them to a country house in Kent where there is a keyboard museum. Being Polish she was fond of Chopin, and Chopin had owned a

Pleyel piano when he lived in France in the 1840s. The museum also had a Pleyel piano from that era, which visitors were permitted to play, so she gave a little impromptu concert of Chopin pieces on the same make and vintage of piano that Chopin himself had used. Nice symmetry, I thought.

We had two further trips that year. My cousin Margaret and her husband George had bought a second property at Mitikas, a seaside village on the west coast of Greece, and they invited us to spend a couple of weeks with them. The village has an attractive little port and is backed by the impressive peaks of the Pindus mountains. Nearby is the 25 kilometre Monolithi beach, named after a single rock in the bay – which no longer exists. It was destroyed by German and Italian mortar fire during the war. There are some interesting historic sites in the area. The naval battle of Actium in 31 B.C. in which Octavian (the Caesar Augustus of the Bible) decisively defeated Mark Anthony and Cleopatra, was fought in Actium Bay just a couple of kilometres to the south. It is rather stirring to imagine the clash between 500 galleys equipped with rams, ballistas, and firebrands, and each carrying 200 marines and archers, and six ballistas. In the end Cleopatra's ships fled the scene, and Mark Anthony's force was overwhelmed. Both he and Cleopatra committed suicide shortly afterwards. In commemoration of the victory Octavian built a city called Nikopolis on the adjacent peninsula, and the ruins are still impressive today. George and Margaret also took us to Zalongo, a cliff surmounted by a striking modern sculpture which commemorates the mass suicide of 22 Greek women and their children who jumped off the cliff in 1803 rather than fall into the hands of the Ottoman Turks.

We hired a car and drove northwards to Gliki to see the Acheron Springs, and then through the impressive gorge of the Acheron river, which according to legend, connects Earth to the Underworld. Eventually we reached the ancient Greek temple of the *Necromanteion*, located at the junction of three rivers which is supposed to mark the gate into the underworld into which Charon ferried the souls of the dead, and where the ancient inhabitants came to communicate with the spirits of the dead. Another day we drove to Dodona near Ioannina, the site of the earliest Greek oracle. The oracle was located in a sacred grove where the priestesses divined the future from the sounds made by objects suspended in the branches – sort of like wind chimes, I suppose.

There has been much dispute about the location of Homer's Ithaca, the home of Odysseus. The small island which currently bears the name, is located east of Kefalonia, but does not match Homer's description in *The Odyssey*. Recent geological investigation has suggested Paliki, the western peninsula of Kefalonia is a more likely site, whilst the island of Levkada is another possible candidate, and contains ruins which have been claimed as the Palace of Odysseus. The white cliffs at the southern tip of Levkada have also been suggested as the site of Sappho's suicide. But this is all the stuff of legend. What is indisputably true is that on the south west coast of Levkada there is an idyllic little cove called Porto Katsiki, with the most stunning turquoise sea, and nubile topless girls frolicking in the waves. Or maybe that was the stuff of legend too.

Our final trip took us to Missolonghi on the Gulf of Patras where Lord Byron died in 1824. From humble beginnings, at the age of ten he inherited great wealth, a title and a large country seat from a distant relative. He was an untiring self-publicist, hedonistic, passionate and wilfully perverse. At Cambridge, the college rules prevented him from keeping a dog or cat, so he kept a bear in his rooms instead. He did the Grand Tour from 1809 to 1811 during the course of which he became enamoured with ancient Greek culture and modern Greek boys. Whilst waiting for clearance for his ship to enter Constantinople he decided to emulate the legendary feat of Leander by swimming the Hellespont from Sestes to Abydos, a distance of about four kilometres, but beset with strong currents. He made it in just over an hour, and referred to the swim in the poem *Don Juan*:

> 'T were hard to say who fared the best,
> Sad mortals thus the gods still plague you.
> He lost his labour. I lost my jest,
> For he was drowned, and I've the ague.

He had numerous affairs with both women and men, and his marriage was of short duration (although his legitimate daughter Ada Lovelace was to become an accomplished mathematician). One of his affairs was with his step-sister and there is evidence to suggest that she had a daughter by him. He left England probably to avoid the scandal, shortly afterwards, never to return. The Greek War of Independence against the Turks began in 1821, largely instigated by the Souliots, a clan of warlike Orthodox Greeks based in Epirus and Albania, whose principal base was at Missolonghi. The town had been unsuccessfully besieged by the Turks in 1821 and 1823 and was still seething with activity when Byron arrived there a month later in January 1824. Byron immediately donated large sums of money to the Souliot leaders from his own funds, selling an estate in Scotland to raise cash, but was soon overwhelmed by armed groups demanding more money. Rivalries and fighting broke out between competing tribal leaders, and Byron quickly became disillusioned by their mendacity and pettiness. He eventually got them to agree a plan to attack the Turkish fort at Lepanto on the Gulf of Corinth, and preparations were well advanced when he fell ill with a fever. The expedition never sailed. His fever became worse, and repeated blood lettings didn't help, and he died in Missolonghi on 19th April. He was 36 years old. His body was embalmed and sent to England, except perhaps for his heart which reputedly was buried in Missolonghi. The following year the Turks again attacked the town and sacked it, slaughtering most of the inhabitants. That provoked the intervention of the Great Powers, which eventually led to Greek independence in 1829. Byron became revered as a Greek hero, particularly in Missolonghi, where a statue was raised in his honour. He was not given a public memorial in England until 1969, and only then after extensive lobbying.

I am not going to detail our trip to Istanbul later in the year, except to say that it is a must visit place. It is a real meeting place of occident and the orient, Europe and Asia, Christian and Moslem, ancient and modern, intrigue and romance. It has wonderful monuments like Hagia Sophia, the Blue Mosque, the Topkapi Palace, the Roman hippodrome and the underground 6th century water cisterns. It has Turkish coffee, Turkish Delight, Turkish baths and Turkish carpets, sellers of pomegranate juice and Turkish street food, oriental bazaars, and the railway terminus of the Orient Express, still pretty much as it was when the service from Paris ceased in 1977. On the Asian side of the Bosphorus is the splendid Haydarpasa railway terminal built in 1909 for the Anatolian Railway, which was designed to allow passengers to continue their journey to Ankara, and points east. During the First World War the Germans attempted to extend the line to Baghdad (the so-called Berlin to Baghdad railway) to threaten Britain's Middle Eastern interests, but they dared not build it along the coastline of the Gulf of Iskanderun for fear of attack from British warships so were forced to route it through the Amanus mountains, which proved to be an insuperable obstacle, and the line was not completed until 1940, by which time Mesopotamia was no longer under Turkish control.

In 1915, with the Turks somewhat reluctantly allied to the Germans, the British, French and Russian navies tried to force the Dardanelles, in order to threaten Istanbul, and remove Turkey from the war. They sent a combined fleet of 94 ships to silence the old-fashioned Turkish gun batteries along the strait, unaware that major improvements had been carried out by German engineers, and that the straits had been heavily mined. Three old battleships were sunk, and three others badly damaged with the loss of about 700 men, and a modern battle-cruiser was also disabled. The losses were serious but not critical, but the Admiralty took fright, unwilling to risk the loss of more ships, without support from land forces. In fact, most of the shore batteries had been silenced, and it was reported by the American ambassador that if the attack had been pressed ahead the following day, the Turks were ready to give up. So, Istanbul was saved, but not so the Turkish Empire. After the war, it was stripped of all its external provinces outside Anatolia and Turkey-in-Europe, the Ottoman caliphate and sultanate were abolished and Turkey became a republic. Under Mustafa Kemal 'the sick man of Europe' was reborn, as a slimmed-down, secular, modern state, which it remains – more or less – to the present day.

Letter 61
My work for Apache ended in the summer, but I continued with the workshops, perhaps the most interesting of which was in Kuwait, where I was able to see the wanton destruction caused during the Iraqi invasion fifteen years before. In the Kuwait Towers there were photographs of the washrooms to show how the Iraqis had smashed even the washbasins and urinals. The scale of destruction was immense. The historic royal palace was flattened and burnt, and the Museum of Islamic Art was trashed and looted – and that by fellow Arabs and Moslems, and most of Kuwait's oil wells were torched. By the time the Iraqis left, Kuwait was a wasteland.

In September I gave a talk on Libya to a conference at Imperial College, and met Danny Clark-Lowes who I had known since Britoil days. Danny had done his PhD on Libya, and had recently marketed a report promoting the potential of the Murzuq Basin. Over a number of weeks, we discussed the possibility of a joint project on oil reserves in Libya. This was an attractive proposition because it had never been attempted before. Published estimates were all based on NOC figures which we knew to be greatly inflated. We planned a much more rigorous approach in which we would calculate the reserves of every field in Libya, the amount of oil already produced and the amount remaining. Unlike NOC we would provide maps and show exactly how our figures had been derived. Of course, this required access to company reports and maps, but it is surprising what you can find. There were a number of fields described in detail in various technical journals, and we had many contacts with companies and individuals who had been involved in Libya, and NOC issued data packages with each licence round. Within three or four months we felt that we had sufficient data to begin the task, but it was a hard grind and it took the best part of two years to complete.

Danny and I were fortunate in the fact that when one project came to an end, another one opened up. We undertook a major study for another Japanese company, headquartered in Tokyo but with an office in London. We conducted numerous studies of blocks, areas, fields and farm-in opportunities, but unlike Apache in Houston, the decision-making process was slow and cumbersome. We would have meetings in London with Japanese technical representatives and agree a set of objectives and priorities, but when these were submitted to Tokyo, as often as not, they would be replaced by something completely different. In the end, the company acquired interests in four concessions and drilled four wells, but unfortunately all four were dry.

Our excursions that year were rather modest, but in May we visited the Italian Lake District which, I have to say, is rather more exotic than the English Lake District. We took a boat from Stresa to Isola Bella, a stunning place of lush floral terraces rising from the water's edge, and then the Villa Taranto, created by a retired Scottish captain, on a barren hillside, and now a place of forest walks with tree ferns, a metasequoia (previously thought to be extinct), a Chinese handkerchief-tree and gorgeous banks of multi-coloured azaleas. Then we followed the last days of Mussolini to the little town of Dongo, only seven kilometres from the Swiss border, but separated from it by a mountain range. He and his mistress, Clara Petacci, were fleeing in a convoy with other fascists and German soldiers. They were stopped by communist partisans of the Garibaldi Brigade and arrested, and later escorted to a secluded farm house in the village of Mezzegra, just across the lake from Bellagio. The following day they were driven down the road, ordered out of the car, stood against a wall and shot. The leader of the brigade at the time was Luigi Longo, an Italian communist who had been imprisoned by the fascists until July 1943, and it seems likely that he ordered the shooting. In an earlier life Longo been an inspector of Republican forces in the Spanish Civil War.

After that we went to see some geology at Lake Iseo, one of the smaller lakes, between Bergamo and Brescia. We zigzagged up into the mountains to the village of Zone to see the remarkable *piramidi*, dozens of tall columns of boulder clay, many capped by a large boulder, which protects the soft clay from erosion. Such features are called hoodoos, named by early explorers who thought the capstones had been deliberately placed by malign spirits. During the Ice Age, the valley was blocked by a large moraine, behind which great thicknesses of boulder clay accumulated. As the moraine was degraded by stream action, walls and totem-pole-shaped columns of boulder clay, some 15 metres high, were left behind, and those capped with a large boulder have survived. It is a most unusual and impressive sight.

Northern Italy is dotted with medieval towns and cities (the setting for five of Shakespeare's plays), like Bergamo, Cremona, Piacenza, Padua, Verona, Mantua and, of course, Venice. On this trip, we explored Mantua, dominated for four hundred years by the Gonzaga family, which is remarkable for two artistic masterpieces. The *Palazzo Ducale* contains the *Camera degli Sposi* (bridal chamber) decorated by Mantegna with wonderful visual illusions and *trompe l'oeil* frescoes and murals, and the *Palazzo del Te* (the summer palace) which contains the *Sala dei Giganti*, with colossal giants covering every inch of the walls, tearing down buildings and columns, with gods and warriors looking down from clouds on the ceiling. In 1630 the city was sacked during the Thirty Years' War, and was then decimated by bubonic plague, from which it never fully recovered, but the murals survived, and still decorate the empty, bare and looted rooms. Most of the splendid Gonzaga art collection was bought by King Charles I of England, but he did not enjoy it for long. After the Civil War and the death of the king the entire royal collection of 1,500 paintings and 500 sculptures was auctioned off by Parliament.

We drove south, over the Apennines, crossing the wartime Gothic Line close to where my father had been killed, and arrived at the spa town of Montecatini. There is a funicular to Montecatini Alto built in 1898, and destroyed by the retreating Nazis on the day after my dad was killed. On we went, to Lucca, with houses still surrounding what had been the Roman amphitheatre, to Monteriggioni, an ancient walled town on a hill, where two of our friends from Brega were living, to Siena with a unique sloping medieval piazza, where the famous *palio* horse race is held, and eventually arrived in Florence where we spent a day at the Pitti Palace and Uffizi Gallery. We drove to the military cemetery where my dad is buried and found two adjacent graves, same regiment and same date, of men who must have died with him. We drove up to Trespiano where he was killed, and were able to locate the spot precisely from a description in the battalion's War Diaries which I had copied. It was very moving to stand there on that bright May morning, children playing, women with their groceries, where the German mortars had been sited. Down below, an olive grove and a farm house and distant views of the mountains. These must have been the last things he ever saw – some corner of a foreign field. I wept.

In summer we were invited to the wedding of Pauline, the daughter of our old friends Masha and Peter. Pauline is a very smart cookie, who made an excellent match with Nick, a landscape garden designer. After the wedding, we took Masha to meet Mike at his 'dacha' at Millfaid near Loch Lomond, and from there the three of us went to Fingal's Cave on the Isle of Staffa. This involves taking a small boat from a tiny harbour on the Ross of Mull (famous for its pink granite, used to build or decorate several lighthouses, three bridges in London, the Albert Memorial, and the docks of Liverpool, Glasgow and New York). The small boat battles the Atlantic swell and local currents for 8 miles across the exposed bay to the famous cave with its vertical columns of basalt, like some gigantic ruined cathedral. Mendelssohn made the trip in 1829 – by sailing boat, which must have been fun – and was inspired to write his *Hebrides Overture*, which perfectly captures the feel of the swell and the blustery wind and the grandeur of the scene. The columns are black – all except one, which is white, because during the war it was damaged by a floating mine, revealing the pristine white coating behind. The boat returned to Iona, one of the earliest Christian settlements in Europe, where St Columba and twelve companions arrived from Ireland in 563.They built the original abbey from where missionaries went out to spread the word. The famous illuminated *Book of Kells*, which I had admired in Dublin, was compiled here in about 800. All of the medieval Scottish kings from 858 to 1097 were buried in the abbey. The community was repeatedly attacked and pillaged by Viking raiders, but it survived until the Scottish Reformation in the 1540s when it was dismantled and abandoned. In the twentieth century, the ruined abbey was rebuilt, almost 1,400 years after its original foundation, but as an ecumenical community. In the bay to the east of Staffa is the island of Inch Kenneth, once owned by Lord Redesdale, father of the celebrated Mitford sisters. Unity Mitford was an ardent Nazi supporter and a confidante of Hitler, who lived in Berlin from 1934 to 1939. When war broke out she attempted suicide, but survived and was repatriated to Britain where her father hid her, out of the way, on Inch Kenneth, where she flew the Nazi flag, played Wagner on the gramophone and decorated her rooms with photographs of Hitler.

Susannah and Alex had their second son, William at the end of August, and you can't help speculating on what the future holds. Will he turn out to be a new Bob Dylan, Bill Gates, Albert Einstein, Lewis Hamilton? Or just Bill, the guy next door, the plumber, the estate agent or the fella in the bakery store? It is like having the great ocean of destiny lying all undiscovered before you. And then a few months later, on a cold, raw day just before Christmas, you were born, not long after your mom and dad had moved to their new apartment in Forest Hills. Your birth coincided with the new moon, which seemed rather propitious, and we came over to see you, Irina and I, like magi, bearing gifts. I remember you had brown hair and long fingers – and we wondered if you were destined to become a pianist. We took you for a stroll in a little woodland park on 98th Street, and then out to West Point – perhaps that too was symbolic of some future career (although I hope not). They called you Saffron, which we thought a very pretty and appropriate name, but someone in the registry office made a mistake and registered you as a boy!

Danny and I continued work on the Libya Reserves Project which was turning out to be a very big undertaking. We collected details on over 450 fields in Libya, made maps of every significant field, plotted the location of every well, planimetered areas, tabulated reservoir parameters, calculated oil in place, recovery factor, recoverable oil, oil already produced and remaining reserves. We produced values for maximum, most likely and minimum amounts, and made estimates of additional potential from enhanced recovery methods. Our conclusions suggested that NOC's estimates were inflated by about 40%. Whether this was a deliberate attempt to mislead, or whether NOC habitually assigned a higher recovery factor, we were unable to discover, but we thought our results were as accurate as the data permitted, and that they were transparent and verifiable to anyone who examined them. We offered the report for sale to a number of companies, and were gratified by the response we received.

We wrapped up our consulting work for the Japanese company, and I conducted one or two workshops, since NOC continued to auction acreage on a fairly regular basis, although companies were now beginning to realise that the terms were extremely harsh and that maybe there were better opportunities elsewhere.

In Spring of 2007 we took another trip to Berlin, a city which I find endlessly fascinating. Things had changed since our last visit ten years before. There was now a Holocaust memorial made of nearly 3,000 concrete slabs resembling chest tombs set on undulating ground and covering an entire city block next to the Tiergarten. The reconstruction of the Reichstag building was completed in 1999 and visitors can now climb to the glass dome and look down on the Bundestag chamber below. Graffiti from the war, many in Cyrillic, have been deliberately left as a reminder of the past. The site of the Führer bunker is now a car park, but at least there is an information board with an outline of its history. The Hotel Adlon on the Pariser Platz, the most famous hotel in pre-war Berlin, reopened in 1997, completely rebuilt and refurbished. The original owner, Lorenz Adlon was hit by a car outside his hotel, and killed in the 1920s, and his son Louis was arrested by the Red Army in 1945, due to his title of 'Generaldirektor', and a few days later was said to have died of 'a coronary embolism'. We had a drink there, but ate at a nearby *kneipe* with bratwurst, mushrooms and fried potatoes – good, solid Berlin-bourgeois food.

I had long wanted to see Bayreuth, a town still living under the shadow of Wagner, so we hired a car and headed south. Wagner was a scandalous opportunist. He had an adulterous affair with Cosima the wife of his friend, and the daughter of Liszt, he fought on the barricades alongside Bakunin in Dresden in 1849, and he unashamedly sponged off anyone gullible enough to fall for his charm. In 1871 Wagner arrived in Bayreuth and declared the town just the place in which his artistic genius could flourish (although he only wrote one opera there). He found the wonderful baroque Opera House, built by the sister of Frederick the Great, too small for his grandiose plans, so he cajoled Ludwig II of Bavaria into bankrolling the construction of a new purpose-built opera house, with seating for 1,900 and devoted exclusively to Wagner's music, plus a villa befitting the composer's perceived status, which he named *Wahnfried* (peaceful delusions). The house was built in the Roman manner with a giant sgraffito panel above the entrance depicting Wotan, King of the Gods being greeted by two women and a child. Inside, the square central space was illuminated by a skylight and decorated by paintings and statues from Wagner's operas. The house and the town were badly damaged by American bombing during the war, but *Wahnfried* was subsequently repaired and eventually became a museum. Wagner is buried in the garden. The directorship of the opera house was jealously guarded, and passed from generation to generation of the Wagner clan from the 1870s until 2008, but their close ties to Hitler during the war, along with some very bizarre directing in recent years, has given the *Festspielhaus* a rather tarnished reputation.

Then we headed to Dresden, following the route that Michael and I had taken as students more than forty years before. The autobahn had been modernised, the region was less dystopian, Karl Marx Stadt had reverted to Chemnitz, and the East German Trabant assembly plant had been taken over by Volkswagen. We stopped off at Meissen to visit the famous State Porcelain Manufactory with its well-known crossed-swords trademark, and then drove beside the Elbe to Dresden. We found the city totally transformed. Gone were the acres of derelict land and the scars of the 1944 bombing. The showpiece buildings of the Zwinger, the Gemäldegalerie, Frauenkirche, Semper Oper, Albertinum, Brühl's Terrace, cathedral and castle have all been meticulously restored. There can surely be no more magnificent baroque city in Europe than Dresden. No wonder they called it the Florence of the Elbe. We wallowed in its architectural luxury, and although I am not a porcelain connoisseur, I had to admit that the collection in the Zwinger Palace is stunning.

But times were changing. Your dad changed jobs to work with an outsourcing company providing computer business solutions to clients in and around New York, and in the UK Susannah moved to a property valuation office in Wimbledon which meant she no longer had to commute into central London. In September 2007, there was a run on Northern Rock, a British building society which had used the deregulation of the banking industry, introduced by Thatcher in 1986, to over-extend credit which it could not recover due to the collapse of the securitised mortage market. Thatcher and Reagan's chickens were coming home to roost; the fallacy of uncontrolled monetarism was exposed. In the US Bear Sterns collapsed, in France BNP Paribas froze their assets. Finance houses fell like dominoes: Fanny Mae, Freddie Mac, Lehmann Brothers, Merrill Lynch, Bradford and Bingley, RBS, Lloyds – all either sold off or bailed out by government. Interest rates plummeted in the UK from 5% in April 2008 to 0.5% in March 2009. It was the start of the most serious recession since the 1930s, and the most protracted. The only solution government could offer was austerity – so long as it did not affect the rich and the multinationals.

The process of rolling-back Britain's public services and welfare state began with Thatcher's privatisation programme of the 1980s, followed by John Major's Private Finance Initiative in 1992 which encouraged 'outsourcing', of public services to the private sector. Tony Blair completed the transformation with his Private Public Partnerships scheme in 2003. This introduced private financing and management into all manner of enterprises, in the belief that private enterprise was inherently more cost effective and efficient than publicly run bodies, so we developed a situation where private companies were awarded contracts to manage prisons, provide 'back-office' services for the police, maintain and improve virtually all of the nation's roads, manage NHS clinical and care services and run local authority housing and schools - to name but a few. Many of the expected benefits of cost reduction, increased employment and improved standards proved illusory, but the companies involved nevertheless racked up impressive profits. The simple expedient of solving the nation's chronic debt crisis by aggressively going after the rich, the overseas tax havens and the multinationals, or of increasing domestic taxation was considered to be unacceptable politically (shades of Gerald Kaufman's 'longest suicide note in history'), so we are left with austerity and a situation where the National Health Service, once the pride of Britain, is being run into the ground, social services have declined to a level not seen in fifty years, police budgets are cut to a point where certain crimes are no longer investigated, and education is fractured into a plethora of different types of school: comprehensive, grammar, faith, public, and academies, each with its own agenda, priorities and funding. It reminds me, quite inconsequentially, of a speech made by a recent British Minister of Education, who said 'it is my aim that the performance of all teachers will be above average'. Good thing she wasn't a maths teacher. So, instead of a brave new world, prosperous, egalitarian and socially integrated, we now have a society more fractured and socially divided than at any time since Margaret Thatcher was elected.

I had enjoyed my two business visits to Japan, but had seen nothing of the country outside Tokyo, and Irina had never been there, so we decided to go and see the famous autumn colours. Unfortunately, the weather was not kind, and several days were spoilt by rain and low cloud, including a trip to the fifth station on Mount Fuji, only to find the mountain totally hidden in mist. They told us it was a typhoon passing through, which was not much comfort. We took a cable car over steaming sulphur springs, with the wind buffeting the cable car, and the mist swirling around, and it seemed like a vision of hell. We spent the night at Hakone, and the next morning the typhoon was gone, the sky was blue, and there was Mount Fuji in all its splendour before us.

We travelled by Bullet Train to Nagoya, and then to see a mountain village with traditional wooden houses with the beams held together by ropes, and open fires in the middle of the house with a slatted roof for the smoke to escape, and a separate rice paddy attached to each house. And that wasn't just for the tourists – all the houses in the village appeared to be occupied. We saw traditional Japanese gardens with immaculately raked gravel instead of grass, and trees very carefully protected with supports for the branches, and covers for the winter. In Kyoto the Shogun's palace has a nightingale floor – a floor which squeaks when walked upon, as warning against possible intruders, and at Nara there is a park with tame deer which you can hand feed with 'tlakkers'.

We left from Osaka's Kansai airport, built on an artificial island in the bay, after the onshore airport became too congested. It has a 35-minute, high-speed rail connection for the 40-kilometre trip to the city centre, and a second runway opened shortly before our visit. (Shanghai has a Maglev train which covers the 35 kilometres from the airport to the city centre in 8 minutes). In London, by contrast, discussions have been ongoing for forty years where to build just one more runway. Osaka's solution would be the perfect answer to London's problem. But so sclerotic is the decision-making process in Westminster that it will probably never happen, and Frankfurt or Paris will become the principal European hub.

Irina's mother, Tatiana, had trained as a nurse before the war, and had served with the Red Army during the war, in Moscow, the Urals, and at the end of the war in Kaliningrad. She had married Ivan Ivanovich in 1947, and continued to work as a staff nurse at a Moscow hospital until she retired. In 2006, she began to show signs of Alzheimer's disease, which rapidly grew worse. Irina's father too was struggling with heart disease and had several scares when he had to be taken to hospital. As a former army officer, he was able to be treated in a military hospital, but early in 2008 he died. Tatiana retreated into a world of her own, sometimes recognising Irina and Yuri, but often not. She fell and broke her hip, and was confined to bed for the last two years of her life. Irina's mother and father had lived almost all of their lives under communism, and were proud of its achievements – the overthrow of the Romanovs, the expropriation of land and factories from the landowners and industrialists, and the performance of the Soviet Union during the war when the Red

Army first halted, and then drove back the Nazis, but at the appalling cost of 8.8 million Russian military deaths, and 22 million military wounded, not to mention the huge numbers of civilian casualties. But they were not blind, they could see that life was more affluent in the west and that people there had more personal freedom. I suspect they regretted that the communist regime had not evolved into something more humane and flexible, that the dead hand of state control had not been gradually lifted, as a reward for all their efforts, but all the same 1989 must have come as a shock, when everything that you had believed in came crashing down like a house of cards. What they thought of that, I never discovered, but I don't think they were enamoured by the likes of Gorbachev and Yeltsin, and they would have been disgusted by the rise of the oligarchs.

Letter 62
One day while trawling the web, I came across an article in a university newsletter about Ralph Fox, who had been a pupil at Heath Grammar School in Halifax before the First World War. I had been fascinated by him as a schoolboy because we were told that he won a scholarship to Oxford, became a communist, and was killed in the Spanish Civil War. That made him a romantic figure to an impressionable school-boy in post-war Halifax. The article quoted Ralph's nephew as the source of the information in the newsletter. I thought this would make a nice subject for an article in the journal of the Halifax local history society, so I contacted the editor and through him Ralph's cousin Gavin, who lives in Scotland, and decided to give it a go. Gavin lived in a quiet village on the Solway Firth with views to the distant hills of the English Lake District. He too had been a pupil at Heath Grammar School, and was a great traveller and cyclist, and on one occasion he had cycled with a friend from England to the place where Ralph was killed in southern Spain. His house had become the repository for the family archives which included a great deal of material on Ralph. He invited me to spend a few days there going through the material and chatting about the family's perceptions of Ralph.

Ralph was an individualist with a mind of his own. He was born in 1900 into a well-off family, attended school from 1908 to 1917 and went up to Oxford just as the First World War was ending, and the Russian Revolution and Civil War getting under way. His sympathies clearly lay with the working classes and he quickly became involved in left wing politics at Oxford, and founded the Oxford branch of the Communist Party of Great Britain. Britain at this time was in a state of flux. The war had opened the eyes of many ordinary people to the inequalities in British society, and there had been widespread support and aid for the Russian Revolution. The government was jittery, and believed that the country was on the brink of revolution, and took a very firm stand against militant strikers in Glasgow in 1919. Ralph became a journalist, writer, poet and playwright, but in 1922 he joined a Quaker charity involved in bringing aid to the starving masses in southern Russia. He learnt Russian and several of the local dialects and became something of an expert on the Mongols and Tatars, and wrote a biography of Genghis Khan. He later worked in Moscow where he married an English secretary, and collected material for another biography, on Lenin. In 1926 he and his wife returned to London just in time for the General Strike, where Ralph demonstrated his talents as an orator. In places like Stanley in the Durham Coalfield, Chatham, home of the Royal Naval Dockyard, and at Colchester in Essex there are to this day streets named for Marx, Lenin and Stalin. Ralph worked as a journalist and continued writing on left-wing issues, and by the 1930s he had become well-known as a communist intellectual. The only work which has stood the test of time is *The Novel and the People*, reissued as recently as 1991, which promotes social realism in literature and pours scorn on mawkish sentimentality and self-indulgent artifice. So he was a fan of D.H. Lawrence, but not of John Galsworthy. There was a curious incident in 1936, which has only recently come to light, when he married Annie Dubowski, a Jewish widow of Polish origin, who was a union convenor in a clothing factory. But he was already married, you might say. I have seen the marriage certificate for Ralph and Annie, but no record of the earlier marriage in Moscow has been found. However, when he went off to Spain a few months later he made a will leaving everything to his 'Moscow' wife, although she is named as Madge Palmer, not Madge Fox. It seems likely that the marriage in Moscow was never officially registered, and that to Annie was a marriage of convenience, so that she would not be deported, but curiously neither wife appeared to know of the other.

In 1936 he signed up with the International Brigades and went off to join the Republicans in Spain, ostensibly as a reporter, but actually as a liaison officer and commissar. The war was going badly for the Republicans who were fighting with poor and antiquated equipment against the Nationalists who were supplied with the latest weapons by Mussolini and Hitler. He was sent to a training camp at Albacete, a barren plain in La Mancha (*al basit* is plain in Arabic) where a British company had been formed, and Fox was appointed assistant political commissar (where he met Luigi Longo, the Italian inspector of the Republican forces, who a few years later played a key role in Mussolini's death). Training was perfunctory, and within a few weeks the company was merged with a French brigade, and rapidly deployed to block the Nationalist advance along the Guadalquivir valley. Fox went with them as an interpreter. They arrived at the disintegrating front on Christmas Day 1936 and were assigned to seize the little town of Lopera, set amid olive groves in the rolling hills south of the river. But they were too late; the town was already occupied by Franco's forces. On the 27th they came under heavy attack from artillery, infantry, and planes, and suffered 50% casualties, including Ralph Fox who was killed whilst attempting, with others, to set up a machine gun post on a small ridge. His body was never recovered, but when the news reached London, Fox's death was front-page news. A memorial service was held at the Quaker Meeting House in London attended by 1,500 people including prominent communist and socialist leaders. Albert Camus wrote 'It was in Spain that men learned that one can be right and yet be beaten, that force can defeat spirit, and there are times when courage is not enough'. Franco triumphed in the end, turning Spain into a military dictatorship, and between 1940 and 1944 200,000 Spaniards died of starvation, revenge, ill-treatment and forced labour. Franco was deeply superstitious and for more than forty years slept with the mummified hand of St Teresa beside his bed. Fox received no official recognition in the town of his birth, and an invitation from the civic authorities in Lopera to their counterparts in Halifax to attend the inauguration of a monument to Ralph Fox went unanswered. My account of Fox's life and death was published in 2009, and I was rather pleased with the result.

In 1964 during my student trip around Eastern Europe with Michael, we spent a day or two in Budapest, which at the time was under rigid communist control, and still recovering from the trauma of the 1956 uprising. Irina and I decided to take a city break to see what difference forty-four years had made. The hotels were totally transformed. We stayed at the Boscolo Budapest, a splendid Art Nouveau building, originally offices of the New York Insurance Company, which fell on hard times between the wars, and even harder times under communism, but is now restored to its original magnificence. It has what has been described as 'the most beautiful café in the world' where the Hungarian *literati* used to meet in the days before the First World War, all gilded stucco columns, frescoes, intricate wooden panelling and Venetian chandeliers. In the castle we came across a string quartet playing in the courtyard which was nice, and in the evening went to the opera to see *Don*

Giovanni sung in Italian with Hungarian surtitles, which was daunting - but opera knows no frontiers, and the performance was good.

The most interesting place we visited was the sculpture park for redundant Soviet statues, an eclectic collection of stakhanovites, heroic soldiers, statues and busts of Marx, Engels, and Lenin and an impressive ensemble in aluminium of a group of revolutionaries being urged on by Bela Kun the Hungarian communist leader, which even includes a lamp-post. The group has been assembled a foot or so off the ground which gives it the effect floating in thin air. But best of all are Stalin's boots. Before the '56 uprising there was a huge statue of Stalin in the main square of Budapest. During the uprising, it was torn down by the mob, leaving only his boots on the plinth. They were removed when the uprising was crushed, but have now been re-erected in the sculpture park, as an ironic reminder that hubris is eventually followed by nemesis. On the banks of the Danube there is a poignant reminder of an earlier dark age, a row of worn shoes and boots by the river's edge, cast in iron. In 1944 the fascist Arrow Cross militia began purging the city of its Jewish population. The victims were led to the river bank, ordered to take off their shoes and then shot so that their bodies fell into the river and were carried away.

In April 2008 the remains of the last two victims of the assassination of the Russian Imperial family, were positively identified by DNA analysis as Alexei and Maria (the most outgoing of the sisters), supposedly bringing to an end the saga of the Romanovs. The other members of the family had already been identified and given a belated state burial in St Petersburg in July 1998 on the 80th anniversary of their execution. The Imperial family had been murdered in Yekaterinburg in the Urals following the Russian Revolution, on the orders (according to Trotsky) of Yakov Sverdlov, Chairman of the Central Committee, with the concurrence of Lenin. It is a fascinating forensic story. The fate of the bodies was known only by rumour until 1979, when a geologist and a film maker, after extensive research, discovered a shallow grave on a forest track, 20 kilometres from Yekaterinburg, containing the remains of nine bodies. Initially the bodies of the Imperial family and four of their retainers were thrown down a mine shaft by Yakov Yurovsky, head of the execution squad, and several drunken assistants. Sulphuric acid was poured over them, but it was discovered the shaft was only three metres deep. They tried to deepen the shaft with hand grenades but to no avail. On reporting to the local Soviet Yurovsky was told to recover the bodies and dump them down a much deeper shaft. The bodies were hauled out but such was the confusion that all kinds of tell-tale clues were left at the site, dentures, buckles, insignia, pearls even the odd diamond. The bodies were loaded onto a truck but travelled only five kilometres before becoming stuck in the mud. Yurovsky ordered a grave to be dug on the track and nine of the eleven bodies were dumped in it, their heads smashed, acid poured over them and then covered in quicklime. The other two bodies (Alexei and Maria) were burnt and smashed with spades and buried in a second pit, 15 metres away, and the two graves covered with railway sleepers (railroad ties) to give the impression of a roadway over a boggy patch of track. And there they remained for sixty-two years. The identity of the remains was established by DNA testing both in Russia and in the USA and the nine remains from the first grave (five family members and four retainers) were interred with great ceremony in the St Peter and Paul Cathedral in St Petersburg. Sverdlov fell victim to the flu pandemic of 1919, Yekaterinburg was renamed Sverdlovsk in his honour in 1924, but regained its former name in 1991. The 'murder house' in Yekaterinburg was demolished on Boris Yelsin's orders in 1977 and replaced by an elaborate 'Cathedral of the Blood', and the Orthodox Church canonized the Imperial family in 2000. But for some mysterious reason the church authorities refused to accept the DNA evidence for the remains of Alexei and Maria, and they remain in a sort of continued purgatory in two small boxes in a vault in Moscow, 'pending further investigation'.

But I digress. Our Libyan Reserves Atlas attracted a number of companies. Danny had a large circle of professional contacts and was very adept at marketing. We exhibited the atlas at several conferences and symposia and made quite a number of sales – payoff for the two years we had spent preparing the report. This was a good example of a report, full of useful information and maps, which individual companies had neither the staff nor the time to compile themselves. We became bolder. Our initial atlas covered only oil; now we decided to enlarge it to include gas fields and gas reserves. We then decided we would produce annual updates, and ultimately we introduced a digital version. Danny lives in Somerset and converted an adjacent barn into a sort of scriptorium where we stored all our reference material and where a small team of graduates helped with the data compilation. Despite the recession this was a good time for us. The four licensing rounds between 2005 and 2007 had stimulated great interest in Libya. Exploration was booming and production rising. I continued with the workshops, and not long after our contract with the Japanese came to an end I was contracted by an

American geophysical company called Ion to advise them on planning a large regional seismic programme in Libya. It made me wish that I had taken up consulting earlier.

Our flat in Warwick Square had been leased to a succession of tenants for around eight years and produced a modest rental income, but this had to be offset against tax on the income, agent's fees, maintenance costs and service charges. The capital value of the flat increased gradually but with the recession we were not optimistic about any likely short-term improvement. At this stage I received an enquiry from a neighbour who asked if I would consider selling the flat. There were some taxation and legal considerations, but Irina and I thought this was a good opportunity to cash in, and be rid of all the hassle of renting. In the end, he made us an offer we couldn't refuse, and we sold the flat at the end of 2008. Big mistake. He was a lot smarter than I was. The recession meant that less and less people could afford to buy, so more and more people were forced to rent. The building of council houses (for people on low incomes to rent) had come to an end with Margaret Thatcher, new builds could not keep pace with demand, the number of new, so-called affordable houses was deliberately kept low, so within a year or two the capital value of properties in central London and rental income far outstripped income available from any other source. We have made several wrong decisions on property, but that was probably the worst.

Letter 63
Since moving to Richmond, Irina had joined the local tennis and bridge clubs. She was a tennis fanatic in the literal sense of the word, playing every weekday, both summer and winter, and she could usually get tickets for Wimbledon, one way or another. On one occasion, we had seats on the front row of Centre Court for the Men's final. It doesn't get much better than that. We had gradually built up a nice circle of friends, amongst whom was Zora, a former professional ice dancer with *Holiday on Ice*, a show with which she toured all around the world. Our Aberdeen friends Tony and Laleh now lived in Oxshott, our Brega friends Liljana and Jovan in Cobham, and our Glasgow friends Mike and Mikiko in Putney, so all were within easy reach. We often went with them to concerts, plays or exhibitions in London, or on country walks, mostly in the Surrey hills – not the rough stuff of Scotland or Yorkshire for sure, but gentle, unthreatening, enfolding countryside - and with plenty of characterful pubs for lunch.

Zora's daughter Christina was working as a journalist in New York, and arranged a surprise birthday party for her mother in Greenwich Village. Zora was expecting a simple family dinner, but Christina had invited many of Zora's friends, including us, to attend. I have never seen anyone so completely taken by surprise. We had been walking with her in Richmond a few days before. It was your dad's birthday a couple of days later, and we met you all downtown and had a celebratory lunch at a French restaurant on 44th Street. You were very bright and lively and the waiters made a fuss of you. But New York in February can be a cold and bleak place, so the next day we headed off to Florida for a few days to catch some sun. Unfortunately, we arrived at the tail-end of a storm and it was dull and wet for the first day or two. We stayed on the Gulf coast at Clearwater in a nice hotel built on an offshore barrier island, over a causeway from the mainland. At breakfast we asked for fresh orange juice – no (and this was Florida in the citrus season), espresso coffee - no, muesli – no, yoghurt, no. So we went to Starbucks instead. We took a ride on a speedboat called *The Screamer* which lived up to its name, fifty-five miles an hour on the open sea is not the place to wear a hat. On the other side of Tampa Bay is the Gamble Mansion, an evocative plantation house, like something from *Gone with the Wind*, on a sugar plantation once worked by 200 slaves. We drove to Homosassa Springs to see the manatees, marine mammals called sea cows, which graze on sea grass, just like real cows, but without the methane emissions. They are slow and ponderous and vulnerable to injury, particularly from speedboats, and are currently protected, and classified as 'vulnerable to extinction'. Along the way, we came across roadside ads like 'Labour of Love' – birth center, 'Keep the gold, dump the digger' – divorce attorney, and 'Good Earth – economical cremations'. They have a sense of humour, those Americans. On the board-walk outside our hotel we found a pianist playing Chopin pieces on a piano on wheels, or we could just sit on the balcony and watch the dolphins enjoying themselves in the lagoon. It was my kind of place.

In September 2008 Susannah had her third child – another boy, who they named Edward, but who has always been known as Eddie. The boys have quite different personalities. James is very self-assured, and did a school project on Russia before the Revolution. His comment afterwards was 'I think I know all there is to know about that period'. William, who has spent most of his life overseas, saw a double-decker bus in London, and asked 'Is the upstairs for first-class?' Eddie had a slight reading problem when he was young, and very proudly proclaimed to his friends 'I'm dyslexic'.

We tend to think of the countries of Western Europe as rather fixed entities identified by common language, traditions, history and religion. But this is a simplistic view. In France, even as late as 1871 only a quarter of the population spoke French as their native language, and in Spain, even at the present time, only 72% speak Spanish as their first language. The regional differences in Spain are profound, and nowhere more so than in Catalonia and the Basque country. Both have powerful aspirations for independence, and a group called ETA, which demanded the separation of the Basque provinces, was long a thorn in the side of the Madrid government. We visited the Basque country on a short break in 2008, and were struck by the strange place-names, like Altzaga, Zabaloetxe, Sodupe and Txorierri. The Basque language is unrelated to any other European language and is considered to be the last remaining vestige of a pre-Indo-European language in Western Europe.

The Basques supported the Republicans during the Spanish Civil War, and paid the penalty. As the Nationalists advanced towards Bilbao their route was blocked by the town of Guernica. They requested air support from the German/Italian Condor Legion, which was under Nationalist command, and on 26th April 1937, the undefended town was bombed by Junkers and Heinkel bombers in the world's first demonstration of *blitzkrieg* terror bombing. The number of deaths was about 300, mostly women and children, as most of the men were fighting at the front, and the town was largely destroyed. News of the event caused an outrage, and Picasso was commissioned by the Republican government to paint a mural of the event for the 1937 World Fair in Paris. The mural has become one of the great icons of art. It was painted in matte house paint in black, white and grey, and is full of symbols and metaphors of the outrage of war. Picasso lived in Paris in World War II (he never set foot in Spain after 1934), and had a photograph of the painting in his studio. A visiting Nazi officer examined the photograph and asked Picasso 'Did you do this?' to which he replied 'No, you did'. The painting was sent to America for safe keeping during and after the war, as Picasso had stipulated that it should not return to Spain until democracy was restored. It finally returned in 1981. In Guernica there is a Peace Museum, where you can vicariously experience the bombing, complete with sound, shaking floors and collapsing buildings.

In Bilbao we stayed in a historic palace which had originally been built 20 kilometres away but had been demolished to make way for a motorway, and subsequently transported stone by stone to the new location. The façade still bore the scars of the Civil War. Bilbao is famous for the Guggenheim Museum, designed by Frank Gehry in a deconstructivist style suggestive of an ocean liner, clad in curved reflective titanium panels like fish scales. It has come to be recognised as one of the most iconic buildings of the twentieth century. There was an exhibition by Richard Serra of large, leaning, weathered steel plates arranged at various angles, and a sinuous snake of curved plates, which were most disorienting to walk through. We passed through San Sebastian (or Donostia, if you are Basque), the tapas capital of the world, where the success of a bar is measured by how many tapas wrappers litter the floor. The city is located on a circular bay, and we were struck by the way the waves broke simultaneously all around the bay.

From there we drove into the heart of Rioja country to a sleepy little village in the mountains called Enciso where there are a dozen or so sites with Cretaceous dinosaur footprints. The most impressive are preserved in a valley near the chapel of *La Virgen del Campo*. Here over 500 footprints have been uncovered on what was originally a sandy shoreline. It is possible to distinguish between prints made by the lumbering four-legged herbivorous dinosaurs (sauropods), the more agile bipedal herbivorous dinosaurs (ornithopods), and the quick and deadly bipedal carnivorous dinosaurs (theropods). Half way up the track the footprints of a theropod show evidence of sudden acceleration to a place where they mix with the tracks of a herbivore. Perhaps the theropod got lucky that day.

We ended our trip at Burgos where our hotel window faced the magnificent façade of the cathedral, built in flamboyant French Gothic style and which architecturally stands comparison with Reims or Amiens or Chartres. It contains the tomb of El Cid, Castile's national hero. He was immortalised in an epic medieval poem which portrays him a chivalric hero of the *Reconquista* against the Moors. The reality is rather different. He was more a soldier of fortune. After being exiled by Alfonso of Leon he fought for the Moors (who gave him the title of El Cid, or lord) in the internecine wars against both Christian and Moorish enemies, before being recalled by Alfonso to fight against Almoravid invaders from North Africa. He was a highly successful military commander and established control of Valencia where he set up an independent principality in which Christians and Moslems were granted equal status. His reputation has generated statues, paintings, novels, plays, operas, movies and even video games. The lavish 1961 movie with Charlton Heston and Sophia Loren ends with the memorable Hollywood fiction of Moorish troops fleeing in terror as the corpse of El Cid, mounted on a horse, advances towards them.

Danny and I made several more sales of our Libyan Reserves Report, and I continued my work with Ion, the American geophysical company, preparing a major regional seismic programme in Libya of deep, basin-wide, high-resolution SPAN lines. My contact at Ion was Vinton Buffenmyer, an astute and experienced geophysicist, who was responsible for planning the survey, and guiding its progress. The programme was designed in association with NOC - which meant that NOC obtained the data, but someone else paid for it. It wasn't Ion who paid however. They operated a system based on pre-commitment where they would promote the planned survey to interested parties and then obtain a proportion of the cost upfront, in exchange for input into where the lines were located. My role was to accompany the Ion team on these promotional visits and present the geological objectives of each line, for which I prepared geological maps and cross-sections to highlight the oil and gas objectives.

Nick Blake, who had accompanied me on my first workshop in Japan, was also a member of the team, and we criss-crossed Europe visiting companies which had indicated an interest – Madrid, Stavanger, Paris, Hamburg, Milan, Kassel, Vienna. Nick became our minder. He knew every city: hotels, transport, eating places; his knowledge was encyclopaedic. He could navigate by train and metro around these cities with the confidence of a native. In Stavanger he took us to a restaurant called the Setra, an exclusive place, but in a simple town house without any exterior indication of its function. I remember we had to ring the bell to gain admittance. They served only Scandinavian products: scallops with salsify, halibut with rowanberries, smoked eel with seaweed, reindeer with chanterelles, plums and chokeberries with juniper sauce, and with each course a matching wine (happily not Scandinavian). Several of the companies we visited signed up for the SPAN programme, so we now had a green light to continue our planning and preparation.

In May I celebrated my 70th birthday, and we decided to make it something special. We hired a marquee and Sue and Alex kindly agreed to let us erect it in their garden in Wimbledon. We invited thirty or so relatives and friends and I sent them invitations on a card with Ben Jonson's poem *Inviting a friend to Supper*, which I thought appropriate:

> *It is the faire acceptance, Sir, creates*
> *The entertaynment perfect : not the cates.*
> *Yet shall you have, to rectifie your palate,*
> *An olive, capers, or some better sallad*
> *Ushring the mutton ; with a short-leg'd hen,*
> *If we can get her, full of eggs, and then,*
> *Limons, and wine for sauce : to these, a coney*
> *Is not to be despair'd of, for our money;*
> *And, though fowle, now, be scarce, yet there are clerkes,*
> *The skie not falling, thinke we may have larkes.*

and so on – a delightful, quaint, archaic poem from the time of Shakespeare. One of Irina's tennis friends was a professional caterer and provided the food. I don't recall the short-legged hen or the larkes, but the coneys were good. We invited people from every period of my life, and in my speech I mentioned them all. Nina came from Moscow, Michel from Paris, Peter and Masha from Scotland, Karen and Richard from Guernsey. It was a delightful gathering, with speeches and music, and a charming song from Imogen, Karen and Richard's daughter. It occurred to me though, that the assembled guests were like ships that pass in the night, and that they would never all meet together again, which I found rather sad.

Letter 64

Fired up by my visit to Norway with Ion, I thought it might be nice to return and see the country in a more leisurely fashion, and we decided to go as a foursome with Mike and Mikiko. I was keen to compare Norway now with the Norway I remembered from 1963, but we were disappointed. The weather was dreadful, the prices astronomical and the hospitality rather underwhelming. However, we did get to see the *Fram*, the ship which took Nansen to the Arctic and Amundsen to the Antarctic, and Thor Heyerdahl's *Kon Tiki*, a balsawood raft which he and five others sailed from Peru half way across the Pacific before smashing into a reef in French Polynesia. We also saw Edvard Munch's *The Scream*, a picture which to me encapsulates the dark side of Scandinavia – the long winters, the gloomy forests, the violent mythology of Thor and Wotan, the anxiety of Ibsen's plays. We discovered that there are actually four versions of *The Scream*, two of which have been stolen at one time or another. To lose one is a misfortune, as Oscar Wilde said, but to lose two is carelessness. Both were recovered, one in a sting worthy of a Henning Mankell novel. Interestingly two British detectives were called in to assist in the hunt for the perpetrators. The thieves were arrested, tried and found guilty, but on appeal the judge released them because the British detectives had travelled using false names. And I thought British law was perverse.

Bergen was the home to two celebrated musicians, Grieg and Olé Bull. Grieg today has a worldwide reputation, and Bull is largely forgotten, but in the 1860s and 70s it was Olé Bull, a virtuoso violinist, the equal of Paganini, who had the higher reputation, and this is reflected in their respective houses. Olé Bull was an extrovert, and built himself an incredible Moorish palace on an island, like a miniature Alhambra, with an elaborate two-storey Moorish entrance, and a bell tower with an onion dome. Inside, he had a music salon which could seat a hundred people, adorned with carved wooden filigree arches, slender columns like twisted rope, crystal chandeliers and oriental carpets. By contrast Grieg had only a modest wooden house overlooking the fjord. In recent years an eco-friendly modern concert hall has been skilfully integrated into the landscape of the garden, with a grass covered roof, and we were treated to a piano and violin recital of music by Grieg, Brahms and Bartok whilst dusk descended over the garden and fjord, a view much as Grieg would have known it.

2009 was a good year for Danny and I. We sold more copies of our Libya Reserves Report, and Danny arranged to have the whole thing digitised and made available as an interactive digitised database to which clients could subscribe, and we decided to offer annual updates, which kept the three graduates busy in Danny's scriptorium. I continued with my workshops and with my consulting for Ion, and after numerous meetings with potential clients, we finally obtained sufficient pre-commitments to make the project viable. In the end, we designed an extensive programme of deep, high-resolution regional seismic lines covering both the onshore and offshore areas of Libya. It was planned as part of a much larger programme to eventually include Egypt, Tunisia and Algeria in addition. It was a very ambitious plan, but Ion had conducted this type of survey in other parts of the world, and was well equipped to handle it.

Our skating friend Zora, with whom we were now firm friends, had flats in both Prague and Nice and she kindly invited us to stay in both. The flat in Prague had belonged to her parents, and was the place where she had grown up. Her father was a photographer, who had owned a photographic shop in *Na Prikope* (The Graben), near the Estates Theatre where Mozart had premiered *Don Giovanni* and *La Clemenza di Tito*. When the communists took over, the shop was nationalised and her father was obliged to become an employee in the store that he had formerly owned. We have been back to Prague several times, and come to know it well. It is one of the few European cities not to have been destroyed during the war, but it was the scene of the assassination of Reich-Protektor Reinhard Heydrich in 1942, as Michael and I had discovered on our trip forty-odd years before. This time I dragged Irina out to the suburbs, by metro and bus to the actual place of the assassination. This kind of thing really fascinates me: to find the exact spot where some famous or notorious event happened – where my father was killed, where Kennedy was assassinated, where Mussolini was shot, where Hitler's bunker was located, but I need to know precisely; hereabouts is not good enough. I knew that Heydrich had been driving in an open-topped Mercedes from his home to his office in Prague Castle. The route took him down Zenklova Street, then round a very tight bend onto Holesovickach Street, and this is where the ambush took place. Photographs taken just after the event show the wrecked car and two trams stopped at the intersection with passengers being questioned by German soldiers, plus one or two onlookers. History caught on camera. The place is easily recognised today, and they have even erected a column to record the event.

Early in 2010 Irina (as usual) came up with a plan to visit South Africa, with me, Zora and another of her tennis friends, Sally-Anne (so it was me and my three chicks). The plan was to fly to Cape Town, drive the Garden Route along the south coast, and then fly to the Kruger National Park. Given the notoriety of the townships we had some misgivings about safety, but we had no trouble at all. The football World Cup was being held in South Africa that year and Cape Town was in festive mood. The views of Table Mountain from our hotel were stunning, but the views from the top (where there is a unique, scrubby vegetation called *fynbos*) were even better, to the Lion's Head, Signal Hill, Robben Island and the west coast. Robben Island, like Alcatraz, was a prison island where Nelson Mandela was incarcerated in a damp cell, initially with only a straw mat on which to sleep. He and the other ANC prisoners spent their days breaking rocks and working in a limestone quarry. They were forced to wear shorts rather than trousers and were denied sunglasses against the glare of the white limestone. He was allowed one visit and one letter every six months. Subsequently conditions were relaxed somewhat, and after a mere 18 years he was transferred to a less severe prison. Robben Island prison was decommissioned in 1991 and can now be visited. It is a grim-looking place of single storey buildings built of large irregular blocks of black sandstone, resembling a concentration camp, with the ambiguous message over the gateway *Ons dien met trots* (We serve with pride). It might as well have been *Arbeit Macht Frei*.

We went to Stellenbosch, a lovely old colonial town, and centre of the South African wine industry, where we took a fancy to Pinotage wine which we sampled in a local winery, a deep-coloured fruity wine, developed from a cross between Pinot Noir and Hermitage vines (hence the name), which thrives on the granite outcrops of the area. We drove to the Cape of Good Hope where you really do feel to be at the end of the world with nothing between you and Antarctica (although it is not quite the southernmost point of Africa). It was hard to imagine the early explorers sailing round here in their primitive caravels, but they did, and at Mossel Bay there is a replica of the *São Cristóvão* in which Bartolomeu Dias sailed to here in 1488, four years before Columbus set sail. Dias sailed on for a further 500 kilometres before his crew mutinied and forced him to return to Portugal. *En route* we overnighted in Swellendam a small, quiet place, and we drove to have a look at the town. I noticed a police car following behind so I drove very carefully and deliberately, but still he pulled me over. There was not another car in sight, but he said my front wheels were over the stop line when I halted at an intersection. He must have been having a boring night. At Knysna I heard about an abandoned gold mine which was open to visitors, but it was disappointing. After a drive of seven kilometres on a dirt road all that was left of the mine was a simple adit 200 metres long which showed precisely nothing, and the local guide could barely speak English. The place had been home to a substantial community in its heyday, but there was virtually nothing left, except a rickety board among the grass indicating 'Main Street'.

We had seen the Kenya game parks, twenty years before, but Kruger was different. In Kenya the parks were widely separated, and each had its own speciality, but the Kruger park contained all of the major animals within reach of the principal lodges. We stayed at the Arethusa Lodge with only a dozen or so individual circular thatched cottages, each equipped with a mosquito net. Dinner was served outside around an open fire. We managed to see all of the big five, but the highlight was when we found a female leopard teaching two cubs how to hunt – or perhaps when our vehicle was chased by an angry elephant. The other day I saw American big game hunters on television shooting elephants in Africa with high powered rifles, just for sport, *at the present day*. It is the hunters who should be shot.

We ended our tour with a stop-over in Mauritius, a volcanic island of which my memories are not great. The principal activity is the production of sugar-cane and rum, and everywhere there were great piles of basalt rocks full of gas holes, collected by the local farmers from the cane fields. We had a nice day at the *Ile aux Cerfs* on the east coast, with an *al fresco* barbecue, and we climbed to see the crater of the dormant volcano *Trou aux Cerfs*, (they are hot on *cerfs* in Mauritius), but on the last day I got seriously sick with food poisoning and had to call the hotel doctor. It was touch and go whether I would make the plane the next day. Not the best end to a spectacular holiday.

I conducted my final workshop on the Petroleum Geology of Libya, appropriately enough in Tripoli for a Chinese company. They had acquired two blocks in the south-west of the country and were about to drill the obligation wells required under the terms of their contract. I gave them a review of the potential of the entire country, but they were obviously most interested to know what I thought about their two blocks, which unfortunately was not much. They were located in an area far from any proven source rock, and where the principal reservoir was thin and rather shallow. I told them it was a high-risk area, with a good many negative indicators, and so it proved. Both of their wells were dry. I told them they should have consulted me (or Nubian) before they took the blocks. That proved to be my last visit to Libya. In less than a year the Arab Spring engulfed the country and exploration came to an abrupt end.

Not long afterwards Alex received a job offer in Dubai, and Sue and the boys flew out to take a look. They liked what they saw, so Alex accepted the offer, Sue resigned her job with the Valuation Office and in May they let their house in Wimbledon and joined the ex-pat community in Dubai, just as we had done in Morocco all those years before. Your mom and dad were on the move too. That year Jonathan joined EMC Solutions LLC, a company which provided Cloud Data Services, and Data Infrastructure Management (whatever they are) as a Solutions Principal (don't even ask), and y'all moved out to Shelton in Connecticut, half-way between Bridgeport and New Haven.

I had long been an admirer of the careful and meticulous paintings of John Constable, a style rather out of fashion these days. I had a book of his paintings which were mostly done in and around his home at East Bergholt in Suffolk, and I conceived the idea of taking a trip there to see whether I could match the paintings to the present landscape. We drove over – it is only 100 miles from Richmond, and found that all the paintings were done within a very small compass around East Bergholt and Dedham Vale. His style is romantic, but painted with freshness and feeling. His cloudscapes are particularly impressive and he spent many hours sketching and painting cloud formations. Whenever we see white, puffy clouds these days we call them Constable skies. We were able to match virtually every painting *(The Hay Wain, Flatford Mill, Dedham Vale, Willie Lott's Cottage,* and even *Boat-building,* and *The Cornfield)* to its location, and it was refreshing to see that many of them had barely changed. The villages around there, like Stoke-by-Nayland, Hadleigh, Kersey and Lavenham, are quintessential, chocolate-box English villages. Interesting too, that the richness of the churches in small Suffolk villages like Lavenham was due to wealth obtained from wool, when this area dominated the English wool trade. It lost this position when the processing and manufacture of woollen cloth became industrialised, and places like Halifax and Huddersfield and Dewsbury took over the trade.

Zora also introduced us to some rather prestigious events. She was a member of the local Polo Club and often invited us to matches, at one of which princes William and Harry participated (and were rather good). She took us to the Guards Polo Club in Windsor Great Park (where the skill of the players was phenomenal), and to the members enclosure at Henley Regatta (her son had rowed there for his school), to Glyndebourne, and to the Royal Enclosure at Ascot. I don't recall how she got us in there, but she had some very influential friends. And so we had a taste of life in the fast lane. It was a long way from the back streets of Halifax.

'But pleasures are like poppies spread, you seize the flower, its bloom is shed'. The following year was not so good. My aunt Eva had reached the age of 96, and we thought she was in line for a Queen's telegram (on reaching the age of 100), but she didn't quite make it. She took an interest in politics and current affairs right up to the end, but she became progressively less mobile, and gradually stopped eating, and in the end 'was ready to go'. We were staying at Zora's flat in Nice when I received a call from my cousin Margaret to say that Eva had died. On that day we were in the mountains and stopped in a village called Gairaut, where we heard the organ playing in the village church. We went in and listened as the organist played *Jesu, Joy of Man's Desiring* and other Bach pieces.

We were the only ones there, and it was like a private requiem for Eva. At the funeral, I gave a eulogy and quoted from *No man is an island*, which ends 'therefore never send to know for whom the bell tolls; it tolls for you', which was certainly apt. She had been a fixed and dependable point throughout my life, and I felt her loss keenly. Then in August, in the middle of the most sweltering summer for years, Irina's mother died in Moscow. She had been unable to walk for some time and suffered from progressive dementia over a period of years, but the heat of that stifling summer proved too much for her. I had liked Tatiana. She had visited us several times in Britain, and had been very welcoming on my trips to Russia, and although we could not communicate easily we got along well together. She was the last of that generation, and I now found myself the oldest member of the family.

Then in November I had a stroke and shortly afterwards I noted the symptoms in my diary as follows. 'Reading; could only see half a page, right side of page not visible, unable to comprehend what I was reading. Breakfast disorganised, searched for bowl and spoon as though I was in a strange house. Forgot to take pills, forgot juice. When watching television could not follow the banner headlines, had to spell them out like a child. Brainstorm. Instead of thoughts coming in a stream, they came in a great rush all on top of one another, and jumbled up.' It was scary. I had a headache for three days, but my reading slowly recovered over the following days. I had a brain scan which showed a 27-mm lesion on the left side of my brain which permanently affected both my balance and my memory, but fortunately I recovered from the other symptoms, and I guess many of the little grey cells eventually reconnected. Since then I have been on a daily cocktail of pills – statins, alpha-blockers, beta-blockers, and probably gamma-blockers as well, if there is any such thing. Without them I probably wouldn't be here.

Letter 65
We celebrated New Year of 2011 with our friends Karen and Richard, their children and their dogs, at their large Victorian house in Guernsey. We poured the drinks (Karen chose red wine), and on the stroke of midnight, as I reached for my glass, I caught Karen's glass with the tip of my finger, knocked it over, and spilt red wine all over the white carpet and antique furniture, and smashed the elegant and expensive glass in the process. We managed to clean up the mess, but not a good start to the new year. We thought perhaps it was an omen, and maybe it was.

In January of 2011 a popular uprising toppled President Ben Ali in Tunisia, in February Egypt followed their example and overthrew Mubarak. There was a rising in Bahrain, and then in Libya. The Arab Spring had arrived; the people had finally had enough, and rose against their corrupt and venal leaders. The Libyans had been under the iron grip of Gaddafi for more than forty years, but he was well prepared for such an event. He had a large, well-equipped army and air force, and stock-piles of conventional weapons including Scud missiles. In the past he had attempted to develop a nuclear capability, but having seen the fate of Saddam Hussein, he abandoned the programme. He also had stocks of mustard gas and sarin which were in the process of being destroyed when the uprising began. What happened to them is anyone's guess. Gaddafi's command centre was in the Aziziya Barracks in Tripoli, from where my uncle Dick had sent us food parcels more than 60 years before.

Gaddafi's wrath was directed mostly against the rebels in Cyrenaica and he launched his forces against them in late February. By 15th March tanks and troop carriers had taken Brega and Ajdabiya and were advancing on Benghazi which Gaddafi threatening to destroy *zenga zenga* – street by street. The UN declared a no-fly zone over Libya on 17th March and on the 19th just as Government tanks entered the suburbs of Benghazi, French jets attacked and halted the advance, and US and British ships fired over a hundred cruise missiles against military targets. All through the spring and summer the front swung back and forth, and Brega, where we had lived for ten years, changed hands six times. The town and its oil installations were very badly damaged, and the airfield, which our friend Branco had helped to build, was obliterated. Gradually, with NATO support, the rebels gained the upper hand and, *shwe shwe*, the Gaddafi regime was toppled. Gaddafi himself was caught hiding in a culvert near Sirt and brutally killed.

Gaddafi had seven sons, the eldest, Muhammad, was a business man and not seriously involved with politics, and now lives in Oman. Number two, Saif al Islam, was a well-known propagandist for the regime. He had a PhD from the London School of Economics, and was a painter of some merit (we once went to an exhibition of his paintings in Hyde Park). He was held for six years by a militia group in Zintan but eventually released and granted an amnesty. The third son, Al Saadi was the footballer who had engineered the demolition of the football stadium in Benghazi out of jealousy. He fled to Niger but was extradited and now faces murder charges in Libya. No 4 was Mutassim, National Security Advisor to

Gaddafi, who was killed alongside his father in Sirt. The fifth son, Hannibal, was a flamboyant playboy who caused scandals in Paris, Geneva, London and Denmark. He ended up in Lebanon where he too is facing charges. The sixth son, Saif al Arab, spent several years living in Munich, but returned to Libya when the revolution began, and was killed in a NATO air strike on Tripoli. The youngest son, Khamis, was a graduate of the Frunze Military Academy in Moscow, but was in the United States when the revolution began. He returned to Libya in time to lead attacks on Zawia and Misurata, but was reported killed in an air strike in August. Gaddafi's wife fled to Algeria and later to Oman, but in 2016 she was allowed to return to Bayda in Cyrenaica, where she had been raised. The fate of Gaddafi's assets is still under investigation. Libya descended into civil war as local militias slugged it out, rival governments were set up in Tripoli and Benghazi, one religious and one secular, Isis became involved, and chaos reigned. The oil industry was badly hit. Foreign companies pulled out their personnel, installations were destroyed or damaged, production plunged and exploration ceased. Ion, the American geophysical company with whom I had been working, had chartered a ship to acquire the offshore SPAN programme in February 2011 and acquisition was just about to begin when the revolution exploded. The programme was immediately terminated. *Hallas*, as the Libyans say, finished.

My stroke had been a bit of a wake-up call. It was a shock since I thought my heart was the problem, not my head. So, we decided to take an extended, but leisurely winter vacation, in case I was unable to do so in future. We were keen to see how Susannah and Alex had settled into life in Dubai, we had long wanted to visit Angkor Wat in Cambodia, Irina had a tennis friend in Australia who invited us to visit, and we thought we should include New Zealand, as we were not likely to come this way again. It was like four different vacations strung together into one.

As you might guess it was a memorable trip. Susannah and Alex and the boys had rented a large villa in Dubai with a lounge like a football pitch, and they had taken to the ex-pat life like ducks to water. They had a live-in maid and an outdoor pool, and life was hunky-dory. Expat life in Dubai is rather different to Libya. Sixty years ago, Dubai had been a sleepy village on a muddy creek which made a living from pearl diving and fishing. Now it is a bustling metropolis of two and a half million people, with dozens of high-rise buildings, including the Burj Khalifah, the world's tallest structure, rising like an inverted icicle (not a very appropriate simile), almost a kilometre high. Not content with building onshore they also constructed an offshore island in the shape of a palm tree on which dozens of houses and hotels have been built, and then, even more ambitiously a grouping of many small islands, which from above look like a map of the world. So, if you don't know what to do with your money, you can buy England, or New York or Libya, and build your pad on it. This has all been made possible by an aggressive policy of attracting foreign investment by offering generous incentives, free trade zones and tax benefits, and money and people have flooded in. Only 15% of the population are native residents, but they are given generous privileges and benefits. It is unquestionably one of the success stories of the second half of the twentieth century.

In Bangkok we visited the house of Jim Thompson, an American entrepreneur who almost single-handedly transformed the Thai silk trade from a cottage industry into a major business. During the war he had served with the OSS in Sri Lanka and Thailand, and afterwards returned to Thailand where he established the Thai Silk Company which farmed-out work to local women working at home, raising hundreds of them out of the direst poverty. He also became an avid collector of south-east Asian arts and crafts. In 1967 he went with a colleague to spend the Easter weekend with friends at their house in the Cameron Highlands in Malaysia. They went to morning service at the local church, then home for lunch and in the afternoon, while the others retired for a siesta, he went out for a walk. He never returned. A search was organised, which grew into the largest manhunt ever mounted in Malaysia, involving police, local residents and British servicemen, extending over 11 days, but no trace of him was ever found. It became a *cause célèbre* and the source of innumerable conspiracy theories.

The highlight of our time in Thailand was a trip to Kanchanaburi where we stayed in a delightful place with a garden overlooking the River Kwai. The town is infamous as the location of the bridge over the River Kwai, located on the Burma-Siam Railway built by the Japanese during the war in order to supply war *matériel* to their troops in Burma. The route ran for 415 kilometres through jungle and mountainous terrain and was built at great speed, in less than 13 months, using a huge workforce of Allied prisoners of war and Chinese and south Asian forced labourers. The conditions were appalling, the prisoners were starved and beaten, and treated with great brutality, and around 90,000 Asians perished and 16,000 Allied POWs. We took a short trip on the railway in antique wooden coaches, with some seats 'reserved for monks', to the Wampoo Viaduct where there are sleepers (railroad ties) dated 1943. The celebrated 1957 movie *The Bridge over the River Kwai* with Alec Guinness, Jack Hawkins and William Holden, was actually filmed in Sri Lanka, and bears no relation to reality. The bridge was bombed by RAF and US planes on three occasions, until finally put out of action in June 1945. We visited the Commonwealth War Cemetery at Kanchanaburi and the first grave we saw was in memory of Leslie Hallett of The Loyal Regiment, aged 24, who died just before Christmas 1943. No relation, I think, but quite a coincidence.

We flew on to Siem Reap in Cambodia (a name which in Cambodian means 'victory over Siam') and spent a couple of days exploring Angkor, the ancient capital of the Khmer Empire. The area is huge, covering twenty by ten kilometres, and containing numerous temples, lakes, secular buildings, roads and villages. The jewel in the crown of course is Angkor Wat, the largest and best preserved of all the temples. Much of Angkor was overgrown by jungle, and was only rediscovered in the 1860s, and not seriously excavated until 1907, when French archaeologists began to open up the site, although considerable areas are still unexplored. The complex flourished under the Khmer Empire from the ninth to the fourteenth centuries, but was largely abandoned following the Thai invasions in the fifteenth century. The temples are a feast of Hindu and Buddhist art, and we saw the mysterious conical spiky towers and terraces of Angkor Wat emerge from the darkness in the early morning sunrise. The temple has a frieze over 800 metres long showing scenes in bas-relief from Hindu mythology and history. Another temple, the Bayon at Angkor Thom is decorated with the smiling faces of dozens of Buddhas, each three or four metres high, cut in bas relief on all four faces of every tower. In eastern Angkor, the monastery complex of Ta Prohm has been left pretty much as it was found (apart from a little essential conservation), covered in strangler figs and silk cotton trees whose roots drape and interfinger the stonework like some malign ectoplasm. This is the temple which was used as the setting for the Angelina Jolie movie *Lara Croft: Tomb Raider*, reviewed by one critic as 'elevating goofiness to an art form'.

On we went to Australia, where we were guests of Sue (another of Irina's tennis friends), and her husband Mark, who had a penthouse apartment in Mosman overlooking Sydney Harbour Bridge and Sydney Opera House. Mark had a little turret on the roof with a flagpole from which he flew different flags each week – the present Australian flag, the pre-1953 flag, the National Colonial flag, the Australian Ensign, the Federation flag, New South Wales, the City of Sydney - he had quite a collection. In the evening, as dusk fell, scores of Flying Foxes (fruit bats) flew over Sydney Harbour from their roosts in the Botanical Gardens to fruit trees on the northern shore.

Irina, being a tennis nut, had arranged that we would be there for the Australian Open, and Sue and Mark just happened to have another apartment in Melbourne, within walking distance of Melbourne Park, and I remember we saw Djokovic beat Federer on his way to winning the championship that year. We were determined to make the most of our trip, because we did not expect to come again, so next day we flew to Uluru, 1,350 kilometres from Sydney and almost in the middle of Australia. It was 46 degrees when we arrived, reminiscent of our visit to Death Valley. You have probably seen pictures of Uluru, a dome of red rock rising abruptly from the flat desert plain. It is sacred to the aboriginal inhabitants who have created all kinds of myths and legends to explain its presence. More prosaically however it is composed of almost vertical Precambrian sandstones, eroded and pitted and channelled, protruding from a thick cover of recent wind-blown sands. The striking red colour is due to the presence of oxidised iron minerals. At that time, you were allowed to scramble up it at certain times, but no longer. It is now off-limits as a sacred site. We walked around part of it, and they insist you carry at least a litre of water with you. The main sense we had was of mystery, timelessness, incongruity, uniqueness. At night we went out into the desert to experience the silence, and the absence of light pollution. In the hotel gift shop there was an exhibition of aboriginal art and I was captivated by one design, an abstract of white dots on a black ground, like closely-packed contour lines picked out on a map. I greatly admired it and regret I didn't buy it.

Then we flew to Christchurch in New Zealand. The town had been damaged by an earthquake just a few months before, and some of the buildings were still closed and cordoned off. We went to a café with a great steel arch in the middle with instructions to take shelter there in case of another one. We stayed at the Heritage Hotel facing Christchurch Cathedral. Just two weeks after we left a much bigger quake occurred which caused 185 deaths and injured 1,500. Damage was extensive, the spire of the cathedral collapsed onto the nave and the tower was so badly damaged it had to be pulled down. The Christchurch TV Building collapsed killing many inside, and the city's largest hotel was tilted to an alarming angle. Many other buildings were judged unsafe and had to be demolished, and present-day satellite images show numerous vacant lots, especially in the older part of the city. New Zealand stands astride the plate boundary between the Pacific Plate, which is being pushed westwards by about 4 centimetres per year, and the static Australian Plate, and earthquakes are not uncommon. Mike Ridd's son Timothy was involved in some of the reconstruction projects.

South Island is very English in appearance, refreshingly uncrowded and scenically attractive. We landed by helicopter high above the snow-line on Mount Cook (in Maori *Ao-raki,* the cloud piercer), but didn't linger, as we heard that the top 30 metres of the mountain had collapsed in a rock avalanche not long before, and had slid more than seven kilometres before coming to rest. We sailed down Milford Sound, almost a carbon copy of a Norwegian fjord, but with tree-fern rain-forests instead of the coniferous forests of northern Europe. I have seen lots of tree-fern fossils in the Carboniferous coal measures, but it was intriguing to see actual living examples (they no longer grow wild in the northern hemisphere). We spent one night at Reefton, a former gold-mining town with a population of just a few hundred. It was like a film set for a Western movie, all boardwalks and false fronts. We sat on the veranda of the hotel sipping our wine and watched a bunch of local hoodies, the modern descendents, I suppose, of the hustlers of the old mining days.

At Kaikoura we took a boat to go whale watching, but they cheat a bit, because they locate the whales by echo-sounder, and then just wait for them to surface. In Wellington, we drove up to a lookout from which you can observe the effects of the 1855 earthquake which raised the ground around the harbour by more than a metre, leaving some areas, like Lambton Quay, high and dry, and more than 300 metres from the sea.

North Island is famous for its hot-springs and hydrothermal activity. Craters of the Moon is a barren area of steaming vents, bubbling mud pools, fumaroles and craters which are constantly changing and reforming, so you need to watch where you walk. It developed as recently as 1958 when a geothermal power station was opened nearby. The volcanic activity is associated with a region of tension called the Taupo Volcanic Zone, along the line of the back-arc basin on the margin of the Australian Plate. Rotorua is a dormant volcano with a collapsed caldera, ten kilometres across, and nearby is the Pohutu geyser which erupts several times a day to a height of 30 metres. George Bernard Shaw visited the area in the 1930s and gave the name Hell's Gate to another geothermal area of steaming cliffs and sulphurous emanations, and the entire area is still highly active.

Earthquakes are a constant hazard in New Zealand and we were reminded of that again in Napier which was flattened by the Hawke's Bay earthquake in 1931, which raised almost 10,000 acres of sea bed above sea level. The town was rebuilt in a most attractive Art Deco style. We were lucky to find a room since we arrived during the annual Art Deco Festival with the locals and waiters and shopkeepers all dressed *a la 1930s*, the streets were full of period cars, and the buildings decorated with bunting and coloured lights. We enjoyed New Zealand – rather far from Europe, and rather prone to sudden shocks, but friendly, easy-going and very geological.

Letter 66
Our consulting activities underwent a bit of a recession following the Arab Spring, but did not entirely collapse. In spite of all the mayhem some oil and gas production continued in Libya, mostly from the offshore fields, and a small number of wells were drilled. There were stories of piracy of oil tankers in the Mediterranean, like modern-day Barbary corsairs, and on land militia groups frequently shut down pipelines, and held them to ransom. One of the more bizarre consequences of the Libyan Revolution was the defection of the former Prime Minister and Minister of Oil, Shukri Ghanem, who fled in May 2011, firstly to Tunisia, then Rome and eventually to Austria. In April 2012 his body was found floating in the Danube in Vienna, with no obvious signs of violence, but his death remains a mystery. Danny and I continued our annual updates on Libyan oil and gas reserves, and had some success with our inter-active, on-line, digitised version. I worked with Ion on an assessment of the oil potential in Tunisia, while Danny did some work for a Japanese company on existing Libyan fields. Eventually we pooled our resources and worked together for both companies. But the boom years were over, at least for the time being.

When I was a student in London I had started to research our family history. I was able to collect information from family members going back two or three generations, but beyond that the main resource was the General Registry Office at Somerset House in London where birth, marriage and death indices were held, going back to 1837. It was a laborious process. The early registers were hand written in beautiful copper-plate handwriting and bound in large, heavy folio volumes. To find a record you had to guess the date of an event, and onerously troll through the volumes until you found the appropriate entry. Then, if you wanted more information, you ordered a certificate which would arrive two or three days later. By this process I slowly traced the Hallett line back to Robert Hallett who was born in York in 1790, and in addition I was able to fill in many of the collateral branches. In the indices I found that the vast majority of Halletts came from the counties of Somerset and Dorset in south-west England, so I assumed our branch had moved north in search of work during the Industrial Revolution.

Gradually, and with the burgeoning of interest in family history, the process has become simpler. The births, marriages and deaths information and the ten-yearly census records, have now been digitised and made available on line, and more recently enormous additional resources have appeared including military records, emigration information, and parish registers. For a long time, I was stuck with Robert Hallett, partly because of spelling changes, but also due to a very misleading entry for his place of birth, but eventually I made contact with a distant cousin who had studied in York and had traced the family back to Matthew Hallett who was born in York in 1582, back in the reign of Queen Elizabeth, so evidently my assumption about the Yorkshire branch moving north in the late eighteenth century was wrong.

I have continued to add information over the years and now have around 3,500 individuals on the tree, with notes and references for many of them. It is an absorbing pastime, full of surprises and dead ends, and occasional, very satisfying breakthroughs. You have the distinction of being the only person on my tree to be born in America. Your dad discovered that a William Hallett arrived in New York, probably in 1630 or 1631, and Hallett's Cove and Hallett's Point on the East River at Astoria mark the place where he settled. Unfortunately for us, he was one of the Dorset Halletts.

When I was a schoolboy our history teacher occupied an otherwise dull afternoon by explaining the origin of some of our surnames. In my case the best he could come up with was 'little Henry', based on *Hal* plus *ette*, which I suppose was quite imaginative, so I thought wow, a French name, we must have come over with William the Conqueror. Quite recently however I had a DNA sample analysed which concluded that my main roots lay in Viking Scandinavia, so you can take your choice. Of course, the Normans were originally Vikings (which is why they were called North-men), so perhaps both are correct. Incidentally you might assume that the human genome is the largest and most complex found in nature. Wrong. The genome of *Paris japonica*, a sub-alpine plant from Japan, has a genome of 150 billion base pairs compared with only 3 billion in humans.

Irina's family was much harder to trace. Many of the records were destroyed during the Revolution or during the Second World War, and records from pre-Soviet times are very fragmentary. Families were torn apart by the upheavals of the twentieth century and the fate of millions of people is simply unknown. Irina's maternal grandfather was an intelligent man who spoke several languages, but was wounded during World War I, and took to drink to alleviate the pain. In the Second World War, he was assigned to a punishment battalion and disappeared without trace. Alla is Irina's sister-in-law. Her father was a sailor on a minesweeper in the Baltic. His ship was blown up by a mine and he and a few others survived by clinging to the wreckage for three days before being picked up. He was taken to hospital in Leningrad, where he was trapped during the 900-day siege. One of her uncles died in Mongolia aged 23 in 1945, another was a Kremlin Guard and KGB officer, and yet another was an Army General. But the real sensation arose when we discovered that Irina's paternal grandmother had had an affair with an English engineer in the town of Ridder in Kazakhstan in 1918 and that Irina's father was the result of that union. The town of Ridder was named for an explorer Philip Ridder (a Russian of German descent), who discovered rich deposits of lead and zinc, manganese and gold in the foothills of the Altai Mountains in 1786. A large mining operation was established there, and foreign concessions were granted, which continued until 1918. The English engineer must have been assigned to one of these concessions. In the chaos of the Revolution and Civil War the English engineer disappeared and her grandmother married a Russian man, who brought up Irina's dad as his own child. Unfortunately, we have not been able to discover the identity of 'the English engineer'. The Altai region rebelled against the communist regime in 1921, but the Red Army quickly quelled the uprising. Ridder was renamed Leninogorsk during Soviet times but has since reverted to its former name. Irina's father's family still live in the town, and after the war her dad was based nearby at Ust Kamenogorsk where Irina was born. Ust Kamenogorsk was a closed city in the post-war period since it produced uranium for nuclear weapons, and in 1990 an explosion released a toxic cloud over the city.

In June 2011, we went to the Wimbledon Tennis Championship and watched three long matches during which I became quite dehydrated. On the climb up the hill to the car I felt woozy and faint and breathless, and the following day I had another mini stroke - similar symptoms to before, blurred vision, loud buzzing, loss of some peripheral vision. This one was not so bad as the first, and subsequent tests showed no further damage. It was assessed as a TIA - a transient ischaemic attack, but I had another angiogram and yet another stent was inserted. My coronary arteries must be like Clapham Junction.

In Libya, following the overthrow of Gaddafi, there was a brief window of opportunity in which it looked as though a smooth transition of power might be achieved. A National Transitional Council organised elections in July 2012, and in August handed over power to the General National Congress which was tasked with establishing a democratic constitution. During this period oil production resumed and a start was made on repairing the damaged oil and gas installations. The Japanese company for which Danny and I were consulting, examined opportunities for improving recovery from some of the older producing fields, and we compiled detailed maps and data sheets to assist them. But the political situation in Libya deteriorated. The GNC was not able to agree a constitution and the window of opportunity closed. New elections were held at which only 20% of people voted, and the country fragmented into rival factions. Two main groups emerged, an Islamist GNC in Tripoli and a secular *majlis*, which for its own protection, established itself in Tobruk in eastern Libya (just as King Idris had done in the 1960s), and the country descended into a second civil war between Islamists and secular groups, but complicated by the presence of local militias, ISIS, Tuaregs and mujahedeen. Libya was fast becoming a failed state.

My mother had been very fond of cruising, but it had never appealed to us. We were not attracted to the idea of being cooped up on a ship, however luxurious, eating too much, being seasick and having limited time to visit the places on the itinerary. It went against my inclination for wandering at will, stopping off to see some curiosity, and not having to meet a deadline. However, in Vancouver we had seen cruise ships heading north to the fjords and glaciers and mountains of Alaska, and we thought it might be an exciting trip, particularly since many of the places are only accessible by sea. Mariola and Artur offered us the use of their apartment in Vancouver, and so we went a-cruising. The ship was a modern cruise liner, twice as large as the *Titanic* (and thankfully with many more lifeboats). We boarded in Vancouver and sailed through the sheltered waters of the Georgia Strait and through the Seymour Narrows, discovered by George Vancouver in 1792, and only 600 yards wide. If he had not found a way through, he would have faced a 200-mile voyage back to the open sea. Incidentally George Vancouver is buried only a couple of miles from where we now live. There is a very strong rip current in the channel which causes severe turbulence, not to mention a submerged rock just nine feet below the surface, which was blown up in 1958, after it had ripped the bottom out of over 100 vessels. Our route wound between the islands and into Alaskan territory just north of Prince Rupert. The place names reflect the early exploration of the region. Many, like Puget Sound, Mount Rainier and Mount St Helens were named by Vancouver, others like Wrangell, Baranof Island, Chichagof Island and Novo-Arkhangelsk were named during the period of Russian America between 1784 to 1867. The Russian population is said never to have exceeded 700, and overhunting drastically reduced the income from fur trading, so in 1867 Russian America was sold to the United States for 2 cents an acre. Three years later gold was found near Juneau, and in the 1890s Skagway became the main access point for the Klondike gold-rush in Yukon Territory. In 1968 the largest oilfield in North America was found at Prudhoe Bay on the Arctic Slope of Alaska. The Russians must be kicking themselves. Part of the evening entertainment on board was provided by a female Russian string trio, which I thought was strangely appropriate.

We put in at a little place called Hoonah on Icy Strait from where we took a small boat to go whale watching. We very quickly encountered a family of humpback whales feeding and blowing air and doing this wonderful thing of raising their flukes out of the water as they dive. We watched them for a long time, and deplored that some countries still practice whale hunting, 'the unspeakable in pursuit of the unpalatable'.

We sailed on to Yakutak Bay, 1,000 miles north of Vancouver, to see the Hubbard Glacier which descends from the permanent snowfields behind Mount Logan (named after a Canadian geologist), a majestic snow-covered peak almost 20,000 feet high. Mount Logan is still believed to be rising, as a result of either tectonic uplift or rebound due to glacial unloading. We were one of the first ships of the season to visit the area and the fjord was littered with floating icebergs. The glacier is 80 miles long and its snout reaches the sea in Disenchantment Bay (so named because it does not link to the North-West Passage). A thousand years ago the snout reached the open sea but has now retreated more than 30 miles up the fjord. Ice which forms at the head of the glacier takes about 400 years to reach the sea. The glacier oscillates however, and in 1986 it advanced to block off a side fjord, behind which a lake formed. The blockage degraded and collapsed during the summer months producing a huge outburst in October, when 1.4 trillion gallons of water burst through the barrier. By way of comparison Niagara Falls has a flow rate of about 750,000 gallons per second, (you can do the math). The snout of the glacier forms an ice cliff in the fjord about two miles long and about 80 feet high. The glacier 'calves' frequently with a crack like thunder as great chunks of ice break off and plunge into the sea. The boats, not surprisingly, stand some way off.

On the return we stopped off at Juneau, the state capital of Alaska, where we took a seaplane flight over the Juneau icefield and Medenhall glacier, and in the tourist shops we found jewellery called ammolite, an attractive, iridescent, opal-like stone, derived from the nacreous shells of fossil ammonites. More than 200 veins of gold-bearing quartz have been found around Juneau, and the region has produced almost 7 million ounces of gold (that is around 200 tons). We visited the Wagner Mine in Gold Creek which opened in 1889 and was one of the principal mines of the area, but it closed in 1930, and now there is very little to see. Most of the coastal settlements have no road connection to the interior, and many of the locals have small sea planes parked outside instead of cars. The tourist shops simply shut down in winter, and the owners decamp to Florida or California.

So, did we enjoy cruising? Well, if you want to see the Hubbard Glacier and the whales and dolphins, taking a boat is the only way to do it, but cruising is for sociable, clubbable people, who like to chat and natter, and we very quickly tired of the same old questions 'where are you from?', 'have you been to…?', 'what's your job?' I don't think we shall be going again.

We spent another week in Mariola's flat and visited Whistler and Tofino, but the highlight was Butchart Gardens near Victoria on Vancouver Island, a fabulous botanical garden, built in and around a former limestone quarry. It makes Kew Gardens look ordinary. It has lakes, fountains, cascades, a Sunken Garden, an Italian Garden, a Japanese Garden and a striking Star Pond. It seems perfectly fitting that the Arabic word *al jana* translates as both garden and paradise.

The Olympic Games were held in London in 2012 and a new Olympic Park was built on a run-down, brown-field site in east London, and for once Team GB did rather well. They were particularly strong in cycling where eight gold medals were won, and in rowing and athletics where we won four each. It may be no coincidence that these three sports received the three largest funding grants from UK Sport. Overall Britain finished third, behind USA and China. The big heroes were Mo Farah, a guy who came to England from Somalia in 1991, who won both the 10,000 and 5,000 metres, Chris Hoy who won two golds in cycling, and Andy Murray who won gold and silver in tennis (he then went on to win the US Open). On one glorious day Team GB won 6 golds, almost like the legendary feat at Gallipoli in 1915 where British servicemen 'won six VC's (Victoria crosses) before breakfast'.

In summer we made a sentimental journey to Scotland to visit Mike and Mikiko and Masha and Peter, and Holmwood House where we had lived in the 1980s. We went to Rosslyn Chapel, made famous by Dan Brown in *The Da Vinci Code*. The chapel was built in the fifteenth century by the Earl of Caithness, but only the choir was completed. It is enriched with lavish carved ornamentation of mythical figures, dragons, Green Men, foliage, and much else, which have generated all kinds of fantastical legends. We returned via Beamish, to see the well-known industrial heritage museum, Durham to see my old college, Halifax to see Westbury Terrace (which was looking a bit down at heel), and Sowerby to see the Georgian church and the grave of Alfred and Mary, my paternal grandparents, which we eventually found after removing a great tangle of brambles. Then we crossed the Pennines to Chester, a lovely old city with well-preserved defensive walls and medieval buildings.

I was keen to revisit Llandrindod Wells in the middle of rural Wales where my dad had been commissioned as an artillery officer in 1943, and try to identify the places that I remembered from that trip, although I was only four years old at the time. I clearly remember watching a display of anti-aircraft guns shooting at sleeves towed behind aircraft, and a farce put on by the cadets which involved an archetypal spy concealed behind a screen. There was a park where the wells and spa were located, and a hotel with an impressive entrance flanked by domed towers which had been commandeered by the army. We found the town sadly run down. Gone were the grand hotels, pump rooms, and band stands. Of the four or five springs only one remained, and that was a pathetic trickle in a wooded glade. The grand Pump House Hotel had been demolished and council offices built in its place, and the Rock Park pump room with its jolly chequered yellow brick façade which had once dispensed four distinct types of water over its marble-topped counter (like a small-scale Marienbad), now stood forlorn and abandoned. The summer 'season', the string orchestras, the bands, the horse races, and the fashionable promenades were a distant memory. The sulphur, saline and chalybeate springs, as I am sure you are dying to know, are derived from Silurian sandstones and shales which underlie the area. We were able to identify the hotel with the grand Edwardian entrance as the former Gwalia Hotel, now converted to yet more council offices, but I was not able to find the place where the ack-ack demonstration had been held, unless it was the Rock Dhole Meadow beside the river – I suspect it probably was.

In August of 2012 archaeologists discovered the remains of King Richard III beneath a carpark in Leicester, positively identified by DNA analysis of some of his descendents. He had been hurriedly buried in the choir of the Greyfriars Friary in Leicester after the battle of Bosworth in 1485, but the friary was demolished by Henry VIII in 1538 and the site of the grave was lost. He was the only king of England whose whereabouts were unknown. This is not quite true as the graves of Henry I, Stephen, Henry II and Richard I and James II have all been subsequently destroyed in various revolutions and upheavals. Richard is reputed to have remained in his mother's womb for eighteen months and been born with hair down to his shoulders, and his skeleton showed clear signs of a deformed spine. Richard III holds a special fascination for historians – was he the monster portrayed by Shakespeare as the hunch-backed murderer of the Princes in the Tower, or was he the unfortunate victim of vicious Tudor propaganda? My schoolboy friend Mark Kirkbright had been a staunch defender of Richard's reputation, and I think myself that Shakespeare's portrait is more concerned with dramatic effect than historical accuracy. There is certainly good documentary evidence that Richard was extremely popular whilst regent of the north during the reign of his brother Edward IV. But we all know that power corrupts, and the Plantagenets were a ruthless bunch.

Letter 67
Spurred on by our visit to Constable country the previous year we decided to do the same for Monet who I had greatly admired since my student days in London, when I had been enormously impressed by his *Autumn at Argenteuil* in the Courtauld Gallery. Many of his paintings (impressions, he called them), were done *en plein air* along the Seine between Le Havre and Paris, so we armed ourselves with a book of his paintings and headed off to France. We started at Mont St Michel, of which Monet produced a pencil sketch, but not I think, a painting. The 'mont' is an island about half a mile offshore reached by a causeway at low tide. The ensemble rises majestically from the fishermen's cottages at sea level, up to the walls and bastions and halls of the castle, to the monastery and abbey church at the top, a perfect microcosm of feudal French society: peasants, knights, nobility, church, God. This impression was reinforced by a group of barefoot pilgrims carrying a cross and walking across the wet sands singing *Ave Maria*.

We stopped off to see the Bayeux Tapestry (which was not made in Bayeux and is not a tapestry), which presents a vivid, almost comic-strip account of the Norman invasion of Britain. The tapestry is full of curiosities. It shows Harold and his nobles wading into the sea without their tights, and a splendid, club-shaped Halley's comet. It shows the hand of God reaching down to receive the soul of the dead King Edward, and there is an enigmatic scene of Aelfgyva, a woman unknown to history, being admonished by a priest, with a naked man in the border below. William's troops are shown apparently having kebabs for breakfast, and later stripping the corpses of the dead. Evidence for Harold being killed by an arrow in the eye remains equivocal. The tapestry image has been modified several times, and in any case, it is not even certain that the figure in question is Harold.

We reached the coast at Arromanches where the rusting remains of the D-Day Mulberry Harbour of 1944 litter the shoreline, and then along the Cote Fleurie (the French are good at attracting tourists with evocative names; Cote d'Opale is another) to elegant and sophisticated Deauville, painted by Boudin, but not apparently by Monet. He preferred its older neighbour Trouville, just across the river, where he painted the famous boardwalk, the *Hotel des Roches Noirs*, and his former mistress, but later his wife Camille, on the beach, fully dressed in a long dress, flowery hat and parasol. They say you can still find grains of sand embedded in the paint of these pictures. We walked along the boardwalk at dusk as thunder clouds gathered, and the dark and shuttered Victorian villas, exactly as shown in Monet's boardwalk picture, looked like something out of Charles Addams.

The small port of Honfleur on the south side of the Seine estuary has a beautiful old harbour surrounded by slate-clad buildings, and the medieval customs house called the Lieutenance. Monet visited the town many times along with Courbet, Boudin and later Seurat and together they established the Honfleur School, painting many scenes in and around the harbour, on the beach and towards the lighthouse. The locations of the paintings are easily recognised

On we went over the recently opened Pont de Normandie to Le Havre, where Monet painted his most famous picture, *Impression, Sunrise*, which gave rise to the term Impressionism. It was painted in 1872 when France was recovering from the defeat of the war with Prussia, and shows a blood red sun rising over a busy industrial port. The port is still busy today, but not quite as picturesque. Le Havre was Monet's home town and he painted several views of the port at different times of day and in different lighting conditions. He was obsessed by the effects of light and began the practice of painting multiple versions of the same subject to capture how a subject could be transformed by different lighting conditions. This can be seen in his studies of Rouen Cathedral, the Thames in London, the coastline at Etretat, in his *Les Meules* (hay-stacks) series, and in his countless water-lilies.

Tucked in behind the port of Le Havre is the commune of Harfleur, once the principal port of the lower Seine, besieged by Henry V in the Hundred Years' War, and the location of one of Shakespeare's best-known speeches 'Once more unto the breach dear friends, once more'. Some of the ramparts are still preserved, but without the English dead. The port eventually silted up and was supplanted by Le Havre. There is little to see these days except for the church, a few half-timbered houses and fragments of the medieval ramparts. As at Mont St Michel, Monet made a sketch of the church in his notebook, but never produced a painting.

On the death of his mother Monet went to live with his aunt at the little sea-side resort of St Adresse at the northern extremity of the Seine estuary. He painted over a dozen pictures there, of his aunt's garden on very edge of the sea, a yachting regatta, beach scenes, and *Pointe de la Hève*, a headland which marks the northern end of the beach. Thirty kilometres north, along the Pays de Caux, are three spectacular natural arches in the Chalk, and a sea stack called L'Aiguille, which attracted not only Monet, but his friends Boudin and Courbet too. Monet painted the scene from almost every angle, in summer, in a storm, in the evening. He particularly enjoyed creating an air of mystery with atmospheric mists and smoky hazes. Next, we drove to Rouen to see the cathedral, of which Monet painted twelve 'impressions' done at different times of day. The architecture is Gothic, but there have been many additions and modifications which give the place a lopsided aspect, with none of the grace and symmetry of Reims. It has an enormous, rather inelegant, cast iron spire which made it the tallest building in the world for several years. Now here is a curious congruency. Reims Cathedral was shelled and badly damaged by the Germans in World War I. The lead on the roof melted and poured onto the cathedral floor. In World War II, Rouen Cathedral was bombed by the British and US, which set fire to the northern tower, causing the bells to melt and pour molten metal onto the floor of the cathedral. Isn't that a remarkable coincidence?

From Rouen we drove up the valley of the Seine, crossing the river at Vernon where Monet painted the strikingly tall and narrow church of Notre Dame from across the river, then surrounded by gardens, but now hemmed in by urban sprawl. From there it is a short drive to Giverny where Monet lived for the last forty years of his life with his extended family. His wife Camille, who had borne him two sons, died at the age of thirty-two (he painted a corybantic portrait of her on her death-bed), and he later lived with the wife of his former patron, who had six children of her own, so it must have been a crowded household. He converted the old barn into a studio and became interested in garden design. Beds of herbaceous plants next to the house were arranged by colour and a nearby water-meadow was developed into the famous water-lily garden with the Japanese bridge, which feature in so many of his paintings. We had lingered too long at the places where he had painted along the lower Seine, and this left no time to visit Vétheuil and Bougival and Argenteuil higher up the river where he had produced some of his most iconic works. We shall just have to go back and complete the pilgrimage.

In October we were invited to the wedding of Anya, the daughter of our Polish friend Zbigy. They still lived in Siedlce near the border with Belarus where we had visited them ten years before. Anya, the girl who had played Chopin polonaises for us back in 2005, was now a dentist, as was the groom. The wedding was a big affair with 200 guests and a banquet which began (surprise, surprise) with a traditional polonaise, and went on all night. It proved a bit too much for the bride who had to be taken off to hospital with tachycardia. Fortunately she recovered in time to participate in the second day of celebrations which began where the first had left off, and ended – you've guessed it – with another polonaise. Zbigy then offered to take us to visit Lithuania where his aunt lived. He had family relations all over the region who seemed to regard the present-day international frontiers simply as temporary inconveniences.

This is a troubled corner of Europe. In the fifteenth century the Grand Duchy of Lithuania extended from the Baltic to the Black Sea, but by the nineteenth century it was firmly under Russian domination. It was overrun by the Germans in 1915, declared itself independent in 1918, but had to struggle against the revanchist policies of Russia and Poland. In the deadly game of musical chairs, it was occupied again by Russia in 1940, then seized by the Nazis in 1941, who carried out systematic ethnic cleansing, especially of the Jews, who were deported *en masse*. In 1944 the Russians forcibly incorporated Lithuania into the USSR, where it languished for the next 45 years, and even after the fall of communism Russia only relinquished its grip with the greatest reluctance. Since then, along with the other Baltic states, it has attempted to protect itself by joining NATO and the EU, but whether this makes it stronger or more vulnerable is a matter of opinion.

Zbigy's aunt lived in the small town of Jonava, home to the largest fertilizer plant in the Baltic states, where a disastrous explosion occurred in 1989, which released large amounts of ammonia, chlorine and nitrous oxide into the atmosphere, forming a toxic cloud 7 kilometres wide and 50 kilometres long and killing and injuring over 50 people. In 1938 the population of the town was 80% Jewish, by 1944 there were none.

On a freezing cold day in early October, Zbigi drove us to see the castle of Trakai, built on an island in one of the innumerable lakes west of Vilnius. The castle was the former home of the Grand Dukes during the glory years of Lithuanian power. Vilnius is reputed to have the largest medieval old-town in eastern Europe, and on the hill overlooking the city Gediminas' Tower, a relic of the former castle, has become a symbol of the county's independence. At the foot of the hill work to restore the former Royal Palace had just been completed, and next door to the palace is the elegant neo-classical cathedral with an incongruous detached bell tower, originally one of the towers of the former city wall. But the most striking sights were in the Russian Orthodox Church where nuns prostrated themselves full length in front of the icons, and in the Ausros Gate to the city, which contains the revered Mother of Mercy icon, where we saw an old woman climbing the steps on her knees. It made us feel rather uncomfortable.

On the return to Siedlce we stopped at the village of Treblinka where the Nazis built an extermination camp in which 800,000 Jews were killed in just 15 months in 1942-43. An uprising in October 1943, in which 200 prisoners escaped, led to the closure and dismantling of the camp. The bodies were burnt, the buildings destroyed and the land ploughed over and an attempt to hide the evidence of genocide. Subsequently a symbolic railway line and platform were rebuilt, and rough-hewn granite stones erected for each town and village from which the victims had been taken. It is a sombre place, but the atmosphere was rather spoilt by the presence of several busloads of chattering Israeli tourists. Death is better remembered in silence.

Several companies were still bold enough to be thinking ahead about opportunities in post-war Libya. Danny and I wrapped up our work on additional oil recovery from mature fields, and continued to market our reserves report, and we undertook further projects for other companies, although the work was fairly low key and did not take up much of our time. We had a meeting in a hotel near Heathrow Airport with a European company which was looking at opportunities in Libya for the first time. They brought a delegation, led by their Exploration Manager, to hear our views and to ask questions, but we had barely started when the Exploration Manager received a phone call from his wife to say that their dog had died, so he immediately packed up his belongings and flew back home, leaving his minions to pick up the pieces. We felt it demonstrated rather clearly where his priorities lay, but were surprised three months later when they purchased a licence for our Libya Reserves study.

Letter 68
By this time your dad had joined a New York-based company which specialised in cloud computing and virtualization services, which was all smoke and mirrors to me, and was finally able to become a man of property in Shelton, CT. It is a long way from New York (and even further from Scunthorpe), but he had a 'work from home' contract which meant that he did not have to regularly commute into the Big Apple, and when he did the company covered his travel costs. We decided to come

and see you all in the new house, and time our visit to see the Fall colours in the Appalachians, along with a tour of the Civil War battlefields of Virginia. We flew to Boston and spent a few days exploring the 'cradle of the American Revolution'. The best way to see the city is on foot and we began on Boston Common, established in 1634, only five years after the Massachusetts Bay Colony was created by English Puritan settlers. The Puritans were a tough, uncompromising bunch who ostensibly fled England because of religious intolerance, only to establish an even more rigorous theocracy in New England, a sort of seventeenth century Taliban. Laws became progressively more draconian, until ultimately capital punishment was prescribed for anyone who attempted to subvert Puritan beliefs. Quakers seem to have been the principal target and between 1659 and 1661 four Quaker women were publicly hanged on Boston Common for defying the Puritan fundamentalists. In the end King Charles II had to intervene and ordered that all the accused must be sent to Britain for trial.

The Granary burial ground, just off the Common, dates from 1660 and contains the graves of Paul Revere, Samuel Adams, John Hancock and several other prominent colonial figures. The Boston 'massacre' took place in 1770, when a group of nine British soldiers was attacked by a mob armed with clubs and stones, who then opened fire to protect themselves, resulting in the deaths of five of the attackers. Interestingly when the soldiers were put on trial they were defended by John Adams, the future United States president, and six of them were acquitted. The location is marked by an elaborate inscription.

The problems between the thirteen American colonies and the British government had begun around 1765 when a tax was imposed on the American settlers for the maintenance of the British army in America (ostensibly to protect them against the French). The situation was exacerbated in 1773 when a further tax was imposed on imported tea. The colonial administrations had no representation in the British parliament, and regarded the arbitrary levy of yet another new tax as unconstitutional. In a gesture of defiance, the colonists took the law into their own hands and dumped a whole shipload of tea into Boston harbour (commemorated today by many a colourful re-enactment). This was the time when the British government should have realised the seriousness of the situation and made some conciliatory gestures, but throughout history colonial governments have resisted independence movements, and this was no exception. The stiff-necked Brits rejected compromise.

We went to see the Old North Church, the oldest building in Boston, famous as the site of the signal sent by the rebels to their colleagues in Charlestown across the river. They hung two lanterns on the church tower to indicate that the British troops were coming by boat. Paul Revere and William Dawes, and others whose names have been lost to history, immediately set out on their famous ride to warn the citizens of Lexington and Concord that the British were on the march. The British troops found and destroyed the arms concealed at Lexington and Concord, but were attacked by local militias (the Minutemen), and harried on their return to Boston. The struggle for independence had begun. The defining song of the war was *Yankee Doodle Dandy*, which has a curious symmetry with that of *Lilli Marlene* in World War II. The modern words to *Yankee Doodle* were written by a British Army surgeon and poked fun at the rough and disorganised colonials who fought with the British regular troops against the French in the Seven Years' War. It was subsequently adopted by the Americans as a badge of pride as they gradually became a force to be reckoned with. Similarly, *Lilli Marlene* was a German wartime song adopted by the Allies in North Africa as the tide of war turned in their favour.

A few months later the Americans learned of British plans to build a fort at Bunker Hill north of Boston, to control the Charlestown peninsula, and decided to forestall them by occupying the hill themselves. They threw up a small redoubt, not on Bunker Hill but on Breed's Hill which was closer to Boston, and assembled a force of 1,500 men to defend it. The British redcoats attacked with about 2,400 men and eventually overran the hill, but sustained considerably higher casualties than the Americans. From this point, there was no turning back.

The struggle rolled on for six more years, swinging first one way, then the other. The 'patriots' declared UDI on 4 July 1776, the British sent more troops and ships, but eventually, with the help of the French navy, which prevented the relief of the British troops in Yorktown, Cornwallis's army was forced to surrender. The relieving force arrived two days too late. A peace treaty was signed in 1783 and the United States was recognised as a sovereign state the following year. Between 80,000 and 100,000 'loyalists' fled the country, about half of them to British North America (Canada), which retained its allegiance to the home country. The war produced a financial crisis in France which, five years later, led to the French Revolution and the overthrow of the French monarchy.

In a dock at the mouth of the Charles River the oldest ship still afloat, the *USS Constitution*, is berthed, just across the river from where she was built in 1797. The ship participated in the wars against the Barbary corsairs in the Mediterranean between 1801 and 1805, but in 1812 she was used to attack British naval ships which were blockading American ports to prevent supplies being sent to Napoleonic France. President Madison declared war on Britain and used the opportunity to attempt to invade Canada, but the Americans were repulsed at Detroit, Niagara and Montreal, and in one of the coastal raids British troops landed and attacked Washington and set fire to the White House, the Capitol and other government buildings. A subsequent attack on Baltimore was repulsed, and the flag seen through the smoke at Fort McHenry was immortalised as *The Star-Spangled Banner*, American words, but to a tune from a bawdy song from an English drinking club. The war was unpopular in the United States (where it was known as Mr Madison's War, and is almost unknown in Britain which was much more concerned with Napoleon at the time), and peace was finally brokered in 1815. Present politicians talk about the special relationship between the United Kingdom and the United States. It is a relationship built on very shaky foundations.

At school, you will no doubt be given a different account of these events, and that version will likely be just as accurate as mine. But that is the interesting thing about history. There is no single absolute truth. Historical events are formed of a complex of interwoven threads; it all depends on which threads you choose.

We spent one more day in Boston visiting Harvard University (founded by John Harvard an English Puritan minister who arrived in Charleston in June 1637, but died just fifteen months later, leaving half his estate to the college), and the Arnold Arboretum created on Jamaica Plain to the north of the city to display of 'all the trees and shrubs which can be grown in the open air of the district', and the Institute of Contemporary Art on the waterfront, an impressive building, but with a less-than-impressive collection. We returned on the Silver Line, part of the underground metro system, but operated by articulated buses, which seemed very odd.

So then we took the train and came to see you and the new house. Your dad collected us from New Haven, and we were able to see for ourselves what all those years of struggle had achieved. The location was delightful, in an area of woodland a few miles north of Shelton. The house was large, spacious and comfortable. You were six years old, and I remember your play area with a big doll's house and a table full of figures, animals, furniture; you name it, you had it. I remember too that you had models and books on dinosaurs, and the following day your dad took us all to the Peabody Museum of Yale University with an enormous bronze *Torosaurus* guarding the entrance, and inside a dinosaur mural covering the entire length of one wall. The museum houses the collection of dinosaurs, including the skeleton of *Brontosaurus*, discovered by Othniel C. Marsh, the nephew of George Peabody, who had risen from humble beginnings through his uncle's patronage. For twenty years he was involved in intense rivalry, known as the Bone Wars, with Edward Drinker Cope, the son of a shipping magnate from Philadelphia. In an effort to outshine his rival, Cope rushed off scores of poorly prepared papers which contained many errors (including placing the head of a plesiosaur at the wrong end). The rivalry was so intense that it led to distinctly unprofessional conduct on both sides, deliberately sabotaging each other's collecting expeditions, and it became a palaeontological *cause célèbre*. We ended the day by visiting the Dinosaur State Park near Hartford where there is an impressive array of over 500 dinosaur footprints (plus thousands more not on display), which date from the early Jurassic.

On our last day we went with you to choose a pumpkin for Thanksgiving at a farm near Shelton, and then to the Maritime Museum at Mystic Seaport, a re-creation of a nineteenth century village with a collection of about twenty vessels ranging from lobster boats to whaling ships and fast, rakish schooners - and inevitably a replica of the *Mayflower*.

After that things were not so good. We headed off to Washington but there was a strike of federal employees so all the museums and galleries (and even the Lincoln Memorial) were closed. The Washington Monument was shrouded in scaffolding due to minor earthquake damage, and it rained continuously. We walked to Foggy Bottom because we liked the name, but found it less romantic than it sounded. Perhaps we had confused it with Sleepy Hollow. From the Lincoln Memorial, we walked to the Capitol and were rewarded with a glorious red sunset, some kind of compensation for an otherwise rain-soaked day.

The Declaration of Independence of 1776 was largely drafted by Thomas Jefferson and contains some lofty and noble sentiments. 'We hold these truths to be self-evident, that all men are created equal, that they are endowed by their Creator of certain unalienable Rights, that among these are Life, Liberty and the pursuit of Happiness' - unless of course you were black and a slave. Washington had 317 slaves at Mount Vernon, Jefferson had 175 at Monticello, and twelve US presidents between 1776 and 1865 were slave owners. At the start of the Civil War Lee, Jackson and even Grant owned slaves (but not Lincoln), which rather tarnishes the image of the founding fathers and their successors. The problems they created are with us still.

We drove down to Harper's Ferry, a tiny settlement at the confluence of the Potomac and Shenandoah rivers, where John Brown and his anti-slavery protestors seized the Federal Armoury in 1859, suppressed by a certain Robert E Lee, an event which ignited the trail which led to the Civil War. The war was a defining episode in the history of the United States, and it was indeed a shocking event, in which modern weaponry was used for the first time, and resulted in around 700,000 deaths either on the battlefield or from disease – more US losses than in all subsequent wars put together, although not all the troops were American; 50,000 British volunteers joined the ranks, mostly on the Union side, of whom 67 were awarded the Medal of Honour.

Next day we drove to Gettysburg, one of the main objectives of our trip, only to find all the federal areas and access roads closed. We got to see some of the main landmarks – Seminary Ridge, Cemetery Ridge, the route of the Confederate advance, Little Round Top where the Confederates were repulsed and then cut down in a bayonet counter-attack, and the site of Pickett's famous charge, a last-ditch, desperate and ill-advised attempt to break the Union line. It failed, as Longstreet had predicted, the battle was lost, and the Confederacy never recovered. The battle was immortalised by Lincoln in his brief but poignant Gettysburg address. The following year Sherman invaded Georgia, and after a four-month siege took Atlanta, then carved his swath of destruction 'sixty miles in latitude, three hundred to the main' all the way to Savannah on the coast.

The Confederate capital was established in Richmond, Virginia, which still has a nostalgic and melancholy feel to it, epitomised by Monument Avenue, a broad tree-lined boulevard with statues of Jefferson Davis, Robert E Lee, Stonewall Jackson, J.E.B Stuart and other Confederate heroes. The Southern White House is a modest little building, actually a villa built for a banker, which was occupied by Davis and his staff until April 1865 when Richmond was finally evacuated by the Confederates, who set fire to many buildings to prevent their use by Union troops. Lee headed west attempting to join the Confederate army of North Carolina, but was intercepted at Appomattox and forced to surrender, bringing the four-year war to an end. The site of the last battle is marked by a small cemetery containing the graves of seventeen Confederates and a lone Union soldier, the last troops to be killed in the war. A week later Lincoln was assassinated in Ford's Theatre in Washington.

We had wanted to see the Fall colours in the Blue Ridge Mountains but Skyline Drive was closed because of the Federal strike, and Blue Ridge Parkway was shrouded in fog and in any case the maples were late that year. They told us we were a week too early. So instead we went to see Thomas Jefferson's mansion at Monticello, a single-storey Palladian house designed by Jefferson himself, with wonderful mountain views, then Mount Vernon, George Washington's rather grander house overlooking the Potomac river, and finally Arlington, General Lee's house with the best views of all, overlooking Washington, but now totally enveloped by the enormous national cemetery, a defiant Confederate house surrounded by the ghosts of thousands of Union troops. The estate had been commandeered during the Civil War by the Union and used as a cemetery (partly to prevent Lee from ever returning there). The earliest burials were in Mrs Lee's former rose garden, and Lee himself was buried in Lexington, Virginia. The historic settlements of Yorktown and Jamestown were both closed due to the strike, and colonial Williamsburg, we thought, had been excessively over-restored, and instead of an authentic seventeenth century town, it now resembled a theme park.

The Civil War produced and popularised many songs and ballads. The Confederates adopted *The Yellow Rose of Texas* from the Mexican wars and *I wish I was in Dixie* from the plantations. They sang *The Bonnie Blue Flag, Maryland, my Maryland*, and *Oh! Susanna*. The Union songs were more martial and less sentimental, like *John Brown's Body*, later recast as the *Battle Hymn of the Republic, The Battle Cry of Freedom, Marching through Georgia, When Johnnie Comes Marching Home*, but on one occasion when the rival armies were camped on opposite banks of the Potomac, someone began to sing *Home, Sweet Home*, which was taken up by both sides in unison.

Letter 69
On the home front things changed dramatically at the end of 2013. Within the space of a week I received an invitation from Elsevier, the publishing

house which had produced my *Petroleum Geology of Libya* book in 2002 to compile a second edition of the book, and then a message from Vinton, my contact at Ion Geophysical, to say that they were resurrecting the SPAN programme for offshore Libya.

The invitation from Elsevier came as a surprise, given the current situation in the country. Initially I was not keen on the idea. It would involve a lot of work, the royalties were very modest, and I knew from the first edition that the cost of drafting over 150 figures would be high. But then I reflected how much had happened in the eleven years since the first edition, how many new discoveries, how much information Danny and I had amassed, the new system of licensing and of course the overthrow of the Gaddafi regime. I asked Danny if he would be interested in becoming a joint author. I told him the first edition had sold twelve-hundred copies, and had brought me some lucrative consulting work, and that a second edition would be easy, as we could simply build on the core of the first edition, and update it with information from our data base. He saw the merits of the proposal, but we decided that we needed to obtain better terms from Elsevier. We said we needed an improvement in the level of royalties, that the figures should all be in colour, and of a good size, and that Elsevier should provide assistance with the cost of drafting. Surprisingly they agreed to all our demands. They more than doubled the royalty fee, they agreed that all the illustrations would be in colour, and they offered to handle all the drafting at no cost to us. They even offered to take care of the proof reading and compile the index. We could hardly believe it. It seemed like a win-win situation. We signed the contract and got down to work.

NOC contracted Ion to acquire the offshore Libya SPAN data in the summer of 2014, three years after it was originally planned to start, and six years after the original proposal. Apart from a few incidents with asylum seekers trying to board their ship, and not being able to work close to the coastline for security reasons, the operation went smoothly. Ion invited Danny and I to act as geological advisors and we had many meetings with their technical staff. We discovered that the Elsevier draftsman for our book was neither as quick nor as accurate as our own draftsman so, as with the first edition, we asked if Ion would like to sponsor our draftsman to produce the figures, in return for acknowledging their help and placing their logo on the title page. They generously agreed, and unlike Talisman back in 2002 they were more than happy to display their logo, believing that it demonstrated their commitment to oil and gas exploration in Libya (which it did). And so the path was clear, and we were able to move forward towards the broad and sunlit uplands - or so we thought.

Irina and I had become *snowbirds*, developing the habit of flying out to visit Susannah and family in Dubai in February each year, when the weather was bad in England but good in Dubai, so that we could attend the Dubai International Tennis Tournament, which attracted many of the big names, and then head off to some other exotic destination. In 2014 we flew to Jordan where we arranged a tour by car with a driver and guide. We started in the north at Umm Qais, a strategically located place overlooking the Yarmouk River, the Sea of Tiberias (aka Sea of Galilee) and the Golan Heights, where Jordan, Syria and Israel meet. The Heights are Syrian territory, seized by the Israelis in the Six Day War of 1967, and effectively annexed in 1981. The annexation was unilateral, and has been condemned internationally and by the UN, but the Israelis, in their inimitable way, have built settlements and fortifications on the heights, and show no sign of leaving. Syrians who fled the area during the 1967 and 1973 wars have not been permitted to return. It was an eerie feeling to look out over this disputed landscape and hear the occasional sound of gunfire. The guide pointed out a couple of farms and several roads on the Jordanian side of the river which he said were under Israeli control, and he should know; he was based there during his national service. In ancient times, Umm Qais was known as Gadara and formed part of the Greek, and later Roman Decapolis, a loose federation of ten cities, distinct from the Nabataean, Aramean and Judean kingdoms further south. It is claimed that Gadara was the location of the story in St Mark's gospel of Jesus casting out demons from a madman into pigs (the Gadarene swine), which then charged down a steep cliff and into the Sea of Tiberias. They must have been very fit pigs; the 'sea' is seven kilometres from Gadara.

We drove on to Ajlun, the location of one of the few castles built by the Moslems during the time of the Crusades. This was the period of Saladin's rise to power. He had first established a base in Cairo, then taken Damascus and Aleppo, and Ajlun Castle had been built to protect the eastern side of the Jordan valley against Crusader attacks. It was not needed. In 1187 Saladin crushed the Christian forces at Hattin, and the Crusader strongholds fell one after another. Within a few months Jerusalem had surrendered, ending Christian control of the city once and for all – or at least until 1917.

We ended the day at Jerash, another wonderfully preserved city of the Roman Decapolis, possibly founded by Alexander the Great in 331 BC, but which reached its peak under Roman rule in the first century AD. A fine triumphal arch, built to commemorate a visit by the Emperor Hadrian, leads to a splendid hippodrome, 300 metres long (where daily re-enactments are staged) and then to the South Gate, beyond which is a stunning oval forum framed by a colonnade of 70 re-erected pillars of red sandstone which seem to glow in the late evening sunlight. The main street is paved with large stone slabs, and it too is flanked by columns, temples, and bath-houses, with a couple of amphitheatres out near the city wall. I suspect that if you had been a successful merchant owning a villa with an enclosed garden and a slave or two, life in Roman Jerash could have been very agreeable.

The following day we drove to Mount Nebo from which Moses saw the Promised Land. Poor old Moses had a really rough deal. After bringing the Israelites out of bondage in Egypt, crossing the Red Sea, and wandering with them for forty years in the desert because they had not believed God's promise to bring them safely into 'the land of milk and honey', he received his final humiliation. The Israelites needed water for themselves and their flocks. God said to Moses 'Speak to the rock and it will pour out its water'. Instead Moses, who was pretty fed up with the whining Israelites by now, struck the rock angrily with his staff, which annoyed God no end. He determined that although Moses should see the Promised Land from the top of Mount Nebo, he would never set foot in it. So, having struggled up the 680 metre mountain at the age of 120, Moses looked out over the land of Canaan and obligingly expired.

I remember as a boy having a magazine with an article on Petra, with the clichéd sub-title 'the rose-red city, half as old as time', a quotation apparently from a poem by the Dean of Chichester, who never actually went there. We had long wanted to see it, and we were not disappointed. We walked down at night when the kilometre-and-a-half route was illuminated by candles within paper lanterns which produced only a feeble glimmer, but created an appropriate air of mystery. The route becomes progressively narrower, and is hemmed in by towering cliffs and ends in the astonishing view of the so-called Treasury building (but actually a mausoleum), carved out of the living rock, surely one of the world's most iconic views. The 'show' unfortunately did not match the setting; it could have been done so much better. If you look carefully you can see that the upper part of the façade is pock-marked by bullet holes around a stone urn, which the Bedouins believed contained treasure. In fact, it is solid sandstone. The next day we walked down in daylight when we were able to see all the features we had missed in the dark. There are dozens of rock-cut tombs, shelters, niches, and temples, all excavated in a soft, reddish sandstone of late Cambrian age called the Umm Ishrine Formation which contains Liesegang banding, nested rings of different coloured sandstone which form striking ceilings in many of the excavations. Beyond the Treasury the route widens out into the town which once held 20,000 inhabitants. It was built by the Nabataeans in the third century BC and was not conquered until 106 AD, after which it declined, and by 700 AD it had been abandoned. It lay unknown to the outside world until 1812 when it was re-discovered by the Swiss traveller and explorer Johann Burckhardt. Today it is one of the Middle East's main tourist attractions.

From there we went on to Wadi Rum, an area of dissected mountains, not far from the border with Saudi Arabia. The area is famous for its desert scenery and has many similarities with the Jabal Akakus in Libya: craggy mountains, sand-filled valleys, natural arches, petroglyphs. There is a prominent outcrop of seven conspicuous crags separated by deep ravines, inevitably dubbed 'the Seven Pillars' after T.E. Lawrence's famous book on the Arab revolt, in which he was a leading figure. We also visited the supposed site of one of his camps, although there is no documentary evidence that he was actually there. The geology is identical to Petra, with most of the reddish sandstones belonging to the late Cambrian period, with an age around 490 million years.

Nearby is Wadi Rum railway station complete with a Japanese steam locomotive built in the 1950s, and a few vintage carriages and trucks sporting the Turkish flag, which are used for re-enactments of the Arab raids on the railway during the First World War. They tell you that this is the Hejaz Railway, but that is being economical with the truth. The Hejaz line, which was the subject of the wartime attacks, lay some distance to the east. The line at Wadi Rum is on a spur line built to transport phosphates to Aqaba, and not actually opened until 1979. It connects to the main Hejaz line at Batn al Ghul (the Belly of the Demon), where the line ascends through a precipitous gorge (almost like the spiral tunnels in Canada) towards Amman. The Hejaz railway was originally built by the Ottomans to carry pilgrims from Damascus to Medina in 1908, but became strategically important during the First World War. The southern half (in what is now southern Jordan and Saudi Arabia) was abandoned after the fall of the Ottoman Empire, and rusting train wrecks are still to be seen on this section of the line.

We spent a pleasant night in Aqaba where you can look across the bay to the lights of Eilat in Israel, and the following day we drove northwards along the rift valley, which extends to the Dead Sea and beyond. It represents a northward extension of the East African-Red Sea rift and is about eight kilometres in width, and is still opening at the present time. The frontier between Jordan and Israel runs along the valley. There were many farms and agricultural projects on the valley floor on both sides of the border, but our guide said that some of the farms on the Jordanian side are leased to Israelis. We saw baskets of tomatoes by the roadside available for anyone to pick up, since there was no ready market for them, which seemed an awful waste.

Eventually we arrived at the Dead Sea, an endorheic (no outlet) lake, in which the salinity is ten times that of normal sea-water. Its surface is 430 metres below sea-level which makes Death Valley (at 86 metres below) appear rather trifling. But even that pales into insignificance when compared with the 3,000 to 5,000-metre drop in sea level in the Mediterranean only 5.7 million years ago. All of these evaporitic basins contain salt deposits which are worked commercially, in the Mediterranean at Realmonte in Sicily, for instance, and in the Dead Sea by evaporation in salt pans around the southern shores. Because of the very high saturation the composition of Dead Sea salt is different to open-ocean seawater, being particularly rich in magnesium chloride and potassium chloride, which give the water an oily feel, and of course it is reputed to have all kinds of therapeutic benefits. Swimming in the Dead Sea is a unique experience, for such is the buoyancy that normal swimming strokes have no relevance. And you do not need to worry about nasty jellyfish or scorpion fish; the sea is completely lifeless.

But all is not well with the Dead Sea. Since the 1950s the size of the sea has shrunk from 80 kilometres in length to 50 kilometres, and the water level has fallen by a metre per year. The river Jordan, which enters the Dead Sea at its northern end, used to be a major river, hundreds of feet wide 'with frequent and fearful rapids', and a flow of 340 billion gallons of fresh water a year. Now it has a flow of 40 billion gallons and contains partially treated sewage, saline water, and water contaminated by agricultural chemicals. This disaster is entirely man-made. In the 1950s the Israelis built a pipeline from Degania, where the Jordan leaves the Sea of Tiberias, to siphon-off huge volumes of water into Israel's National Water Carrier System. According to a recent study, agricultural schemes in Israel use half of the country's water, but yield only 3% of its GNP. In addition, the Syrians extract 55 billion gallons, and the Jordanians 71 billion gallons a year from the Yarmouk River, the principal tributary of the Jordan. As a result, where the Jordan enters the Dead Sea, it is now a sluggish, polluted stream about ten metres wide. This spot is the traditional site of Christ's baptism, and facilities have been constructed on both the left and right banks for pilgrims to be re-baptised. Most of the people in our group donned a white robe, waded in, and fully immersed themselves three times in the muddy polluted water. It seemed to us akin to the Hindu ritual in the Ganges at Varanasi.

The surrounding countryside is rich in archaeological interest. The cities of Sodom and Gomorrah are reputed to have been located at the southern end of the Dead Sea, but the exact location is unknown. In the biblical account, Lot's wife was turned into a pillar of salt for simply turning her head to look back at the burning city of Sodom, contrary to God's command. That Old Testament God really had a short fuse. Both the Israelis and the Jordanians will show you rock-and-salt pillars purporting to mark the spot. The great rock fortress of Masada, where nearly a thousand Jews committed suicide rather than fall into the hands to the Romans in 73 AD is close to the southern end of the Dead Sea on the Israeli side, and the Qumran caves, where Bedouin shepherds found the original Dead Sea scrolls in 1947 while searching for a stray goat, are located in the occupied West Bank just across the water from our hotel.

Letter 70
Over the years Danny and I had conducted all manner of studies for our clients – evaluation of open acreage, regional reviews, identification of favourable areas, migration fairways, unexplored areas, neglected and overlooked areas, the exploitation potential of the many undeveloped fields, the development of secondary reservoirs in producing fields, farm-in opportunities with other companies, joint ventures with state companies, fields suitable for enhanced oil recovery or for cluster developments. And in the background, we maintained our Libyan oil and gas reserves database and sold the occasional licence. Usually Danny and I worked these projects together, but occasionally we needed to bring in specialists such as seismic interpreters or petroleum engineers, but our collaboration was very harmonious with barely a disagreement between us. The only problem was that following the revolution very few

companies were willing to commit funds to Libya until their security could be guaranteed, and that did not appear to be imminent. Early in 2014 we received one more commission and, given the current turmoil in the country, considered ourselves lucky that we still had projects to work on. We could not help wondering how much longer the golden goose would quack.

Danny was a keen climber and a member of the Alpine Club, and he had climbed in both the Alps and Himalayas. I was very impressed when he told me he had climbed the Matterhorn, and since neither Irina or I had even seen the Matterhorn, we decided it was time to pay a visit. We came up with four main objectives: to see the north face of the Eiger, a classic rock wall where many climbers have met their deaths, the great white tooth of the Matterhorn, thirdly the Aletsch glacier, the largest in the Alps, shown in almost every geology text-book, and finally the Reichenbach Falls where Sherlock Holmes met his Waterloo - until being resurrected to satisfy a disbelieving public. We flew to Zurich and made our way to Grindelwald from where there is an awesome view of the Mönch, Eiger and Jungfrau. A rack railway takes you up to Kleine Scheidegg, from where you can continue on the Jungfraujoch railway, an astonishing achievement, which tunnels through the Eiger and Mönch mountains and includes two stops inside the tunnel from which visitors can admire the view through windows cut into the mountainside. The terminus is at an altitude of 3,454 metres on a saddle between the two mountains. At the top you can take a walk through an ice tunnel, where we both suffered from claustrophobia, a condition we had never experienced before, perhaps exacerbated by the altitude and the lack of ventilation. Once out in the open air we quickly recovered.

The North Face of the Eiger is one of the most demanding climbs in the world. It rises as a sheer, north-facing rock wall, 1,800 metres from base to top, and the early attempts to climb it became legends of daring, heroism and folly, inflamed by intense national pride during the 1930s. The first attempt in 1935, by two German climbers, ended in tragedy when they were caught in bad weather and literally froze to death in their bivouac half way up the face. The following year it was attempted by two Germans and two Austrians, the result was even more calamitous. On the ascent they had made the very difficult Hinterstoisser traverse, but once across they removed the fixed rope for later use. One of their party was injured by falling rocks so they were forced to retreat, but on reaching the traverse they found it covered in ice, and impassable. Three of the party were swept off the face by rock falls, and one remained suspended on a rope. Heroic efforts were made to reach him from the window at Eigerwand station, and they got close enough to touch his crampons with an ice axe. They passed him a rope, but he was exhausted, and with frost-bitten hands he could not get the knot through his carabiner. After struggling for hours he finally lost consciousness and died. The face was finally conquered in 1938 by a German-Austrian team who claimed it as a victory for the German Reich and 'the Ostmark'.

We then headed for our rendezvous with Sherlock Holmes at the Reichenbach Falls. The description in *The Final Problem* refers to 'the torrent, swollen by melting snow, plunging into a tremendous abyss, making a man giddy with its constant whirl and clamour. The spray rises like smoke from a boiling pit of incalculable depth within an immense chasm'. Well, the description is a little overblown. The stream is quite modest (although we were there in summer whereas Holmes' visit we are told was in April), and the waterfall cascades down a rocky gully before making its final plunge, but the viewing platform is just as Dr Watson described it, at the end of a narrow path which ends abruptly so the visitor has to return the way he came. This is the spot where Dr Watson was tricked into returning to the guest house to treat a sick patient, leaving Holmes to his final encounter with Professor Moriarty. When Watson returned, after realising he had been duped, all he found was Holmes' alpenstock, cigarette case and inevitably, a brief note of farewell.

From there we drove to Zermatt, or at least to Täsch where you have to leave your car and take the train. We took the Gornergrat railway up to 3,089 metres from where there are wonderful views, not only of the magnificent Matterhorn but, we learned, of 28 other peaks over 4,000 metres high. On the way down, we stopped at Riffelalp with more stunning views and had an idyllic picnic among the alpine flowers and gentians.

God made a little Gentian –
It tried – to be a Rose –
And failed – and all the Summer laughed –
But just before the Snows

There rose a Purple Creature –
That ravished all the Hill –
And Summer hid her Forehead –
And Mockery – was still –

Finally, we ascended the Rhone valley and took a funicular up to Bettmeralp from where there is a panoramic view of the Aletsch glacier, descending majestically from Concordia Platz, where three glaciers converge. It is a text-book example of an Alpine glacier: U-shaped valley, hanging valleys on the flanks, lateral moraines much higher than the surface of the glacier, indicating substantial retreat of the ice in recent years, and, most prominently, median or supraglacial moraines which form long trails of debris on the surface of the glacier. It was a lesson in physical geology.

We continued eastwards to the fashionable resorts of Andermatt, St Moritz, Davos and Klosters in the footsteps of Conan Doyle, who introduced skiing to Davos, Robert Louis Stevenson who completed *Treasure Island* there, and Thomas Mann who wrote *The Magic Mountain*. But during our visit most of the shops were boarded up for the summer, and the place had a rather dejected air. On the way back, we stopped to see the Rhine Falls at Schaffhausen, a Swiss city built in a small enclave of Swiss territory on the northern bank of the river, completely surrounded by German territory. In 1944 it was bombed in error by US planes which mistook Schaffhausen for Ludwigshafen, more than 200 kilometres away. An enquiry put the blame on tail winds which greatly increased the air-speed of the planes. The falls are impressive, not for their height but for the sheer volume of water cascading over them. They were formed as a result of changes in the course of the river during the Ice Age.

Irina had a friend who invited us to use her holiday home in Tuscany for a week or so. It was located in the mountains behind Sarzana, not far from the port of La Spezia. What made it special was that it was formerly a tiny chapel capable of seating at most twenty people. It had fallen out of use, been deconsecrated and left as a roofless ruin, but Irina's friend had restored it in a most imaginative fashion. The chapel had no windows, but a sliding roof had been installed so you could lie in bed, open the roof and gaze at the stars. The top of the former altar could be raised to reveal a cooker, dishwasher and fridge, and the tiny sacristy was now a bathroom. The 'capella' was surrounded by olive groves with views to the Gulf of La Spezia and the Ligurian Sea.

We took a boat from Lerici to Porto Venere across the Bay of La Spezia. It was here in 1822 that Lord Byron undertook one of his legendary swims (along with the Hellespont, the Venice lagoon, and the the Grand Canal).

He swam from a little cove called the *Grotta del'Arpaia* across to San Terenzo near Lerici, a distance of about six kilometres, to visit his friend Shelley who had rented a house there. (There is now a two-kilometre breakwater which marks the approximate line of the swim). Shelley bought a boat which he named *Don Juan* as a compliment to Lord Byron, but just a month or two after Byron's epic swim, Shelley and two friends were returning from Leghorn to Lerici, when the boat was caught in a sudden squall and sank, and Shelley and his companions were drowned. Due to quarantine laws his body could not be buried, so it was cremated on the beach at Viareggio in the presence of Byron and Leigh Hunt and a few friends. He was 29 years old.

Porto Venere is the gateway to five picturesque villages, known as the *cinque terre*, which cling to the rugged Ligurian coastline. Cars are prohibited, so most people arrive by boat, but disembarking at some of them can be a bit hazardous. They are rather spoilt by the hordes of tourists and the tatty souvenir shops, but their steep cobbled streets and pastel coloured houses, dark mysterious alleys and baroque churches make them attractive places to visit.

Another day we drove over the mountains to Carrara to see the famous marble quarries. They have been worked for over two thousand years from more than 600 quarries extending over an area of 70 square kilometres, although about half the quarries are now abandoned. They have provided the raw material for some of the world's most famous sculptures and structures: Trajan's Column in Rome, Donatello's *St Mark*, Michelangelo's *David, Pietà* and *Dying Slave*, Canova's *Three Graces*, not to mention the Marble Arch in London. Many sculptors would personally select what they regarded as the most suitable marble for their work. The marble was originally a fine-grained, white Jurassic limestone, which was subsequently metamorphosed into a finely-crystalline sucrosic marble. There are several different grades, the best is known as *Statuario* but comprised only 5% of the total, and is now worked out. Other grades went by names such as *Calacatta* (ivory colour with stripes of grey), *Arabesque* (with net-like veins of grey), *Zebrino* (white and grey stripes), and *Bardiglio* (dark grey with blueish spots). It is an awe-inspiring sight to stand on the rim and look over a white landscape, the hillsides torn apart, decapitated, tunnelled and quarried, with huge trucks and excavators roaring up and down the zigzag roads to the quarries. Some of the marble is now mined, and we went into one of the mines where huge pieces of marble are cut in situ by precision drills and huge band saws, like a giant cheese cutter, ultimately leaving a space reminiscent of a cathedral - or a bunker.

The culinary speciality of the area is *lardo*, herbed, pure-white, back-fat from pigs, which provided much-needed protein for the miners. It is stored in marble vats and served in thin slices on toast. It comes as a bit of a shock to health-conscious Brits, but is reputedly much healthier than butter. Curiously (or maybe not) this is a popular dish in Russia too – and probably for the same reason.

The spiky limestone mountains behind Sarzana and Carrara are part of the Apuan Alps and contain ancient villages like Paghezzana, Gragnana, Castelpoggio and Fosdinovo, hardly touched by the passage of time, where olives are harvested by shaking the tree and crushed in an olive press in the village, hay is stacked by hand in the fields, and old women dressed in black cross themselves as they head into church. The church for them is almost a second home, and still the centre of the community. We climbed up through beech and pine forests to Campocecina at a height of 1,320 metres where there are fantastic views over the Carrara quarries, and far out to sea, and we had lunch in a little Alpine Club hut full of climbing equipment and memorabilia.

Letter 71
'The book' proved a good deal more demanding than we had anticipated. Danny and I decided on the chapter headings, and added a couple of new ones which had not been covered in the first edition. We drafted an outline of the contents for each chapter, and then decided which chapters each of us would tackle. It turned out that about 500 technical papers had been published since the first edition, which all needed to be checked to determine whether they were useful or not. Thanks to Ion's financial assistance our draftsman Bob Needham was able to produce the numerous illustrations in colour, and the result was very pleasing. It is a well-known fact that most people first check the illustrations before reading the text, so it was important to make them look impressive. We had several more meetings with Ion to clarify the objectives of the Libyan offshore SPAN survey and to tie the survey to wells where we knew the geology, and we produced a series of maps and geoseismic sections, as requested by our other remaining client.

In 2015 we extended our Dubai trip by flying on to the Seychelles. We stayed in a garden hotel within a nature preserve, which was delightful, but all did not go well. There was a little island about 300 metres offshore, and at low tide the water was only knee-high so lots of people waded out to the island. I did the same. The first half was over coarse sand which was not too bad, but the second half was over a hard, stony bottom with sharp coral debris. I should have turned back, but no I carried on, and cut my feet quite badly. The return was worse, and to add insult to injury I was bitten by a fish on the ankle as I struggled back on my bleeding feet. The following day we decided to take it easy with a leisurely kayak trip across the bay and up a creek into a mangrove swamp. Our minder had his own kayak and Irina and I had one each. I don't know if you have ever tried it but if not, my advice is don't bother. It is extremely strenuous, not just on the arms but on the pectoral and lumbar muscles too. I suppose we travelled about three or four kilometres there and back in about an hour and a half. By the time we returned we realised how unfit we were. We were so fatigued that we skipped dinner that night.

The highlight of the trip however was pure serendipity. Walking back to the hotel one evening, we came across a guy unloading about thirty big tuna to sell by the roadside. We showed an interest, but of course had no means of cooking one. However, a young lad of about eighteen offered to cook it for us. We bought a two-kilo tuna for 100 rupees (about £5), the boy went home and brought bread and beer and a few supplies and set up a barbecue beside the sea. He filleted the fish, made a fire, covered the grill with banana leaves and cooked us the best, freshest, most tasty tuna of our entire trip.

By the time of our flight back to London my ankle had swollen, probably because of the fish-bite, and I could barely hobble, and I had to request a wheelchair on arrival at Heathrow. But worse was to follow. Four days after arriving home I began to feel nauseous and developed abdominal pains. My doctor referred me to the local hospital for investigation, where they diagnosed a gall-bladder infection. I was admitted into a men's surgical ward with about eight or nine others, and started on a heavy course of antibiotics. There were some serious cases in there; one had terminal bowel cancer and kept shouting in pain, and repeating the same phrases over and over again, and abusing the nurses for not giving him morphine, two others had had strokes and were comatose. There was also a young chap who walked around carrying a drain pot which was collecting fluid from his abdomen. He questioned every decision the doctor made, because he thought his condition was worsening, so he pulled out the drain tube, the stitches and the dressings, which caused a real *furore*. Next to me was an interesting chap, 99 years old but with all his marbles. He had served on the HMS *London* during the war, in pursuit of the *Bismarck*, on Arctic convoys, including the infamous PQ17, and after the war in the *Amethyst* incident on the Yangtze river. He had later been a steward on the Royal Train. I told him he should write his memoirs. The nights were bad with shouting and yelling and lights on and off all night long. It demonstrated very clearly to me that although the NHS staff were very professional, caring and kind, they were almost overwhelmed by the sheer number of patients they had to treat. On the sixth night I had severe pain and difficulty with breathing, and had to have morphine. My gall bladder had burst.

The daily blood tests showed a marked deterioration, my stomach swelled and I was not responding to the massive doses of antibiotics. They inserted a drain tube into the gall bladder and drained out the accumulated fluid, after which I transferred to a private room where at least I could get some peace and quiet. By now there was fluid on my lungs too, so another drain was inserted, and I had to stagger around with two drain pots. I was astonished to see how much fluid was drained from the pleural cavity. I won't bore you with the gory details, but I was in and out of hospital for three months, until eventually the surgeon removed the gall bladder. I never felt that my life was in danger, but I think my condition was worse than I realised. I got an inkling of this when he warned me about the potential dangers of the operation in rather stark terms. Fortunately, due to his skill, I survived, but I am convinced that the problem was triggered by the kayaking trip in Seychelles. I shall not be taking any more kayaking trips.

Danny and I finally finished work on the book in February 2016. In the end we researched, compiled, checked and drafted 187 figures, and assembled a text of 200,000 words. It had a bibliography of 1,200 items, almost all of which we had examined and evaluated. The project took us 27 months to complete, and instead of 25% of new material, as we had originally envisaged, we ended up with 90%. It wasn't just a second edition, but a total re-write.

Unlike the first edition which I simply submitted in page-ready format, the second edition had a whole team assigned to it. There were copy editors, designers, typesetters, printers, indexers, hyper-linkers and no doubt widget-engine operators as well. Elsevier produced both a print version and an electronic version, and the latter had a neat system of linking a text reference to the bibliography and to the original publication on the web. The project management was based in California but the production team was in Mumbai, and there were the inevitable problems of too many cooks. At one stage, we had Indians correcting our technical English and telling us how to spell Libyan names, and they rather spoilt the bibliography by attempting to standardize every item. Fortunately, we were able to get them to desist and leave well alone. The final result however, was very impressive. The design, the layout, the cover, the figures were all first-class, and we were delighted with the finished product. It was published in June 2016 and we felt it was well worth all the effort we had put into it.

Unfortunately, Elsevier's marketing techniques fell short of their production expertise. They enthusiastically embraced every modern marketing gimmick and cliché, and loved to 'reach out' and 'touch base' with people, without any actual action plan. They had no concept that in order to bring the book to the attention of the people who mattered, it was necessary to arrange book reviews and get them published, promote the book at conferences, and advertise it in a meaningful way. We found that although it appeared on Amazon's website, it was shown below, and subordinate to the out-of-date first edition. Furthermore, they kept sending us suggestions on how to 'maximise impact' with potential customers and advocated 'innovative' ways for us to promote it. We rather felt that was their job. We gave a number of copies to Ion, proudly displaying their logo on the title page, in genuine gratitude for their help in funding the drafting.

Ion now had the preliminary results from the offshore Libya SPAN survey and we had several meetings with them at their office in Chertsey to identify some of the rather surprising features which the survey revealed. We were impressed to see how quickly seismic reflectors could be picked and maps produced, compared with the old days when this had to be done laboriously by hand on paper. Danny and I had obtained data on some key wells whilst researching the book and Ion had data on others, which we were then able to tie to the seismic reflectors. One of the features of the SPAN lines was the depth of penetration, so deep in fact that it was possible to identify the Moho, the discontinuity between the earth's crust and mantle. It also produced much new information on the thickness of salt deposited during the 'Messinian crisis' when the Mediterranean shrank to a fraction of its present size. Ion was then able to complete processing the lines and begin marketing the survey. We made our final presentations to their Chief Geologist and a number of clients in May of 2016, eight years after I had first begun work on the project. My contact throughout this time had been Vinton Buffenmyer, an excellent geophysicist and a delightful colleague – and an impressive sportsman. He ran marathons, and thought nothing of cycling a hundred miles through the Surrey countryside. I once took him on a hike to Leith Hill in Surrey, and it took about an hour to get there by car. He said he could have done it quicker by bike.

Our three projects, consulting, Ion and the book had now come to an end, and there seemed little prospect that the political problems in Libya would be solved anytime soon. Libya had provided me with fascinating work and a livelihood for thirty years. At the age of seventy-eight it seemed like an appropriate time to hang up my boots.

In June 2016 David Cameron, the Tory Prime Minister, held a referendum asking the British public whether they wished to remain in the European Union, or leave, fully expecting that there would be a comfortable majority for the remain camp which would strengthen his position in Parliament. It was a calamitous miscalculation. For forty-three years Britain had been a member of the union. Initially it had simply been a Common Market with free trade among its nine members, but it had grown into something much larger, a union of twenty-eight countries, some rich, some poor, governed and administered by a Commission, Council and Parliament. The general aim was to create a free association of countries sharing common values, freedom of movement, freedom to work anywhere within the union, no tariff barriers, a common currency, a common agricultural policy and a supreme European Court of Justice. Member states contributed to a central budget and funds were then redistributed to members to cover agricultural subsidies to farmers, infrastructure projects, environmental issues, scientific research and so on. In general, the rich countries were net contributors, and the poorer countries net beneficiaries.

This system had brought much needed stability to Europe. It created a counterbalance to the power blocs of the United States, Russia and China, and it brought added security to the member states. But it was not without its problems. The organisation was perceived as being too bureaucratic and unresponsive to criticism. Many of its rules and regulations seemed unnecessarily complicated and intrusive, and some countries seemed to believe that they could take advantage of the EU by operating cavalier financial policies and then expect the EU to bail them out when things went wrong. Britain had been a willing member of the Common Market and in general terms supported the principles of the EU. It benefitted from agricultural subsidies to farmers and from a supply of cheap labour from Eastern Europe, but it was very wary about the perceived trend towards a 'United States of Europe', the flood of immigrants coming into the country, the 'faceless bureaucrats of Brussels' and the supremacy of the European Court of Justice, which on occasion seemed to deliver some quite perverse judgements.

It could be said that Britain had never been a whole-hearted, enthusiastic member of the club. It had not signed up to the concept of an ever-more integrated union. It had not joined the common currency, nor the Schengen Agreement for the abandonment of border controls, and it resented the fact that it could not restrict the enormous flow of European migrants into Britain (around 300,000 per year), the perception that 'foreigners' were taking 'British' jobs, that 'foreign' boats had free access to 'British' fishing grounds, and that decisions of British courts could be overturned by the European Supreme Court. Unfortunately for Cameron, the British referendum coincided with the arrival in Europe of millions of displaced persons from the troubled areas of the Middle East and Africa, all expecting to be sympathetically received and supported, and the EU made it clear that it expected Britain to take its share. It was in this climate that the British public said 'no' to Cameron's referendum, a momentous decision, which will inevitably have grave and long-term consequences for Britain. Cameron resigned immediately, washing his hands of the whole sorry business.

> *I could not dig: I dared not to rob:*
> *Therefore I lied to please the mob.*

It seemed to us (and most of our friends) that it was a wrong decision, taken primarily for emotional rather than practical reasons, which would exclude Britain from Europe, jeopardise our trade, result in London's position as Europe's financial centre being threatened, and in the reduction of the value of the pound. And the irony was that the referendum need never have been held. Decisions in Britain are made by Parliament, not referenda. It was a calamity symptomatic of Britain's declining status in the world. We were now in danger of becoming even more marginalised.

Letter 72

Irina and I celebrated the end of 'the book' with a trip to Germany (while we could still travel without a visa!). We were keen to visit the scenic Rhine Valley between Koblenz and Bingen. Many people do this trip by boat, but I was keen to visit the castles and villages which dot the valley and see the Lorelei rock and its association with the Rheingold legends used by Wagner in his Ring cycle of operas, so we hired a car to criss-cross the area. We started in Heidelberg which is dominated by the castle on the hill, destroyed during the Thirty Years' War and slighted by Louis XIV in 1689. The remaining ruins however are still impressive. During our trip we were surprised to see how much damage was done to the area by French troops during the Nine Years' War under the rallying cry *'Brûlez le Palatinat!'* We stopped to see Worms Cathedral, a Romanesque building with four unusual round towers where Luther was condemned as a heretic by (as every schoolboy knows) the Diet of Worms, and then spent a delightful day at the spa of Bad Kreuznach where there are elegant warm-water saline pools, a vapour spray, a graduation wall (on which bundles of blackthorn twigs concentrate salt by evaporation), and a radon tunnel. Why can't they produce attractive spas like that in England, where there are dozens of neglected and abandoned spas? We crossed the Hunsrück slate massif (which contains amazing Devonian fossils, 400 million years old, a so-called *lagerstätte* in which even the soft parts have been preserved), and spent a night in Trier, a town with a Roman amphitheatre and basilica and a splendid four-storey Roman gateway called the Porta Nigra, later used as a church, until it was forcibly closed by Napoleon and restored to its original function. Trier is the birthplace of Karl Marx. His ancestors were Jewish, but his father converted to Lutherism, and the family lived in comfortable circumstances in an elegant house in a fashionable neighbourhood of the city, on his father's income as a lawyer and vineyard owner. At the age of thirty-one Marx came to London, where he spent the rest of his life, writing and sponging off his long-suffering friend Friedrich Engels. The house is a shrine for communists, especially Chinese, as is his grave in Highgate Cemetery in London.

We drove down the Moselle valley past hectares of vineyards, which spread along the steep valley sides, growing mostly Riesling grapes, which produce light, flowery wines with slight mineral characteristics derived from the Lower Devonian slates through which the valley passes. The slopes are so steep that many of the vineyards have to be harvested by hand. We passed through some of the well-known producing regions like

Piesport, Bernkastel, Zell and Cochem, and sampled one or two on the way down the valley. We ended the day at Koblenz, and walked to Deutsches Eck (German Corner), a spur of land where the rivers Moselle and Rhine meet, where there is a colossal equestrian statue of Kaiser Wilhelm I. A cable car crosses the Rhine to the Ehrenbreitstein Fortress on the right bank, rebuilt by the Prussians after the earlier fortress was blown up by the French. From the fortress there are spectacular views of the confluence of the two rivers, one grey and the other a muddy brown.

From there we drove north-westwards over the Eifel mountains to Aachen, stopping off at delightfully picturesque towns like Bad Munstereifel and Monschau, full of half-timbered houses and sparkling rivers, and waitresses carrying enormous steins of beer. Real, authentic, bucolic Germany. Aachen by contrast was a disappointment, full of blacks and refugees loitering in threatening-looking groups, and with the famous cathedral covered in scaffolding and plastic sheeting, which is a pity because this is where Charlemagne was crowned Emperor of the Romans in 800 AD (in the fiction of re-founding the Roman Empire), and where every subsequent German monarch was crowned until the 16th Century. In Bonn we paid homage at the birthplace of Beethoven, a modest apartment in the courtyard of a house belonging to the court musician, where he spent the first fifteen years of his life. It is the only house in which Beethoven lived, to have survived.

Then we saw Bad Ems on the map and, being fond of spas, we decided to pay a visit. Bad Ems is a fashionable spot on the river Lahn, where emperors and princes used to gather to take the waters during 'the season' - and to start wars. The Franco-Prussian War was provoked by the Ems Telegram, sent from here in 1870 by Kaiser Wilhelm to Bismarck in Berlin. Bad Ems was a spa town to rival Baden Baden and Marienbad, but the emperors and their entourages are long gone, and the clientele these days is much more *hoi polloi*. Apparently, the modern preference is for hedonistic places like *Therme Erding* near Munich (discovered by accident by Texaco in 1983) developed into a giant spa complex, with an indoor water park and dozens of chutes and rapids, and where naked bathing is *de rigueur*.

From there we began our exploration of the Rhine Gorge, which cuts through the Devonian slate formations which we had seen earlier in the Hunsrück and Eifel mountains. We travelled up the western bank and down the eastern. Between Koblenz and Bingen there are more than sixty castles and fortresses commanding the most strategic points on this vital trade route and frontier area. They range from Carolingian ruins to late Neo-Gothic residences of industrial grandees, and include some of the most romantic, photogenic, fairy-tale castles you will encounter anywhere. Some are on the commanding heights, others on eminences closer to the river, and one on an island in the river. The Rhine is about three-hundred metres wide, but resembles a motorway with a constant stream of barges and tourist boats going up and down. We visited about six castles during our two-day perambulation, and on the return leg climbed up to the top of the Lorelei rock, a massive triangular outcrop of Devonian slate, whose name translates as 'murmuring rock' because in former times there were rapids here. According to legend Lorelei was a siren who distracted sailors and lured them to their deaths on the rocks. Wagner may have been influenced by this legend when he conceived the three Rhinemaidens as guardians of the Rhinegold, who introduce (and end) his *Ring* tetralogy. Germany is a rich country, cultured, inventive, civilized, the very last place you would expect to fall for the extremes of fascism, but they did, and within living memory.

Irina had never been to Ireland so we decided to visit Dublin for a mini-break. I wanted to show her Trinity College where my boyhood friend Michael had studied, particularly its wonderful library containing the medieval illuminated manuscript called the *Book of Kells*, and the elegant Georgian architecture for which the city is famous. I was also interested to visit some of the sites associated with the Easter Rising of 1916. Your English grand-mother, Wendy, my first wife, had a cousin who married an Irishman called Con Colbert. His grandfather was an Irish nationalist who participated in the Easter Rising and was executed by the British for his role in the rebellion.

There is not much love lost between the Irish and the English, and with good reason. The Brits have treated the Irish very badly for centuries. Beginning with the Normans, but intensifying under the Tudors, successive English governments have attempted to Anglicise the Irish and force them into the Protestant fold, but with spectacular lack of success. The Celtic Irish (Catholic) refused to submit to the will of English 'invaders' (Protestant), and the English, who invariable had superior resources, treated them ruthlessly. James I (Protestant) began the 'plantation' of Ireland with English and Scottish settlers in 1607 (James I was the first king to rule over all three kingdoms of England, Scotland and Ireland). After the Civil War Irish Catholics rose against the newly-established English Republic, only to be savagely put down by Cromwell (Protestant) at Drogheda in 1649, and when James II (Catholic) tried to regain his throne with the help of Irish Catholics, he was routed by William III (Protestant) at the Battle of the Boyne in 1690. All clear? Religion has caused more wars than you can shake a stick at.

During the eighteenth century, power in Ireland was in the hands of a small Protestant elite, but a further revolt in 1798 proved a step too far. Ireland was forced into a Union with England and Scotland, and the Irish Parliament was dissolved. Great Britain thus became the United Kingdom. Incidentally the 'Great' in Great Britain does not mean illustrious, it was introduced in classical times to distinguish England and Scotland (Megale Bretannia) from Ireland (Mikra Bretannia). Worse was to come. The native Irish peasants were grindingly poor, often tenants of rapacious English landlords who could be turned out of their hovels without notice. The Great Hunger of 1845 to 1849, when the potato crop failed, led to the death by starvation or disease of close to a million Irish men and women, and the emigration (mostly to the USA) of a further million. It was a disaster on an unprecedented scale, far worse than the Highland Clearances in Scotland (when about 70,000 people were ejected). The greatest tragedy of all is the government in London looked the other way, and did nothing.

But the English Liberals began to have a conscience about Ireland and Gladstone made two separate attempts to give the Irish some form of home rule, but was voted down each time. Finally, on the eve of the First World War parliament agreed to the principle of Home Rule, but with the advent of war the process was temporally shelved, and many Irish regiments fought alongside English and Scottish regiments in the war. But there was a hard core of Irish nationalists who were determined to get rid of the British presence in Ireland. Planning began within a month of the beginning of the war and the Irish nationalists began to organise an uprising for which they attempted to enlist German support. A retired British diplomat who had become sympathetic to the Irish cause, Sir Roger Casement, was sent to Germany to request help, and whilst there he attempted to suborn Irish prisoners-of-war to join the cause, with minimal success. The Germans offered weaponry and ammunition but no German troops. The rising was planned for Easter 1916 and an arms consignment was despatched. Then things went badly wrong. The arms were intercepted by the Royal Navy and Casement was arrested shortly after landing from a German submarine. Nevertheless, the rising went ahead.

On Easter Monday 1,200 nationalists, armed with antiquated weapons, seized a number of locations in Dublin city centre. The General Post Office was occupied and the republican flag raised, and beneath the grand portico Patrick Pearce read out the Proclamation of the Irish Republic. Most Irish citizens, however were less than enthused. There was a war on, their men-folk were out in France fighting the Germans, and the English had promised home rule after the war. Why raise a rebellion now?

It was quickly put down. The rebels failed to take the key strategic locations - Dublin Castle, the Telephone Exchange, Trinity College, the railway stations, the port. The British brought in a gunboat, and thousands of reinforcements, and set up artillery in Trinity College. Rebel positions were shelled and dozens of buildings, including the General Post Office, were left as smoking ruins. After six days of resistance the republicans surrendered, and the leaders were arrested and taken to an army barracks, where they were court-martialled in secret, and without defence counsel. If they had simply been sentenced to a spell of imprisonment all would have been well, but ninety rebels were sentenced to death and taken to Kilmainham Jail, and within four days the first executions took place. This provoked outrage within the Irish population, and the British Prime Minister quickly called a halt, but by then fourteen rebels had already been shot, and the damage was done. The victims, for whom there had been little sympathy, became instant martyrs and the struggle for Irish independence was revitalised. Its legacy is with us still.

We visited some of the key sites in Dublin, the Post Office, Trinity College, the Four Courts, but the most chilling was Kilmainham Jail, a grim, intimidating Georgian prison, decommissioned in 1924 and now a museum. It has a central hall, a panopticon, with cells on three floors reached by iron staircases linked by iron bridges, and has been used in several movies, most notably the original version of *The Italian Job*. The cells each housed five inmates, and there was a death row for those awaiting execution. The yard where the 1916 rebels were shot is surrounded by high blank walls thirty feet high which reminded me of the execution wall at Auschwitz. This is where our distant relation, Con Colbert, was shot. He was a member of the Irish Republican Brotherhood and held the rank of captain. He was the bodyguard of Tom Clarke, one of the chief architects of the rising, and was one of four rebels shot on 8th May. Sir Roger Casement was hanged for treason at Pentonville Jail in London three months later.

We had seen the Rift Valley in East Africa, the Dead Sea Rift, the meeting of the Pacific and Australian plates in New Zealand, the collision zones of the Alps and Himalayas. That left the mid-Atlantic Ridge, and the only place to see that is Iceland. We visited one cold December with our friends Tony and Laleh, hoping to see the Northern Lights, but my main interest was the geology. Iceland is situated on a hot spot where the Earth's crust is thin, and a mantle plume rises to the surface forming a great lava pile, with lots of active volcanoes and geysers. The hot spot, and Iceland itself, are located astride the mid-Atlantic Ridge, where new oceanic crust is continuously being generated, and it acts like two giant conveyor belts, tearing Iceland apart. The ridge marks the boundary between the European Plate and the North American Plate, and I was very interested to see it. One of the best places to examine the rift is at Thingvellir, 40 kilometres from Reykjavik where there is a 7-kilometre-wide graben, containing a lake, which is opening at a rate of about two centimetres per year, and has subsided by 40 metres within the last 9,000 years. Within the lake there is a chasm just a few feet wide, called the Silfra Rift, which is one of the active pull-apart zones at the present time, and the edge of the rift on the American side forms a very conspicuous escarpment. It is a curious thing to stand there, looking over the snow-covered hills and the ice-covered lake, and think what is happening beneath your feet. And yes, we were treated to a great display of the Northern Lights.

Letter 73

My previous letter would probably have been a good place to end this inconsequential history, but I am conscious that most of my letters have been about places and events, and very little about interests and beliefs, so at the risk of disappointing you or embarrassing you with some of my views, this letter will be a sort of coda. Everyone eventually arrives at his own personal decalogue and, however facile, this is mine.

If you have read this far I think you will have realised that politically I would call myself a liberal socialist (some might say a champagne socialist, but not of the Tony Blair variety). I am very conscious that I owe a great debt to the benefits of free education, equality of opportunity, the welfare state and the freedom to act and think as I like, which were features of the socialist years following World War II. It was these benefits which allowed me to advance from a working-class background to a professional career, and to profit from the enrichment which an education brings. I believe the state has a responsibility to care for its citizens and to nourish their potential (since the state also benefits). I deplore the attitude of 'for whosoever hath, to him shall be given, and he shall have abundance: but whosoever hath not, from him shall be taken away even that which he hath'. I hate unrestrained capitalism which allows the few to enrich themselves, and for the gap between rich and poor to grow ever wider. I have no reason to think that things will change, but some ideals worth striving for. The extreme policies of fascism and communism were tried in the last century and produced monsters. What most people yearn for is a moderate, caring, peaceful society in which people are treated with fairness, equality and dignity, and where no group is in a position to exploit the rest. Unfortunately life is not like that. There will always be greedy individuals who cheat and exploit the system for their own advantage.

In Britain we have a system of democracy in which Parliament is supreme. Members are elected by the people to approve or reject laws and policies proposed by the government. It is far from perfect, but at least members of Parliament can be held to account, and we have the safeguard of elections every five years when an unpopular government can be kicked out. The second chamber, the House of Lords, is a historical anachronism which has no executive power and would not be missed if it was abolished. Similarly, the monarchy has gradually been stripped of its political powers – the last time a monarch refused the royal assent was in 1707 – and its function is purely ceremonial. In times past the church formed an important element of government, but many countries, for example France and the United States, deliberately chose to separate church and state. Over here, England and Scotland still retain an established church, but Ireland and Wales do not. In this day and age, I think it is inappropriate that in Britain (or in any other country) the church should be formally connected to the state.

There obviously needs to be a balance between the power of the state and the power of the individual. The state has its responsibilities but so do its citizens. The question is how far the state control should extend. In principle the answer should be as little as possible, but in practice the state has to take responsibility for collective defence, law enforcement, taxation, foreign affairs, education and so on, but I believe there are other areas, like infrastructure, energy, health and environment where government also needs to be involved. In many countries, like France, Germany, Spain, Switzerland and Italy, railways are largely state controlled. Aeroflot, Emirates, Thai and Singapore Airlines are similarly operated by the state. Other countries have state-owned oil and gas sectors, others have state control over water supply, electricity generation and power distribution, nuclear power and mining. Britain has none of these. Privatization under Thatcher and her successors transferred virtually all state-owned enterprises in Britain into private ownership. It went further, as I outlined in an earlier letter, and now private companies are involved in education, hospitals, prisons and police. I profoundly disagree with this policy. The argument that we tried public ownership before and it failed is not valid, since the railways, the energy industry and nuclear power in Britain have all experienced dire problems whilst under private ownership, and even among those who think that multinationalism is irreversible, there are some who are not happy with the idea of having nuclear power stations in Britain built by a Chinese-led consortium, electricity supplied by a French-dominated energy company, or water by a Canadian investment group. There is no reason at all why public ownership should be any less efficient than private ownership – there is no such concern in Germany or France or Norway; all that is required is effective financing, leadership and management. Indeed, Crossrail, one of the largest and most prestigious civil engineering projects in London, is largely a state-run operation, and that other great icon of British and French cooperation, *Concorde*, was built by state owned companies. I believe that renationalisation of some of Britain's basic industries, particularly railways, water and energy, would bring significant benefits to Britain, not least by re-establishing political control over essential national assets.

Let's move on. We have a situation in Britain where national expenditure has been higher than national income for decades and austerity has been the name of the game since the financial crisis of 2008. Budgets have been drastically cut for the military, the police, the National Health Service and education, leading to a critical decline in all these services. The solution is surely obvious. The government should increase taxation on the more affluent members of society, shut down overseas tax havens and tax avoidance schemes, and force multinational companies to pay their fair share of taxes. Personally I would go further and stop government subsidies to all schools other than state schools. But of course, it won't happen, the government is too interested in clinging onto power at all costs, and unfortunately it is not in the hands of philanthropic idealists, but largely of self-serving politicians, susceptible to lobbying from powerful interest groups.

Now let's have a go at religion. I am an atheist, and I make no apology for that. We are all products of our age, and like most of my contemporaries I was brought up as a Christian, but in my 'teens I began to doubt, then to disbelieve, and finally to vehemently oppose all religions. I agree with Christopher Marlowe who said 'I count religion but a childish toy, and hold that there is no sin but ignorance', and with Nietzsche's more succinct 'God is dead'. To see people shouting *Allahu Akhbar* as they fire their bazookas at fellow human beings, or Orangemen flaunting their Protestant symbols through Catholic areas of Northern Ireland, or Buddists chasing Moslems out of Burma, is an affront to humanity. Think of the Dominicans (the hounds of God), *De heritico comburendo*, the Inquisition, the Thirty Years' War, the persecution of Tyndale, Luther, Galileo and Spinoza, and the scores of shameful things done in the name of religion. And then ask yourself whether Hezbollah and the Taliban are not simply their modern-day equivalents. Then consider the peculiar and unappealing doctrine of original sin. Why accept the dogma that mankind is born in sin, and that there is no escape from its perpetual legacy? Not only that, but for centuries Christians have been indoctrinated to believe that sex is for one purpose only, and that outside procreation it is sinful, immoral, shameful, unacceptable. What kind of religion is that? Where is the joy, the wonder, the pleasure of life? Why do we need to feel guilty for enjoying life? Is it not puzzling that many well-educated, civilised, perceptive people can simply suspend rational thought when it comes to religion? The dead do not rise, water does not turn into wine, people do not walk on water. Yet many people, even those who are not regular worshippers, continue to believe in life after death, to think that God is keeping a benevolent eye on them. Indeed, in some places – like Russia – religion is undergoing a resurgence, with Putin an enthusiastic supporter – could it be because it gives him the support of a considerable proportion of the population? As Lenin said 'Religion is the opium of the people'. People naturally yearn for security and reassurance, but it is not in God that we must trust, but in ourselves. We alone can improve society, we should not expect divine intervention. For my money science holds out far more promise for the future of mankind than any religious mumbo-jumbo.

Consider some of the benefits of science. When Victoria came to the throne 75% of the population of England worked as labourers on the land. Life was 'nasty brutish and short', marred by disease, poverty, malnutrition. The fastest means of travel was the horse, travel by sea was at the whim of wind and storms, communication was by letter, illumination was by candles, toilets… well, let's not go into that. People thought that the earth was created in 4004 BC (on the 23rd of October), that each species of plant and animal was created separately by God, that the world had been covered by a universal flood, and that the only creatures to survive were those in Noah's Ark. And then judge how far we have come. As a trivial but fascinating example of science in action, the Crossrail project in London involved tunnelling through a burial ground of victims of the Great Plague of 1665, and DNA studies have identified the presence of the *Yersinia pestis* bacteria responsible for the disease. This can now be compared with DNA from the victims of the medieval Black Death to determine whether the same bug was to blame.

People like to think there is a purpose in life. But the life of a human being has no more purpose than that of an ant (although ants are pretty smart). It is governed, just like every other species, by two overriding factors, staying alive and reproducing. Man may be a sentient creature, he may be able to improve his material condition, and quite literally move mountains, and impose his will on the natural world (or spoil it), but in the end it makes no difference whether you become a musician, housewife, nuclear scientist, supermarket cashier or poet. The main objective should be to find satisfaction, fulfilment and enjoyment in life – 'life liberty and the pursuit of happiness', at least they got that bit right. Some find this by charitable or creative activities, others by exploring new horizons, or by trying to advance the frontiers of science, or simply by going to the pub and having a night out with the lads. If that sounds trite I disagree; making yourself happy does not mean making another person unhappy.

In science, as in life, the main motivation is curiosity, a wish to dig deeper, to see further, to ask one more question. Science advances irregularly, in fits and starts, accelerating in one area, only to retreat in another. It is a process of refinement, nibbling away at the edges, like revealing a mosaic by slowly removing the covering of sand. And who knows where it will lead? But we still have far to go. The holy grail in physics these days is to discover a theory of everything, an elusive goal which aims to provide a unified explanation of physics at the cosmic, mundane and subatomic levels, where at present the laws which apply at one level do not appear to apply at another.

Then consider art. Art fulfils a deep need in many people, and I find it fascinating to speculate which has the greatest emotional impact – poetry, painting, literature, music, architecture, theatre, opera, film. For me poetry has the greatest emotional appeal, followed by music and then film. I discovered the fascination of books at an early age; there are always new discoveries waiting to be made. We had a small collection of books at home when I was young, mostly inherited from my granddad. It wasn't just the contents, but the feel of them, the smell, the look that seduced me.

> *A precious – mouldering pleasure – 'tis –*
> *To meet an Antique Book –*
> *In just the Dress his Century wore –*
> *A privilege – I think –*
>
> *His venerable Hand to take –*
> *And warming in our own –*
> *A passage back – or two – to make –*
> *To Times when he – was young –*
>
> *His quaint opinions – to inspect –*
> *His thought to ascertain*
> *On Themes concern our mutual mind –*
> *The Literature of Man –*

A book allows you to enter the world of the author and share his or her thoughts, whether it is Aristotle, Dickens, Macaulay, Bernard Shaw or Emily Dickinson. But books have a life of their own, and once published they are, for better or worse, outside the control of the author. I gradually began to acquire books and later became addicted. To this day browsing an old bookshop is one of my great pleasures, and finding something unusual is pure serendipity.

You will have noticed my fascination with history. Actually there is no such thing as history, it is simply collective biography, combined and interwoven strand by strand, set in the context of place and time. It has appeal on many levels; the fact that it appears differently depending on the viewpoint, that it has unexpected twists of fortune, that it is tied to a particular location, and that it shows human frailty and fallibility, even among the greatest, and the satisfying fact that hubris is often followed by nemesis.

Poetry I like because it allows thoughts to be expressed in an original, precise, vigorous way. Poetry has the power to intensify emotion; it sucks you in and envelops you. Take an example or two:

Milton: *Avenge, O Lord, thy slaughtered saints, whose bones*
Lie scattered on the Alpine mountains cold.

Thomas: *No one left and no one came*
On the bare platform. What I saw
Was Adlestrop – only the name
And willows, willow-herb, and grass

Lawrence: *A snake came to my water-trough*
On a hot, hot day, and I in my pyjamas for the heat,
To drink there.

Anon: *For the pace is hot, and the points are near,*
And Sleep hath deadened the driver's ear:
And signals flash through the night in vain.
Death is in charge of the clattering train.

The arts, like science, have evolved dramatically over the past hundred years. Think of the progression from Monet to Picasso, Warhol, Rothko and Hockney; Housman to Dylan Thomas, Eliot, Auden, Betjeman and Ted Hughes; Stravinsky to Shostakovich, Schoenberg, Stockhausen and Gorecki; Proust to Joyce, Lawrence, Hemingway and Orwell; Frank Lloyd Wright to Gropius, Gehry, and Norman Foster; Bernard Shaw to Osborne, Pinter, Stoppard and Alan Bennett. The kind of art which satisfied our parents no longer satisfies the present generation. It has been replaced by art which is experimental, daring, challenging and complex. As an old-timer, born before the Second World War, I find that I can absorb new trends in graphic art and architecture more easily than new trends in music and poetry. I find it hard get my head around most modern classical music written since, let's say, the 1970s. Serialism or twelve-note music has no resonance with me. Is it me, or has classical music become wilfully obscure? Is it inherently more difficult to appreciate music than graphic art? I don't know, but it is interesting to speculate.

I also believe that enjoyment of the arts can be increased by knowing something about the artist or composer. Handel composed the complex, three-hour oratorio *Messiah* in just twenty-four days. Mozart was commissioned to write the *Requiem* by a mysterious, anonymous patron who sent his servant as a go-between. It turned out to be his last composition, and in a sense became a requiem for Mozart himself. Shostakovich was given a rough ride by the Soviet authorities, but he found ways of introducing coded messages into his music which cocked a snook at his philistine critics. Enrique Granados was an eminent Spanish pianist and composer who wrote an opera called *Goyescas*. Due to the First World War it was premiered in New York in 1916, following which he was invited to give a recital at the White House for President Wilson. This resulted in him missing his ship back to Spain, so he sailed to the UK instead and took the channel ferry from Folkestone to Dieppe with his wife. The ferry was torpedoed by a German U-boat and broke in two. Granados jumped into the sea to help his wife, but neither could swim, and both drowned, along with about 50 others. Ironically the part of the ship in which their cabin was located did not sink, and was towed back into port. It adds a certain piquancy to the music.

I guess we are all collectors of one sort or another. When I was a kid it was stamps and coins and cigarette cards, later it was fossils and books, and I find myself these days collecting words and quotations. When I come across a word that I cannot precisely define I add it to my collection, along with its definition and etymology. Did you know for instance that honeydew is a sweet substance produced by aphids which are 'farmed' and 'milked' by ants (I told you they were smart), or that a hoodoo is a tall rock pinnacle with a capstone, so-called by fur-trappers who thought they had been placed there by malign spirits, or that Dr Johnson misunderstood the word internecine, or that Browning had an embarrassing encounter with a hat, or that taboo is a Tongan word, or that the word egregious, which now means outstandingly bad, originally meant exceptionally good? Well, there you go. You see the rewards of curiosity. And think of things blue: blue bind, blue blood, Blue Book, Blue Hawaii, Blue John, Blue Max, blue men, blue moon, blue murder, blue note, blue on blue, Blue Peter, blue screen, blue stocking, blue vitriol – fascinating, isn't it?

As another trivial pursuit, I keep a timeline of events that have happened during my lifetime – family events, career highlights, travel destinations, and more general events like political crises, sporting triumphs, shipwrecks, earthquakes, wars, revolutions, anything which interests me, really. I find it intriguing to juxtapose disparate events – cricket victories with a volcanic eruption, the death of a world leader with a tsunami in Indonesia. They are incongruous, and apparently unrelated, yet they are related in the sense that they all occurred in the same narrow time frame. It is a view of the world that people would have had at that point in time.

In the autumn of 2016 your dad came over to England on one of his flying visits, and kindly fired up a new Apple computer for me. My work on

Libya was finished. What I needed now was a new project. My friend Mike Ridd had given me a copy of a book he had written about his life and travels, and he urged me to do the same. I wondered who could possibly be interested, but he made the point that few people leave a record of their passing, and within a generation or so all is forgotten. 'And some there be, which have no memorial, who are perished as though they had never been born'. I thought of you, and how little you know about my life and times, and I thought that, just maybe, when I am gone, you might like to know a bit more about your English grandfather; after all we carry many of the same genes, you and I. And so I settled down and began to write *'Dear Saffron'*.

Acknowledgements

I am grateful for permission to quote from the following works:

'*But though kind time*' from '*Perhaps*' by Vera Brittain, by permission of Mark Bostridge and T.J. Brittain-Catlin, Literary Executors for the Estate of Vera Brittain 1970.
  '*No churns, no porters, no cat on a seat*' from '*The Slow Train*', by permission of the Estates of Michael Flanders & Donald Swann 2017.
  '*Two roads diverged in a wood*' from '*The Road Not Taken*' from '*Collected Poems*' by Robert Frost, by permission of the Random House Group Ltd. © copyright 1930.
  '*And still of a winter's night*', from '*The Highwayman*' by Alfred Noyes, by permission of The Society of Authors as the Literary Representative of the Estate of Alfred Noyes.
  '*Pharaoh Story*' from '*Joseph and the Amazing Technicolor Dreamcoat*', lyrics by Tim Rice, music by Andrew Lloyd Webber, © copyright 1975. All rights reserved, by permission of The Really Useful Group Ltd.
  '*He's a cheery old card*' from '*The General*' by Siegfried Sassoon. Copyright Siegfried Sassoon, by kind permission of the Estate of George Sassoon.
No copyright holders were identified for '*When you look around the mountains*' from '*D-Day Dodgers*' by Harry Pynn, or for '*I Teach them from a Bloodless Book*' from '*Geography*' by G.D. Martineau.

Other quotations are as follows:

'*What is life?*' from Gluck's *Orfeo ed Euridice*, '*When lovely woman stoops to folly*' by Oliver Goldsmith, '*Lizzie Borden took an axe*', anon, '*There rolls the deep*' from '*In Memoriam*' by Alfred Tennyson, '*Half house of God, half castle 'gainst the Scots*' from '*Harold the Dauntless*' by Walter Scott. '*I have a rendezvous with death*' by Alan Seeger. '*When I was a child*', 1 Corinthians 13, '*Some love too little*', from '*The Ballad of Reading Gaol*' by Oscar Wilde, '*Oh, what a tangled web we weave*', from '*Marmion*' by Walter Scott, '*Ye who behold perchance this simple urn*' from '*A Memorial to Bosun*' by George Byron, '*I met a traveller from an antique land*' from '*Ozymandias*' by Percy Shelley, '*Something hidden, go and find it*' from '*The Explorer*' by Rudyard Kipling. '*Yet each man kills the thing he loves*' from '*The Ballad of Reading Gaol*' by Oscar Wilde. '*Fear no more, the heat of the Sun*' from '*Cymbeline*', by William Shakespeare. '*In silks and satins the ladies went*' from '*Chemin des Dames*' by Crosbie Garstin, '*T were hard to say who fared the best*' from '*Swimming from Sestos to Abydos*' by George Byron, '*It is the faire acceptance, sir*', from '*Inviting a Friend to Supper*' by Ben Jonson, '*God made a little Gentian*', by Emily Dickinson, '*I could not dig, I dared not rob*' from '*The Dead Statesman*' by Rudyard Kipling, '*A precious mouldering pleasure*', by Emily Dickinson, '*Avenge O Lord, thy Slaughtered Saints*', from '*On the late massacre in Piedmont*', by John Milton, '*No one left and no one came*' from '*Adlestrop*' by Edward Thomas, '*A snake came to my water-trough*' from '*The Snake*' by D.H. Lawrence. '*For the pace is hot*' from '*Death and his brother Sleep*' anon, Punch 1890. '*And some there be which have no memorial*' from Ecclesiasticus 44:9.

Picture credits
Malham Cove by kind permission of Dr Tony Waltham, Atlantic II semisubmersible and Thistle Production Platform, photos from BNOC PR Department, 1985, Great Man-Made River, Libya, photo by Jaap Berk 1987, made available via Wikimedia Commons. Fozzigiaren Arch, Libya, photo by J-P Malavialle, made available via Wikimedia Commons.

Design and cover
Particular thanks to Cathy Hickey of *Could You Just...?* of Kingswood, Kent, (who also expertly prepared an earlier book of mine for publication), for her invaluable help in again assembling the text and figures, and designing the cover.

Note
Imperial and metric units are used in accordance with the custom of the countries concerned.